Sustainable Infrastructure:
Principles into Practice

Sustainable Infrastructure: Principles into Practice

Charles Ainger
Centre for Sustainable Development,
Cambridge University Engineering Department, Cambridge, UK
Cambridge Programme for Sustainability Leadership
MWH UK

Richard Fenner
Centre for Sustainable Development,
Cambridge University Engineering Department, Cambridge, UK

Sponsored by

BUILDING A BETTER WORLD

Published by ICE Publishing, One Great George Street, Westminster, London SW1P 3AA.

Full details of ICE Publishing sales representatives and distributors can be found at:
www.icevirtuallibrary.com/info/printbooksales

Other titles by ICE Publishing:

Environmental Geotechnics, Second edition.
R. Sarsby. ISBN 978-0-7277-4187-5
Intelligent Buildings, Second edition.
D. Clements-Croome. ISBN 978-0-7277-5734-0
Environmental Impact Assessment Handbook, Second edition.
B. Carroll and T. Turpin. ISBN 978-0-7277-3509-6

www.icevirtuallibrary.com

A catalogue record for this book is available from the British Library

ISBN 978-0-7277-5754-8

© Thomas Telford Limited 2014

ICE Publishing is a division of Thomas Telford Ltd, a wholly-owned subsidiary of the Institution of Civil Engineers (ICE).

Commissioning Editor: Rachel Gerlis
Production Editor: Imran Mirza
Market Specialist: Catherine de Gatacre

FSC
www.fsc.org
MIX
Paper from
responsible sources
FSC® C013604

Typeset by Academic + Technical, Bristol
Index created by Dr Laurence Errington
Printed and bound by CPI Group (UK) Ltd, Croydon CR0 4YY

Contents

Foreword

Infrastructure organisations, through the projects that we deliver, are at the heart of how humanity interacts with the environment and the natural world that we rely on. We must always be driven to serve social and economic needs for a better quality of life. So, even if we have not thought of ourselves in that way, we are all in the sustainability business – and all our projects matter.

This truth was brought home to the then global CEO and President of MWH, Bob Uhler, in late 2006 when he attended a Clinton Global Initiative conference. After listening to expert talks on global challenges, including climate change, his working table brainstormed responses. On Bob's table was the CEO of a major global information technology company, many times larger than MWH. This CEO asked Bob what MWH did and he replied: 'our sole business is to plan, design and build environmental and water infrastructure ... we are water around the world'. An hour or so later the man turned to Bob and said: "I envy you – with your geographic platform, scale and focus – you've got a chance to make a real difference in the world." Bob later described this as 'a eureka moment' – he had made sense of climate change, in direct relation to the MWH business. He subsequently made a 'Climate Change Commitment' for the global company, based on three key areas.

1 Work with our clients on measures to manage their greenhouse gas emissions.
2 Reduce the impacts of our business by minimising our own greenhouse gas emissions.
3 Help to educate our local communities.

These commitments are real, as is the determination of MWH to work to the principles of sustainability. How to actually implement them is the real challenge in the time, cost and efficiency driven world of infrastructure provision – to *do more with less*. These pressures don't just apply to the physical materials and equipment in projects; they also apply to our people and their planning, management, design and construction time – the 'engineering overhead' on any project. So, if we are to effectively add sustainability thinking into what has to be considered on a project, we must deal with it in just as efficient and integrated a way as all the other issues.

Without a strong and logical framework for this, and under those same pressures, it is tempting to compromise on answering the hard questions that sustainability asks of us, or to miss opportunities for more creative sustainable solutions, because we forgot to ask the right question at the right time. However, we cannot leave sustainability questions to be thought of as 'nice to haves', we must embed them into standard engineering practice.

Such frameworks have been hard to find in the sustainability literature. There are books that have a strong sustainability science foundation, but give little help on making the proposals practical enough. There are others that, in being practical, tend to dumb down or compromise on the issues. Also, there have been almost too many principles and measurements that different authors have set out for sustainability, with no one set of principles that matches well with the needs of infrastructure.

This is why we welcome and have sponsored this book. It marries Charles' years of practitioner experience, gained with MWH, of what actually happens during the project delivery sequence, with Richard's deep knowledge of the academic and science foundations. The principles that they have derived focus particularly on the impacts that infrastructure has and the new kinds of thinking that we need to apply. Their relating of 'what engineers can do' directly to each stage of project delivery helps us chose the right time to ask each question. The result is a framework that is practical enough for us to use as a basis for efficient standard practice, but that continues to remind us of the hard science that we must respond to.

All of us who work in infrastructure have a responsibility, and a wonderful opportunity, to make a difference for sustainability. This book can help us all take that opportunity. I am delighted to commend it to you all.

Garry Sanderson, MWH UK Managing Director

www.mwhglobal.com

About this book

Background and purpose

The authors bring together 43 years of front-line engineering practice, and 35 years of engineering research and teaching experience, in both infrastructure and sustainability. We find a considerable gap in understanding and language between the concept of sustainability, and the practice for actually achieving it. This can be seen in the different language used by 'environmentalists', 'engineers' and 'policy-makers', and by academic papers compared with engineering manuals and methods. While many books have been published about sustainability in all its aspects, few of them have approached the subject from the time-limited, output-driven perspective of the practical infrastructure engineer/practitioner. We aim to fill that gap.

So this book, and the others in this series, are 'engineering books' about sustainability. They are for practitioners – mainly engineers, but also all the other built environment professionals who play vital roles in creating and maintaining infrastructure services – who know generally of the challenges of sustainability, but have not studied the subject.

The aim is to help you to understand sustainability; to identify and apply the critical changes needed to provide more sustainable solutions; and to embed these changes into our tradition of excellence – delivering infrastructure to provide a good service of the right quality, on time, at lowest cost – as straightforwardly as possible.

The book series

This first book, *Sustainable Infrastructure: Principles into Practice*, identifies the key sustainability issues for infrastructure, and their common principles, and shows how you can apply them in practice, across all types of infrastructure. Following this overarching volume, there will be sector-specific books, which will apply the common set of sustainability principles to practice in different infrastructure sectors. These will cover: 'buildings', 'transport', 'municipal water and wastewater', and 'waste management', and more may be added. Each sector-specific book will take a similar approach to applying the common set of sustainability principles.

We have written the books in a practical style, with 'we' as authors directly addressing 'you' as readers. They are full of references and web links. For simplicity, and with apologies, we have used 'engineer' or 'engineering' as shorthand for all infrastructure practitioners, including the many who are not engineers.

The books cover the *sustainability* aspects of infrastructure, so use them alongside your standard texts and methods. We do not aim for complete technical coverage, nor to offer a 'balanced' point of view. We focus on things that most need changing, and will make the most difference, using latest best practice – to help you deliver sustainable infrastructure.

Book structure

This book is divided into four parts.

Part I: Principles introduces the key issues and concepts needed to develop sustainable infrastructure (Chapter 1). It then describes the core principles that can guide engineering decision-making, and frames these in four groups (Chapter 2).

Part II: Practice is structured to match the typical stages of project delivery (the names of the project stages are taken from Chapter 2 of the *ICE Client Best Practice Guide*), to help you ask the right questions at the right time. Chapters match the following stages:

- *Planning*: Chapter 3 (Business strategy), Chapter 4 (Project scoping) and Chapter 5 (Stakeholder engagement)
- *Development*: Chapter 6 (Procurement) and Chapter 7 (Outline design)
- *Implementation*: Chapter 8 (Detailed design) and Chapter 9 (Construction)
- *In-use and end of life*: Chapter 10.

In each stage within Part II we ask 'What can engineers do?' Some of the proposed actions may seem difficult to accomplish, given present roles and practices; but they are where the opportunities lie, and they are already being done as best practice. You have more capacity to change things than perhaps you think.

Part III: Change helps you to see *how* you can take those actions to transform things; to understand the opportunities and constraints given by your organisation or project (Chapter 11); and how to be persuasive, and take care of yourself, as an individual (Chapter 12).

Finally, **Part IV: Tools** offers reference advice on some of the tools and approaches that are available to help you define, test and measure sustainability in infrastructure. It describes tools that are specifically relevant to infrastructure (Chapter 13). It then provides some final thoughts as an Afterword (Chapter 14), and briefly summarises some of the better known general sustainability principles already in use.

The structuring of Part II around the stages of project delivery will allow those of you with very limited time to focus on those chapters that are most immediately relevant to your work. We hope most of you will read the whole book. At least start by reading Chapter 1, and the framework for principles in Chapter 2, to make sense of everything that follows.

If you know *what* questions to ask, and *when*, then innovating towards sustainability is easier, and much more doable than you think, even if you do not see yourself as a 'change agent'. We can all, in effect, incorporate sustainability into our definition of 'good engineering'.

This book is to help you choose 'What?' and 'When?' We hope you will use this book, and enjoy it, and feed your own innovative experience into the common store of knowledge.

Reference

ICE (2009) *Client Best Practice Guide*. Institution of Civil Engineers, London.

Acknowledgements

The collaboration on this book grew out of Charles' visiting role in the Centre for Sustainable Development in Cambridge University's Engineering Department, working with Dick, the MPhil programme Director. We thank the Royal Academy of Engineering and Robert Mair for their support, and all our colleagues at the Department, particularly Peter Guthrie and Heather Cruickshank, for their stimulating discussions and contributions.

Many of the concepts included reflect Dick's work in developing professional practice postgraduate Master's programmes, designed to develop sustainability concepts in young early career engineers, and encouraging them to lead the necessary changes within the engineering industry. Thanks go to all of the 376 students who have so far graduated from the MPhil in Engineering for Sustainable Development, for their challenging ideas, their passion and their determination to embrace new approaches in their subsequent careers. Their encouragement has been immense, as have the support and constructive critique of our endeavours in this field from many academic colleagues, including David Fisk at Imperial College, Adisa Azapagic at the University of Manchester, Nick Ashford, Eric Adams, John Ochsendorf and Susan Murcott at MIT, David Laws at the University of Amsterdam, and Charles Kennel at the University of California San Diego.

The 'practitioner' engineering and sustainability experience, distilled into the book, comes from Charles' career with international consultant MWH. Thank you to very many MWH colleagues and clients, in the UK and worldwide, who are too many to name, and to the company. We thank directly Ian Davies, Peter Ratcliffe and Andy Timms, for their feedback on the first draft, which has improved it greatly; Adrian Johnson, Alison Bradley, Dave Rendall, Gavin Gilchrist and Alan Crossman for examples, and Garry Sanderson for MWH's book sponsorship, and the Preface. Charles started working on change, from the stimulating discussion and support of teachers and colleagues at the Master's course at the University of Bath; thank you particularly to Judi Marshall, Peter Reason, Gill Coleman, Chris Seeley and Karen Karp.

For stimulating conversations, contacts, case studies and references we thank Ron Watermeyer, Paul Jowitt, Mary Lou Masko, Azad Camyab, John Summers,

Jeremy Purseglove, Quentin Leiper, Roger Duffell, Dale Evans, Joanne Fielding-Cooke and Sarah Pinkerton. Finally, for general guidance on the structure and style of the book, we acknowledge the early comments from Bill Addis (Buro Happold), Tony Parry (Nottingham University) and Alan Yates (Building Research Establishment) which helped shape its development.

We thank Victoria Thompson and our commissioning editor Rachel Gerlis at ICE Publishing, for their support and patience throughout the book's development and completion.

Writing the book took time and attention away from our families, and we thank our wives Judy and Gill for their forbearance of missed hours – more than expected. The book could not have happened without their love and support.

Part I

Principles

This part sets the scene for the rest of the book.

Chapter 1 answers the question 'Why sustainability?' and introduces the key issues and concepts needed to develop sustainable infrastructure.

Chapter 2 sets out core principles for sustainable infrastructure, to guide engineering practice and decision-making.

> An army of principles can penetrate where an army of soldiers cannot.
>
> Thomas Paine

Sustainable Infrastructure: Principles into Practice
ISBN 978-0-7277-5754-8

ICE Publishing: All rights reserved
http://dx.doi.org/10.1680/sipp.57548.003

Chapter 1
Why sustainability?

We need an industrial revolution for sustainability, starting now.

John Schellenburger

1.1. Sustainable infrastructure: new challenges

The world is rapidly changing. We face a process of global warming, which has already started, and threatens our future; and at the same time rising human population is increasing the strain on the Earth and its resources (IPCC, 2007a). A loss of biodiversity in the natural environment and the adverse impacts of rapid urbanisation on individuals in the built environment are other consequences, as societies rapidly industrialise out of poverty (Butchart *et al.*, 2010).

Others have comprehensively catalogued and described this range of global problems, which have become apparent in the last few decades (Worldwatch Institute, 2010). Recent events have shown infrastructure systems to be vulnerable to natural events; such as air transport to the Icelandic volcano's ash cloud, the Fukushima nuclear reactor to the Japanese tsunami, or even basic infrastructure and shelter to earthquakes, floods and hurricanes, from China to Haiti, and from the USA to East Africa. The most striking impact of Hurricane Sandy in October 2012 was the disruption it caused to New York's critical infrastructure. But sometimes infrastructure itself can be the cause of the problem, such as the oil spill in the Gulf of Mexico from the Deepwater Horizon drilling platform, or the contribution that the building and operation of energy-intensive facilities makes to the build up of greenhouse gases (GHGs) in the atmosphere, and global warming.

Box 1.1 summarises the scale and range of issues involved. These issues are the drivers for providing sustainable infrastructure.

1.2. Characteristics of infrastructure

Infrastructure assets, such as roads, bridges, railways, power plants, residential and commercial buildings, pipelines, and water and energy utilities, have several things in common. They provide the basic services that allow modern communities and global society to function. They are specific to the geographic location within which they must operate, and represent capital goods that are typically long lasting (sometimes for centuries). Some types of infrastructure, such as transportation, power and water systems, rely on complex networks to deliver their functions, and can benefit from economies of scale and of scope (Cleveland, 2012).

Box 1.1 The global situation: some trends (Worldwatch Institute, 2009) with 2012 updates

---

Population	The world's population surpassed 7.0 billion in 2012, with no significant slowing, yet.
	The number of people on the move involuntarily worldwide may be as high as 184 million – roughly equivalent to the entire population of Brazil, or one out of every 36 people on Earth.
	Each day, 200 000 people are forced to move to a big city because their previous environment is no longer capable of supporting them. Since 2008 more people live in cities than in rural areas.
Energy	World production of fossil fuels – oil, coal and natural gas – increased by 2.9% in 2008 to reach 27.4 million tons of oil equivalent (Mtoe) per day.
	In just 8 years, it is projected that the world will be consuming nearly 50 000 gallons of oil every second. By that time, the world will not be able to meet the projected demand, for one simple reason – we are using up oil at breakneck speed.
	In 2006, coal accounted for 25% of the world's primary energy supply.
Climate	In 2007, worldwide carbon emissions from fossil-fuel combustion reached an estimated 8.2 billion tons, which was 2.8% more than in 2006 – and 22% above the total in 2000.
	In 2007, there were 874 weather-related disasters worldwide, a 13% increase over 2006 and the highest number since the systematic recording of natural perils began in 1974.
	Temperatures in central England have increased by 1°C since the 1970s; total summer rainfall has decreased in most parts of the UK; and sea levels around the UK have risen by 10 cm since 1900.
	If global temperatures rise by only 2°C, 20–30% of species could face extinction.
	Around 1.5 billion people currently live in water-stressed regions. Climate change and population growth could increase this number to nearly 7 billion by the 2050s. Latest predictions are that we are heading for +4°C by 2100 – twice the +2°C 'safe' target.
Resources	The wealthiest 20% of the world's population account for 76.6% of total private consumption. The poorest fifth just 1.5%.
	The world's annual consumption of plastic materials has increased from around 5 million tonnes in the 1950s to nearly 100 million tonnes today.
	Calculations show that the planet has available 1.9 ha of biologically productive land per person to supply resources and absorb wastes – yet the average person on Earth already uses 2.3 ha worth. These 'ecological footprints' range from the 9.7 ha claimed by the average American to the 0.47 ha used by the average Mozambican.

Box 1.1 Continued

Biodiversity	About one-fifth of the world's coral reefs have already been lost or severely damaged, while another 35% could be lost within 10–40 years, according to the latest review by the Global Coral Reef Monitoring Network (2008). The areas of the world that are officially protected – national parks and the like – grew by some 26% between 1997 and 2007, roughly one-third as fast as during the preceding 10 years, when the rate topped 75%. 2000 trees, or the equivalent of seven football fields, are cut down every minute in the Amazon rainforest.
Water	By 2020, water use is expected to increase by 40%, and 17% more will be required for food production to meet the needs of the world's growing population. According to the United Nations, by 2025, 1.8 billion people will be living in regions of absolute water scarcity, and two out of three people could be living under conditions of water stress. Humans already appropriate over 50% of all renewable and accessible freshwater flows. South Australia is the driest state in the driest continent in the world, yet its water consumption is 445 l/day per person (2001/2002).
Pollution	Pollution affects over 1 billion people around the world, with millions poisoned and killed each year. The World Health Organisation estimates that 25% of all deaths in the developing world are directly attributable to environmental factors. The presence of lead in children lowers IQ by an estimated 4–7 points for each increase of 10 μg/dl. In Cairo, breathing daily air pollution is equivalent to smoking 20 cigarettes a day. According to the Environment Agency in the UK, in 2008 there were 126 pollution incidents that had a serious impact on air quality, 199 pollution incidents that had a serious effect on land quality, and 442 pollution incidents that had a serious impact on water quality.

In many parts of the world, infrastructure systems are reaching the end of their intended design life, and in some cases exceeding it. This can lead to unsafe bridges, inadequate flood defences, decaying pipelines, tunnels and airports, or water treatment facilities and power supply systems operating beyond their intended capacity. Much is being done to address these ageing assets, and many opportunities exist in developing refurbishments and replacements to provide smarter, more efficient and lower carbon ways of delivering improved services.

Despite these difficulties, infrastructure services remain at the heart of meeting human needs. The infrastructure we develop now, if done within the increasing range of new constraints and higher expectations of society, has the potential to provide safety, security and the adaptations that will be necessary in this century to deal with climate change, more mouths to feed, and how the support systems of the planet are protected. This presents a critical challenge to how we design, develop and implement adaptable – sustainable – infrastructure, capable of dealing with the political, economic, techno-logical, climatic, demographic and other changes we face. All aspects of human activity have to adapt to meet these challenges. The ways in which infrastructure services are defined, planned, procured, designed, built, operated and disposed of requires, in response, our innovation and improvement.

1.3.   Understanding sustainable development

It is often said that the notion of sustainable development is vague, contradictory and even uncomfortable when it poses challenges to previously acceptable solutions. The term 'sustainable development' is progressively more and more (over-) used by both professionals and an increasingly aware public, but many engineers think that the concept remains difficult to deliver at the operational level:

> If no-one really knows what it is, who's accountable for it, and how you measure it, then how do you know if you've achieved it?

> (Cole, 2012)

Definitions of the term abound. From the most quoted Bruntland definition,

> Sustainable development is development which meets the needs of the present without compromising the ability of future generations to meet their own needs.
> (Bruntland, 1987)

To other, less human-centred, attempts to capture the essential principle, such as the one offered by Forum for the Future:

> Sustainable development is a dynamic process, which enables all people to realise their potential and improve their quality of life in ways that simultaneously protect and enhance the Earth's life support systems.
> (Chambers et al., 2008)

There are many more definitions of sustainability, and much has been written about their merits (Mawhinney, 2002), but we believe we can sum up the key meanings in the following way.

First, the impacts of *all* humans' development must be **sustainable within environmental limits**: that is, must be able to continue indefinitely within the environmental 'carrying capacity' of our one planet. A good measure of this is our 'ecological footprint', which is a standardised measure of the demand for natural capital, expressed as an equiv-alent land area per person, expressed in 'global hectares' (Ewing et al., 2010). It represents the amount of biologically productive land and sea area necessary to supply the resources the human population consumes, and to assimilate the associated waste. This highlights the obvious constraint that we must work within the environment limits of the natural world, its supporting systems, and its overall system integrity. Limits are critical globally, but may also be critical regionally or locally.

Second, within these limits, all people have the right to **social and economic development**, to have an acceptable quality of life. A good measure of this is the UN's human development index (HDI), which combines information on health and education (society) and gross domestic product (GDP) per capita (economy) performance into a single score out of 1.000. (The HDI is a composite index measuring average achievement in three basic dimensions of human development – a long and healthy life, access to knowledge and a decent standard of living (see HDI, n.d.).) This applies country by country, but is also a regional and local issue.

Figure 1.1 shows the combined assessment of many countries against these two measures, and the scale of the improvements needed for the world's sustainable development. The sustainable 'target' for all countries is the box in the lower right-hand corner. The poorest countries (on the lower left) need quality of life improvements of up to 250% to get to the target HDI of 0.8, without increasing their ecological footprint beyond 1.9 ha/person. The most developed and environmentally damaging countries (on the upper right) need 'eco-efficiency' improvements of up to 80% without unacceptable losses in their HDI, to reduce their ecological footprints to the global average capacity of 1.9 ha/person.

But it is not just about tackling the problems of the present. The decisions and actions we make now must not compromise the ability of **future generations** to respond to their own needs. This recognises that the future bears much uncertainty, and that 'progress' is not

Figure 1.1 Human development index and ecological footprint, 2009. (© Global Footprint Network, 2013. *National Footprint Accounts*, 2012 edition)

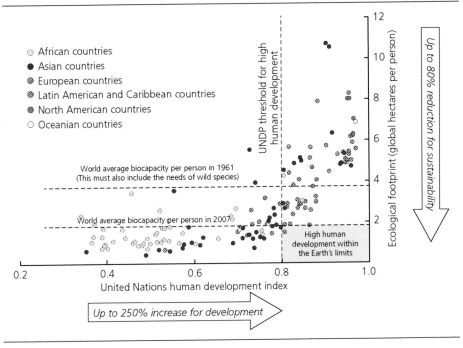

guaranteed to always be positive. We have a 'stewardship' responsibility – 'We do not inherit the Earth from our ancestors; we bequeath it to our children' (Anon.).

Finally, we live in a highly interconnected natural and man-made world, so actions in one sphere can have unintended and unforeseen consequences elsewhere. Infrastructure solutions, therefore, have to operate within highly **complex systems**, and be flexible and adaptable to deal with the uncertainties this implies. ('Complex' systems are not just the same as 'complicated' systems, with which engineers are familiar. Complexity involves working with 'wicked' problems. A wicked problem is a complex issue that defies complete definition, and for which there can be no final solution, because any resolution generates further issues, and solutions are not true or false or good or bad but the best that can be done at the time. Such problems are not morally wicked, but diabolical in that they resist all that can be done at the time (Rittel and Webber, 1973).)

These components of sustainable development follow either from the laws of science or from the moral and ethical responsibilities of civilised human society. They are absolute, and so form the basis of the sustainability principles that must guide infrastructure decision-making at all levels. The components will be developed in Chapter 2. A timeline of how these sustainability concepts have evolved is given in Box 1.2.

Box 1.2 Development of sustainable development concepts

Early environmentalism
Nineteenth century
Henry David Thoreau, Ralph Waldo Emerson, John Muir
Significance of nature as a mystery full of symbols and spirituality.
The importance of preserving wilderness and the beginning of national parks.
Twentieth century
Aldo Leopold
Ecosystems tied directly to human survival. 'A thing is right when it tends to preserve the integrity, beauty and stability of the biotic community. It is wrong when it tends otherwise.' (Leopold, 1949)
Rachel Carson
In *Silent Spring* (1962), Carson described the dangers of agricultural pesticides for animals and humans, and is often credited as the founder of the modern environmental movement and the setting up of the US Clean Air Acts, Clean Water Act and the Environmental Protection Agency.

Modern environmentalism
1972   **The Limits to Growth, Club of Rome Report (Meadows et al., 1972)**
Systems modelling applied to the entire world showed trends of highly damaging resource depletion and the accumulation of pollution.
1974   **James Lovelock's 'Gaia' hypothesis**
The Earth seen as a single self-regulating organism, with life creating the conditions for life; the beginnings of Earth systems science (Lovelock and Margulis, 1974).

Box 1.2 Continued

Roots of sustainability

1972 **United Nations Conference on the Human Environment, Stockholm, Sweden**
Focused on regional pollution, e.g. acid rain in Northern Europe. First steps to find positive links between environmental concerns and economic issues such as development, growth and employment.

1972 **United Nations Environment Programme (UNEP)**
Mission to 'provide leadership and encourage partnerships in caring for the environment by inspiring, informing and enabling nations and people to improve their quality of life without compromising that of future generations'.

1980s **Robert Allen's *How to Save the World* and Lester Brown's *Building a Sustainable Society***
Allen's book (1980) was based on the World Conservation Strategy prepared for the United Nations Environment Programme and the World Wildlife Fund. Brown (1981) analysed the economic predicament facing the world because of a fundamental disregard for ecological limitations, outlining a strategy for integrating economic and environmental issues.

Emergence of sustainability

1983 **The World Commission on Environment and Development**
Asked to formulate 'A global agenda for change'.

1984 **The Worldwatch Institute published its first *State of the World Annual Report***
Provides a global perspective on the relationship between the world's resource base and the dynamics of economic development. Helps create a global consciousness about the interconnection of ecological, economic and social issues.

1987 **The Bruntland Report, *Our Common Future* (Bruntland, 1987)**
Helped define sustainability by creating the first framework for concerted action to protect the Earth's life support systems, while promoting both economic and social justice goals. Seven critical actions are identified.

1992 **United Nations Conference on the Environment and Development, Rio, Brazil**
Known as the Earth Summit, it agreed 27 principles and adopted a global programme for action on sustainable development through Agenda 21.
Set up Framework Convention on Climate Change, the Convention on Biological Diversity, and the Forest Principles.

2002 **World Summit on Sustainable Development, Johannesburg, South Africa**
Dealt with poverty eradication, consumption and production issues, and health concerns.

2009 **UN Climate Summit, Copenhagen, Denmark**
The Copenhagen Accord recognises the scientific case for keeping temperature rises below 2°C, but does not contain commitments for reduced emissions that would be necessary to achieve that aim.

2012 **Rio +20 United Nations Conference on Sustainable Development, Rio, Brazil**
Heads of state and ministers from more than 190 nations signed off on a plan to strengthen global environmental management, tighten protection of the oceans, improve food security and promote a 'green economy', but was criticised by environmentalists and anti-poverty campaigners for lacking the detail and ambition needed.

Box 1.2 Continued

2012   **UN Climate Change Conference, Doha, Qatar**
Extends the 1997 Kyoto protocol limiting greenhouse gas emissions to 2020. Fails to agree any new global framework, and plans to achieve one only by 2015.

2012   **World Bank warns that the trend is to a +4°C world by 2100**
'We should avoid +4°C at all costs' (World Bank, 2012).

1.4.   Infrastructure for sustainable development

The familiar 'triple bottom line' Venn diagram for sustainable development implies that the economy, society and environment are of equal priority (Figure 1.2(a)). This fails to reflect the fundamental ecological basis of all our society's and economy's assets, within the real limits of our single planet.

So the actual relationship is more like three nested components (Figure 1.2(b)). Within the Earth's single planet limit, the *environment* nurtures our human *society*, which has invented the *economy*, to serve its needs. The combination of the climate-change trends, and our re-thinking following the 2008–2009 financial crisis, perhaps usefully reminds us of this dependence: it is the laws of nature that are unchangeable, while society's behaviour can be changed, albeit with difficulty; and economic rules can be re-invented, if we wish, better to fit our objectives.

As Figure 1.2(b) shows, engineering **infrastructure** provides the critical interface between society and the environment. It is only infrastructure that allows humans to live in the large numbers and concentrations that urbanisation dictates. It works both ways: infrastructure shelters humans from the environment's risks, and allows us to exploit it for our benefit, but it also protects the environment from our wastes.

A useful representation by Oxfam, combines the first two fundamental principles – sustainable environmental limits and minimum socio-economic development standards – in one 'donut' radar plot (Figure 1.2(c)), to define the 'safe and just space on our planet' for humanity. The inner boundary represents the 'social foundation' target for development – satisfying 11 needs of any community, in terms of minimum quantities of food, water and energy, as well as the right to education, gender and social equality, and employment. The upper boundary is the 'environmental ceiling', representing the limits of the nine planetary boundaries (Rockström *et al.*, 2009) that we must work within to maintain the essential natural support systems of the planet.

These evolving ways of representing sustainability show how the understanding of the concept is continually being refined, while retaining the fundamental core components. The Natural Step Framework's (see Appendix A) metaphor of a physical funnel (Box 1.3) may appeal to engineers, and help visualise how these combined pressures

Figure 1.2 Representations of sustainable development: (b) Parkin (2003); (c) Raworth (2012). © Oxfam 2012

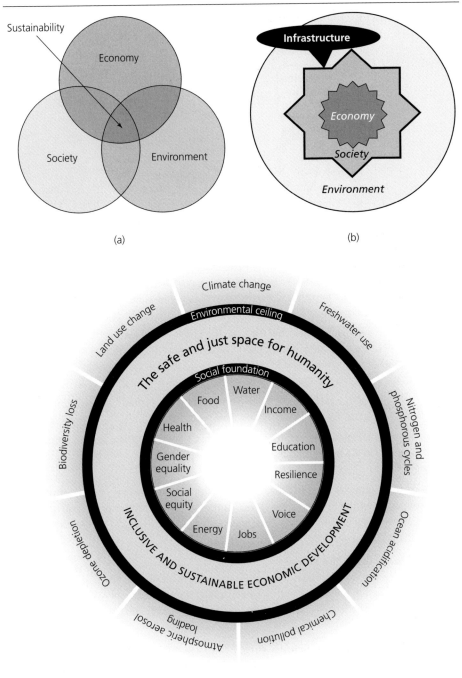

(a)

(b)

(c)

Box 1.3 The Natural Step resource funnel

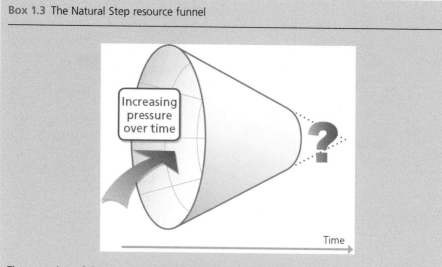

The metaphor of the tunnel (© The Natural Step, 2013)

We use the funnel as a metaphor to help visualise the economic, social and environmental pressures on society that are growing as natural resources and ecosystem services diminish and the size of the population and its consumption grows.

Imagine looking at a giant funnel from the side. The upper wall is the availability of resources and the ability of the ecosystem to continue to provide them. The lower wall is our demand for these resources, which we need to make clothes, shelter, food, transportation and other items, and the ecosystems that create them.

The things we need to survive – food, clean air and water, productive topsoil and others – are in decline. So is nature's ability to regenerate them.

But, at the same time, our demand for these resources is growing. There are more than 6 billion people on the planet, and the population is increasing. Our level of consumption is increasing.

As our demand increases and the capacity to meet this demand declines, society moves into a narrower portion of the funnel. As the funnel narrows there are fewer options and less room to manoeuvre. Organisations that continue business as usual are likely to hit the walls of the funnel, and fail.

Opening the walls of the funnel

Every one of us lives and works in this funnel, and every one of us has the opportunity to be more strategic about our choices and long-term plans. Through innovation, creativity and the unlimited potential for change, we can shift toward sustainability and begin to open up the walls of the funnel.

Forward-looking organisations can position themselves to avoid the squeeze of the funnel, and invest toward opening the walls and creating a truly sustainable and rewarding future. Chambers et al. (2008)

impinge on us, as natural resources diminish, our climate warms and the population grows.

Infrastructure directly impacts all measures of sustainable development, and it is essential for achieving sustainable development goals. All our projects help 'development' either by providing infrastructure serving basic needs and improving quality of life, or by making such services more 'sustainable' environmentally, within the natural limits of the Earth, and often both. Infrastructure practitioners cannot be neutral with respect to sustainable development, as every project makes things better – or worse.

One keen debate around sustainable development and infrastructure increasingly cannot be avoided. It is about which objectives comes first. It assumes that 'if the three pillars (Figure 1.2(a)) are actually incompatible … there is always the chance that we are prioritising one sustainability pillar over another, to suit our own agenda' (Cole, 2012). As the main 'agenda' for infrastructure provision is generally meeting social and economic need, staying within absolute global environmental limits has tended to take second place. So, for instance, we can rightly applaud the clever reductions in environmental impact – compared with past practice – achieved in building for the London 2012 Olympics, while ignoring the real unsustainable extra impact of the whole, very short-term and travel-intensive, idea.

The metric measure used to compare the emissions from various GHGs based on their global warming potential (GWP) relative to that of CO_2 is the carbon dioxide equivalent (CO_2e). For a gas this is derived by multiplying the tons of the gas by the associated GWP. The most recent trends (late 2012) in CO_2e emissions are driving us towards a +4°C world by 2100 (World Bank, 2012). This is a place we do *not* want to arrive at:

> There is a widespread view [among experts] that a 4°C future is incompatible with an organised global community; is likely to be beyond 'adaptation'; is devastating to the majority of eco-systems; and has a high probability of not being stable (i.e. 4°C would be an interim temperature on the way to a much higher equilibrium level). *We should avoid +4°C at all costs.*
>
> (Anderson, 2012)

As we come to accept the 'nested' view of sustainable development (see Figures 1.2(b) and 1.2(c)), we realise that staying within *absolute* environmental limits, and changing this global warming trend becomes more and more imperative. The challenge, then, for us as infrastructure engineers is to accept a new norm: that we will *only* build infrastructure, for any socio-economic purpose, so that it can at the same time *reduce* 'whole-life' environmental limits impacts, compared with building nothing. Encouragingly, we can see increasing evidence that the 'three pillars' are not necessarily incompatible, and that this can be achieved. For example, recent thinking on flood defence now focuses on creating and enhancing natural environments such as storage in wetlands, which are often cheaper than hard defences, and provide amenity value as well as a wide range of ecosystem service functions, together with carbon sequestration and mitigation of urban heat-island effects.

1.5.   What can engineers do? – Be a key part of the solution

Some of the public have long regarded engineers as part of the problem, not the solution. In 2001, the magazine *New Civil Engineer* published a debate around the question: 'Major engineering projects are by their very nature bad for the environment. Can they be justified?' (NCE, 2001).The environmental activist's answer was challenging:

> NO; these firms' quest to boost economic growth, it seems, knows no bounds. ... Almost no claim is too grand for the engineers' vision of the world where Nature is bound in concrete and milked for as much cash as it can yield. And it seems that no ecological price is too high. ... If the Earth is to survive ... then those who shape the process of development must radically change their approach. ... Are there some new civil engineers ready to rise to that challenge – or only old civil engineers with different public relations?
>
> (NCE, 2001)

Note that his challenge goes beyond concern for environmental damage; it reflects a distrust of the socio-economic purpose of the typical project. In the writer's mind, we fail against those first two fundamental principles.

Understanding the total dependence that modern society has on infrastructure, our answer must be that engineers *have* to be a key part of the solution. Our professional ethos, and approach, must incorporate a commitment to achieving sustainable development.

Since ancient times infrastructure engineers have always had a strong sense of public service in their professional ethos. Within the definitions of 'good engineering practice', or of engineering 'value', 'quality' or 'excellence', we have responded to the evolving needs of society, and have developed and incorporated the key innovations and ideas needed to respond to the challenges of the time. For example, ancient world and medieval builders applied local, empirical and experiential 'know-how' to produce structures designed for *strength*, and then for the *permanence* manifest in achievements such as the Pyramids and the great gothic European cathedrals of the Middle Ages.

The setting out of *Newtonian mechanics* facilitated the industrial revolution and enabled a rational 'know-what' approach, leading to a greater *economy of design* using new materials, such as iron in the famous bridge across the River Severn at Ironbridge, UK. In the twentieth century, a creative force influenced engineering, focused on the *aesthetics* of structures, such as the Sydney Opera House, the design of which itself drove early innovation in the use of *computerised structural analysis*. And in the last 30 years a much stronger focus on *health and safety* in engineering and construction, so as not to kill and injure workers, has become an essential part of good practice. During the same period, *environmental issues* have come to be addressed to some extent in most projects, partly through the need to meet compliance with regulations and legislative goals.

As our definition of 'good engineering' has evolved to include new criteria, the tension in design and construction has remained between providing 'safe service performance' and cost-effectiveness. Usually, the service requirement – human safety, or environmental protection, say – has been seen as an overriding driver, to be met at least possible cost. Because there had been no challenge to reduce materials or energy use (except as reflected in cost), our design standards and manuals for 'safe service provision' have tended (in general) to be cautious, somewhat risk-averse and conservative.

In the twenty-first century the new driving criteria in 'good engineering' are to incorporate this broader concept of 'sustainable development' – the need to serve a proper **socio-economic purpose**, and to work within our single planet's **environmental limits** – as a fundamental part of 'safe service performance'. In one sense, this next phase of engineering will not differ from those of the past; we must continue to deliver vital infrastructure services to society within prevailing constraints.

However, in other ways these new drivers make a game-changing challenge to our approach to 'good engineering' practice, for three reasons.

■ The **vicious cycle**: our usual infrastructure solutions, using materials and energy, and thus emitting GHGs, now do not just add to costs; they also warm the planet and actually make the problem we are trying to solve *worse*. Figure 1.3 shows such a cycle, applied to action to reduce local water pollution.
■ The **scale**: to get back within one-planet environmental limits in 'developed' countries we need to achieve up to 80% improvement in our overall resources 'eco-footprint' (Ewing *et al.*, 2010), and an 80–90% reduction in GHG emissions, or 'carbon footprint'. This requires a *transformation* in our approach to

Figure 1.3 Local water quality improvements can increase CO_2 emissions

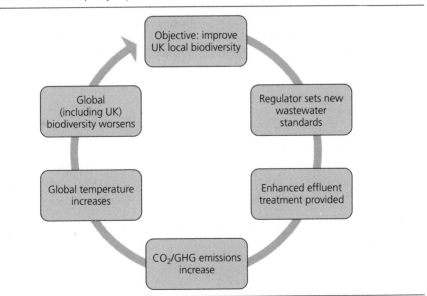

infrastructure engineering, which must achieve more than the underlying steady incremental improvement (Head, 2008).

■ The **complexity**: we cannot simply specify 'sustainability' as an extra engineering requirement, as the comparison in Box 1.4 shows. It can only be defined in the *context* of local, regional and global constraints, and requires satisfying the multiple needs of many diverse stakeholders.

These three characteristics challenge our previous ideas of 'efficiency' and 'service performance', and some of our old ways of solving engineering problems; but they also offer exciting opportunities for innovation. To take these opportunities requires us to examine how to ask the right question at the right time, by examining the opportunities at each stage of project delivery.

1.6. Asking the right question at the right time

Figure 1.4 shows a diagram of the typical stages of infrastructure project planning and delivery. The language used in different sectors, disciplines and contract formats for infrastructure differs (i.e. civil, building, architecture, etc.), but the main generic stages are common. The planning, development. implementation, operation and decommissioning stages shown in Figure 1.4 are as defined (Kershaw and Hutchison, 2009) by the Institution of Civil Engineers (ICE), UK. Within any one of these stages, sustainability can be improved by what you do, and particularly by what questions you ask. In that way, we can all take some responsibility, because *every* infrastructure project contributes to, or detracts from, sustainability (Section 1.4).

Whatever stage of project delivery you are working in, there are opportunities to deliver more sustainable solutions, but, as the value management curve in Figure 1.4 makes clear, the biggest opportunities are often in the earlier stages of the process, and also at the very start of each stage. This is because the output of each stage (e.g. the project brief, say, emerging from scoping) has to be worked within at the next stage, in this case outline design. Therefore the scope for innovation has been constrained and narrowed. However, do not underestimate the difference that can be made even in later stages – the best opportunities to deliver more sustainable infrastructure projects come by asking radical questions right at the start of each stage, before the pressures of 'just get it done' production take over.

Most engineers are not usually involved in the early planning stages, which have the most scope for innovation; often, our first opportunity on a project comes at the start of the outline design stage. This time is critical, because in 'clarifying the scope' it is the *last opportunity* to make many of the, possibly unsustainable, decisions and when opportunities missed earlier in project planning can be questioned.

The case study in Box 1.5 is an example from the water sector, in which early discussions with the client at this point redefined the *scope* of the project, and led to a solution that was 30% of the expected capital cost, and effectively zero carbon emissions. So, asking the right question at the right time enabled innovation that saved money, energy and carbon.

Box 1.4 What is sustainable engineering? Inca vs. Roman bridges

Question: Which bridge is the most sustainable?

| Roman arch bridge (still standing) | Inca grass bridge (in use for 700 years) |

(Source: http://upload.wikimedia.org/ wikipedia/commons/d/d5/ Bridge_Alcantara.JPG. Reproduced under GNU Free Documentation License, Version 1.2, available at http://commons.wikimedia. org/wiki/Commons:GNU_Free_ Documentation_License_1.2)

(Source: http://upload.wikimedia.org/ wikipedia/commons/6/69/ IRBSideViewClip.jpg. Reproduced under Creative Commons Attribution 3.0 Unported license, available at http://creativecommons. org/licenses/by/3.0/deed.en)

Engineering characteristics
- A high load capacity (wagons and legions)
- Design life: lasts for centuries
- Uses locally available materials (stone)
- Low stresses, low maintenance
- High initial cost, reusable materials

The Roman bridge is perfectly suited to an expanding empire, in which the bridge demonstrates permanence. It does not need regular maintenance and so has survived, suggesting to local populations that the Roman Empire is here to stay.

Engineering characteristics
- Low load capacity (people and mules)
- Design life: very short – rebuilt annually
- Uses locally available materials (grass ropes)
- High stresses, high maintenance
- Low initial cost, renewable materials

The Inca grass bridge is reconstructed every year as part of an important festival. Each community member makes a contribution to the effort involved, thus adding to the glue of social cohesion. The maintenance plan is fundamentally tied to the social fabric of the community and would fail if the society ended.

Answer: Both! Because both perform well in the *context* in which they operate. The context of both bridges extends beyond merely the engineering brief of crossing a river or ravine, to the surrounding society for which the service is provided.

Based on concepts developed by Ochsendorf (2003).

Figure 1.4 Opportunities for more sustainable outcomes at typical stages of infrastructure project delivery

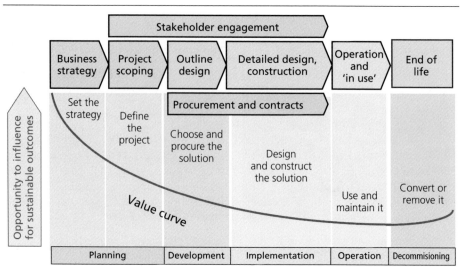

1.7. Dealing with complexity, long timescales and future uncertainty – a mindset change

When we consider 'science' as the basis for engineering we tend to assume we mean Newtonian mechanics. This linear, predictive science is vital to allow us to design and build structures that do not fall down, and infrastructure that operates as expected. It is the key to providing 'stuff that works'. We rightly apply this thinking to the

---

Box 1.5 Redefining the problem – an example

Aim: treatment of spring source water to control cryptosporidium.

Original scope: add extra treatment; chose between membrane or activated carbon options.

Redefined the problem: to manage the risk by protection and improvement of spring source; no extra treatment then needed. Residual risks were evaluated and shown to be acceptable and manageable.

- £1 million capital expenditure; saved 70% of the estimated £3.5 million for extra treatment.
- Near-zero operating expenditure, energy and CO_2e emissions (extra treatment would have led to increases).
- 'Natural' stream course replaced the concrete spring basin; added to habitat benefit.

Result: Saved money and energy, reduced CO_2e emissions.

(Source: MWH)

'middle' stages of project delivery, encompassing design, procurement and construction. It is absolutely essential for this aspect of engineering activity.

With this approach, exercised through clever tools such as building information modelling (BIM) (Crotty, 2011), we can now handle very *complicated* problems – such as the thousands of internal interconnections in a modern building between foundations, structure, cladding, office layouts, furniture, services, energy use and IT systems – with an ability to predict the results, the performance and the way the whole will work.

However, when we are planning and scoping that building, earlier in the delivery sequence, and in its use, at the end, we are working out how we want that building to interface externally. That is, how we want the building to interface with its surroundings, including all the other components of urban areas, and its users and their behaviour. This is the realm of 'wicked' (Brown *et al.*, 2010) problems (see Section 1.3), which have to be examined within the twentieth century science of complex adaptive systems – of how everything is interconnected, in unpredictable ways.

Where unexpected consequences have arisen from infrastructure projects, this has often been because of a failure to think at these early stages in terms of the whole (interactive) system. As former President of the ICE Paul Jowitt observed:

> In the era of technical rationality – which has dominated the last two or three centuries or so – economic and technical progress has generally been embedded in narrow technical disciplines which, despite our scientific understanding, have not anticipated the wider physical and non-physical consequences at the system level.
>
> (Jowitt, 2004)

Examples of unexpected user behaviour include new roads that just generate more traffic; and assistance with household energy saving that saves money, which people then spend on more energy-using appliances. One amusing example (at least in hindsight!) that illustrates the point about hidden interconnections with surrounding systems is given in Box 1.6.

So, when we are at the stage of project planning, scoping and outline design, and of examining performance in use, we need to adopt a systems thinking approach, and follow the recommendation of the many senior engineers who agree with this view, such as Paul Brown, who said in 2008:

> to achieve sustainable urban development we need to do more than improve the efficiency of the component parts of the infrastructure – we need to do so at a systems level
>
> (Brown, 2008)

One response to dealing with the user behaviour interaction is the development of complex behavioural modelling tools, capable of running long-term simulations of how users might respond to different choices of infrastructure design. The example in

> **Box 1.6** Complexity – Operation Cat Drop
>
> ---
>
> 'In Borneo in the 1950s, many Dayak villagers had malaria, and the World Health Organisation (WHO) had a solution that was simple and direct. Spraying DDT seemed to work: mosquitoes died, and malaria declined.
>
> But then an expanding web of side effects started to appear. The roofs of people's houses began to collapse, because the DDT had also killed tiny parasitic wasps that had previously controlled thatch-eating caterpillars. The colonial government issued sheet-metal replacement roofs, but people couldn't sleep when tropical rains turned the tin roofs into drums.
>
> Meanwhile, the DDT-poisoned bugs were being eaten by geckoes, which were eaten by cats. The DDT invisibly built up in the food chain and began to kill the cats. Without the cats, the rats multiplied.
>
> The WHO, threatened by potential outbreaks of typhus and sylvatic plague, which it had itself created, was obliged to parachute fourteen thousand live cats into Borneo. Thus occurred Operation Cat Drop, one of the odder missions of the British Royal Air Force.'
>
> (Source: Conway GR (1969) 'Ecological Aspects of Pest Control in Malaysia',
> quoted in *Natural Capitalism* (Hawken *et al.* (1999))

Box 1.7 shows that optimising at the systems level may involve the use of complex design tools to design and construct very simple technical components.

Another challenge to our traditional mindset is posed by the long timescales, and future uncertainty, that are implicit in the idea of taking 'stewardship' for the interests of future generations. Our 'predict and provide' approach to infrastructure needs to adapt in order to cope with the slow, inevitable but uncertain impacts of, for instance, global warming on infrastructure, resources and urban areas, where forecasting performance on an understanding of past behaviours may no longer be valid.

As one example, some climate scientists think that we have already passed the increased temperature tipping point that triggers the eventual complete melting of the Greenland ice cap (see, e.g., the credible work summarising experts' opinions and published data in *Climate Wars* (Dyer, 2011).) This would mean that, over a few hundred years, a global sea-level rise of about 7 m is inevitable – far too much to cope with in our many large coastal cities around the globe by 'sea defences' or barriers. How would we decide when to stop investing in new infrastructure in those city areas under such a threat, and start planning new or adapted urban areas, with possibly very large population movements involved?

We are only just beginning to grapple with such questions, but they have already led to new concepts that we can apply. These include the ideas that, to deal with multiple interacting and uncertain risks, we should aim for *resilience* in infrastructure; and that to cope with uncertainty over long, slow timescales we should adopt *adaptive management* in our infrastructure planning. (Resilience is the ability of a social or ecological system to absorb

---

Box 1.7 Pedestrian modelling of Oxford Circus

Oxford Circus in London is a busy intersection with more than 200 million visitors a year. Work involved a £5 million project to pedestrianise part of Oxford Circus. Based on crossings in Tokyo, the new design stops all traffic in all directions, and allows people to cross diagonally as well as straight ahead. Street clutter and barriers at the junction of Oxford Street and Regent Street have been removed. The design is based on modelling.

Two-dimensional traffic and pedestrian models were coupled with 3D Studio Max in such a way it is difficult to distinguish the model from real video footage. This makes the finished simulation particularly compelling as a simulation testing various designs.

The block model was used to check the proposed CAD plan design issues in 3D, with vehicles from all angles. (© Crown copyright 2013, courtesy of Atkins.)

---

disturbances while retaining the same basic structure and ways of functioning, the capacity for self-organisation, and the capacity to adapt to stress and change (IPCC, 2007b: 880). Adaptive management is a decision process that promotes flexible decision-making that can be adjusted in the face of uncertainties as outcomes from management actions and other events become better (Walters, 1986).)

1.8.   Am I acting sustainably?

One regular challenge to engineers who want to be 'part of the solution', is to be able to answer this question, and to satisfy not just fellow engineers, or project sponsors, but the wider views of other project stakeholders. In doing so, issues beyond the familiar engineering criteria of time, cost and quality need to be considered, and this involves thinking about the project scoping and actions over wider horizons. These might include issues of scale and boundaries, ethics, context, socio-economic purpose and

Figure 1.5 Wider project horizons for engineers (Fenner *et al.*, 2006)

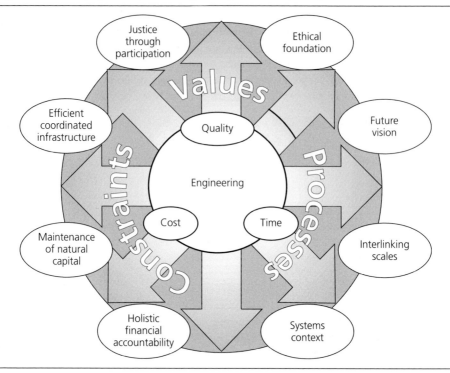

impacts, and future timeframes. These represent an expanded set of choice criteria by which a project is defined and evaluated as 'sustainable'.

In Figure 1.5, these wider horizons are represented by eight elements, named deliberately to avoid labels such as 'environment' or 'stakeholder engagement', which might be dismissed as versions of sustainability that are already being extensively addressed. This wider framework has been developed to stimulate a list of better questions to be asked at every stage of project delivery (Fenner *et al.*, 2006).

They might include the following, for instance.

- How has the engineering process shown respect for people and the environment?
- Has an extended range of options been examined? How have these been documented?
- How is careful and informed material selection ensured and overspecification avoided?

Answers to such questions, asked at the right time, will help more creative and sustainable solutions to emerge. At the very least, even if they do not change the project scope, specifications or procurement, they contribute to a more explicit set of answers to the stakeholder's question: 'How sustainable is this project?'. Furthermore, most infrastructure practitioners belong to a professional institution that has set out codes of practice for ethics and sustainability. These contain good general principles, but it is hard to demonstrate performance against them; these answers may also contribute to that. They could be recorded in a standard 'sustainability ethics' record, which might be attached to each project information document.

It is important at this point to understand that achieving a sustainable infrastructure project is not achieved by following a set of prescriptive rules, or merely running through checklists that reflect generic issues that may have no local relevance. Instead, it is no more than thinking more widely about how solutions are conceived, the impacts they will have, and whether they are fit for that particular place. It simply requires a new set of questions to be asked at the right time.

We hope that the ideas, approaches and questions in this book will help infrastructure engineers give increasingly positive answers to those questions, project by project.

Why sustainability?: summary

Global warming, increasing population, rapid urbanisation and growing environ-mental losses set large new challenges: to provide infrastructure services in innovative new ways.

The four fundamental concepts of sustainable development are: living within Earth's finite limits, helping everyone to achieve an acceptable quality of life, 'stewardship' for future generations, and dealing with complexity.

Infrastructure provides the interface between humans and the environment, so all our projects are 'sustainable development' projects. Each one makes things better or worse – we cannot be 'neutral'. We are a key part of the solution, by embedding the sustainable development approach in our 'good engineering'.

This requires game-changing innovation. We know how to do this, but it needs us to ask new questions, at the right time at each stage of project delivery. The earlier we ask, the better, but we can improve things at every stage.

Dealing with complex systems, long timescales and future uncertainty requires some additions to our mindsets and skill sets, too.

Suggested Further Reading is listed after References.

REFERENCES

Allen R (1980) *How to Save the World: Strategy for World Conservation*. Kogen Page, New York.

Anderson K (2012) *Real Clothes for the Emperor: Facing the Challenges of Climate Change*. Cabot Institute Annual Lecture, University of Bristol. Available at: http://www.bristol.ac.uk/cabot/events/2012/194.html (accessed 22 July 2013).

Brown LR (1981) *Building a Sustainable Society*. WW Norton, New York.

Brown P (2008) *Cities of the Future*. Keynote address, IWA World Water Congress, Vienna.

Brown VA, Deane PM, Harris JA and Russell JY (2010) *Tackling Wicked Problems Through Transdisciplinary Imagination*. Earthscan, London.

Bruntland GH (ed.) (1987) *Our Common Future: The World Commission on Environment and Development*: Oxford University Press, Oxford.

Butchart SH, Walpole M, Collen B, van Strien A, Scharlemann JP *et al.* (2010) Global biodiversity: indicators of recent declines. *Science* **328(5982)**: 1164–1168.

Carson R (1962) *Silent Spring*. Houghton Mifflin, Boston, MA.

Chambers T, Porritt J and Price-Thomas P (2008) *Sustainable Wealth Creation within Environmental Limits*. Forum for the Future, London.

Cleveland AB (2012) *Sustaining Infrastructure*. A Bentley White Paper. Available at: http://ftp2.bentley.com/dist/collateral/whitepaper/Whitepaper_Sustaining_Infrastructure_eng.pdf (accessed 22 July 2013).

Cole M (2012) Deconstructing the case for sustainability. *New Civil Engineer* 12 July: 10–11.

Crotty R (2011) The Impact of Building Information Modelling: Transforming Construction. Routledge, New York.

Dyer G (2011) *Climate Wars: The Fight for Survival as the World Overheats*. Oneworld, London.

Ewing B, Moore D, Goldfinger S, Oursler A, Reed A and Wackernagel M (2010) *Ecological Footprint Atlas 2010*. Global Footprint Network, Oakland, CA. Available at: http://www.footprintnetwork.org/images/uploads/Ecological_Footprint_Atlas_2010.pdf (accessed 22 July 2013).

Fenner RA, Ainger C, Guthrie P and Cruickshank HJ (2006) Widening horizons for civil engineers – addressing the complexity of sustainable development. *Proceedings of the ICE – Engineering Sustainability* **159(ES 4)**: 145–154.

Global Coral Reef Monitoring Network (2008) *Status of Coral Reefs of the World 2008*. Available at: http://gcrmn.org/gcrmn-publication/status-of-coral-reefs-of-the-world-2008 (accessed 26 July 2013).

Global Footprint Network (2013) *The National Footprint Accounts*, 2012 edition. Global Footprint Network, Oakland, CA, USA. See 2013 graph at: http://www.footprintnetwork.org/en/index.php/GFN/blog/human_development_and_the_ecological_footprint, accessed 7th October 2013.

Hawken P, Lovins A and Lovins H (1999) *Natural Capitalism: Creating the Next Industrial Revolution*. Little, Brown and Company, Boston, MA.

HDI (n.d.) *Human Development Indices*. Available at: http://hdr.undp.org/en/media/HDI_2008_EN_Tables.pdf (accessed 22 July 2013).

Head P (2008) *Entering the Ecological Age – The Engineer's Role* – Brunel Report 2008. Available at: http://www.ice.org.uk/Information-resources/Document-Library/Entering-the-ecological-age–the-engineer-s-ro-(1) (accessed 22 July 2013).

IPCC (2007a) *Summary for Policymakers. Climate Change 2007: The Physical Science Basis* (Solomon S, Qin D, Manning M, Chen Z, Marquis M, Averyt AB, *et al.* (eds)). Contribution of Working Group I to the Fourth Assessment Report of the Intergovernmental Panel on Climate Change. Cambridge University Press, Cambridge.

IPCC (2007b) *Climate Change 2007: Impacts, Adaptation and Vulnerability* (Parry ML, Canziani OF, Palutikof JP, van der Linden PJ and Hanson CE (eds)). Contribution of Working Group II to the Fourth Assessment Report of the Intergovernmental Panel on Climate Change. Cambridge University Press, Cambridge.

Jowitt P (2004) Sustainability and the formation of the civil engineer. *Proceedings of the ICE – Engineering Sustainability* **157(ES 2)**: 79–88.

Jowitt P (2009) *Now is the Time. Presidential Address 2009.* Available at: http://www.ice.org.uk/Information-resources/Document-Library/Presidential-Address-2009 (accessed 22 July 2013).

Kershaw S and Hutchison D (eds) (2009) *Client Best Practice Guide.* Thomas Telford, London.

Leopold A (1949) *A Sand County Almanac.* Republished in 2006, Ballantine Books, New York.

Lovelock JE and Margulis L (1974) Atmospheric homeostasis by and for the biosphere: the Gaia hypothesis. *Tellus, Series A* **26(1–2)**: 2–10.

Mawhinney M (2002) *Sustainable Development: Understanding the Green Debates.* Blackwell Science, New York.

Meadows DH, Meadows D, Randers J and Behrens III WW (1972) *The Limits to Growth.* Universe Books, New York.

NCE (2001) Major engineering projects are by their very nature bad for the environment. Can they be justified? *New Civil Engineer*, 6 December 2001.

Ochsendorf J (2003) The role of bridges in the Inca Empire. Harvard Department of Anthropology, Graduate Seminar in Andean Studies.

Parkin S (2003) Sustainable development: the concept and practical challenges. *Proceedings of the ICE – Civil Engineering* **138(6)**: 3–8.

Raworth K (2012) *A Safe and Just Space for Humanity: Can we Live Within the Doughnut?* Oxfam International, Oxford.

Rittel H and Webber M (1973) Dilemmas in a general theory of planning. *Policy Sciences* **4**: 155–169. [Reprinted in N. Cross (ed.) (1984) *Developments in Design Methodology.* Wiley, Chichester, pp. 135–144.]

Rockström J, Steffen W, Noone K, Persson Å, Chapin III FS, Lambin E *et al.* (2009) Planetary boundaries: Exploring the safe operating space for humanity. *Ecology and Society* **14(2)**: 32.

The Natural Step (2013) *The Funnel – Society is Being Squeezed.* Available at: http://www.naturalstep.org/en/the-funnel (accessed 22 July 2013).

Walters CJ (1986) *Adaptive Management of Renewable Resources.* McGraw Hill. New York.

Worldwatch Institute (2009) *State of the World 2009: Into A Warming World.* Worldwatch Institute, Washington, DC.

Worldwatch Institute (2010) *Vital Signs.* Worldwatch Institute, Washington, DC.

World Bank (2012) *New Report Examines Risks of 4 Degree Hotter World by End of Century.* Available at: http://www.worldbank.org/en/news/2012/11/18/new-report-examines-risks-of-degree-hotter-world-by-end-of-century (accessed 22 July 2013).

FURTHER READING

Allenby B (2011) *The Theory and Practice of Sustainable Engineering.* Pearson Education, London.

Novotny V and Brown P (2007) *Cities of the Future, Towards Integrated Sustainable Water and Landscape Management.* IWA Publishing, London.

Pollalis SN, Georgoulias A, Ramos SJ and Schodek D (eds) (2012) *Infrastructure Sustainability and Design.* Routledge, New York.

Randers J (2012) *2052: A Global Forecast for the Next 40 Years.* Chelsea Green Publishing, White River Junction, VT.

Visser W, on behalf of Cambridge University Programme for Sustainability Leadership (2009) *Landmarks for Sustainability Events and Initiatives that Have Changed our World.* Greenleaf Publishing, Sheffield.

Sustainable Infrastructure: Principles into Practice
ISBN 978-0-7277-5754-8

ICE Publishing: All rights reserved
http://dx.doi.org/10.1680/sipp.57548.027

Chapter 2
Key principles

Important principles may, and must, be inflexible.

Abraham Lincoln

2.1. The role of principles

Engineers understand the concept of physical principles, such as conservation of mass, or Newton's laws of motion. They provide the ideas, rules, or concepts that you need to keep in mind when solving an engineering problem. Designers also adopt principles that help guide their thinking, from 'keep it simple stupid' to 'keep the target user in mind'. Similarly, in order to test that you are acting sustainably, you need to be able to set your choices and engineering decisions against guiding principles for sustainability. These provide a touchstone against which you can test your actions. A good example of this is the principle within Aldo Leopold's famous land ethic:

A thing is right when it tends to preserve the integrity, stability, and beauty of the biotic community. It is wrong when it tends otherwise.

(Leopold, 1949)

Many sustainability principles and measures have been suggested. There are ongoing debates, and sometimes esoteric arguments, surrounding the focus that different principles choose to emphasise: 'deep green' or 'light green', weak or strong sustainability, ethics and/or economics. Their history and a philosophical perspective is given by Simon Dresner (2008); and they have been usefully reviewed and summarised by Edwards (2005), who sees strong similarities in the values these principles express. He summarises the key themes they embody simply as

1   stewardship
2   respect for limits
3   interdependence
4   economic restructuring
5   fair distribution
6   intergenerational perspective
7   nature as a model and teacher.

He points out that such principles play a key role in setting the context for the choices that organisations make, and that, while some focus strictly on values, others include a defined methodology or standard for implementation. If these can be attached to the key sustainability themes, you can then use them to help and guide sustainable decision-making at all stages of infrastructure project delivery.

Through the United Nations Environment Programme (UNEP) and the International Organisation for Standardisation (ISO) there is an active process 'aiming at the formulation of international and European standards for the assessment and declaration methods of sustainability aspects of buildings and building products' (Hakkinen, 2009). So far this has developed ISO 15392 : 2008, which identifies 'general principles for sustainability in building construction ... as it applies to the life cycle of buildings and other construction works, from their inception to the end of life'. This makes a start but, as with other proposals, it only provides some rather general cornerstones for sustainable decision-making; the challenge we face is how to 'operationalise' these for practice.

An example of a principle at the operational level is the often-referred to 'precautionary principle'. One expression of this is from the Rio Conference in 1992:

> Where there are threats of serious irreversible damage, lack of full scientific
> certainty shall not be used as a reason for postponing cost-effective measures to
> prevent environmental degradation.
>
> Rio Declaration on Environment and Development (UN, 1992)

Another example of a practical sustainability principle, very relevant to infrastructure, is 'resource efficiency'. One operational application of this is to waste management and disposal, through the well-known 'waste hierarchy', as shown in Box 2.1.

So our search here is for a set of *principles* from which we can derive some practical, realistic *practice* to apply to infrastructure planning and delivery. We develop these in the next section.

2.2. A structure for sustainability principles relevant to infrastructure

The range of different sustainability principles that abound can be confusing, rather than helpful. Some well-known ones have little relevance to infrastructure, while some less quoted ones are critical to it. So, to establish the principles that are important in as straightforward and clear a way as possible, we have set out our own structure, but have also considered others' well-established principles in relation to it.

This structure adopts three levels on which to set the principles you need to apply to good engineering practice for infrastructure. The levels are described below.

■ **Absolute principles**: these are incontrovertible and are consequences of natural science laws and basic humanity. They express the four definitions already introduced (see Section 1.3), which are at the heart of driving sustainability. In Figure 1.5 they represent the *constraints* within which you need to deliver infrastructure services.

Box 2.1 The waste hierarchy – Sydney example

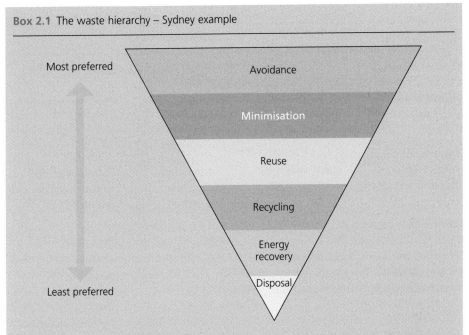

Waste hierarchy (City of Sydney, 2012). (Reproduced courtesy of the City of Sydney.)

'The waste hierarchy has been adopted as a core principle of environmental protection by most Australian states. It ranks ways of dealing with waste, with avoidance being the most preferred and disposal the least.

The City's resource recovery approach also follows the waste hierarchy. The most preferred outcome is to avoid waste being created. The City does not have direct control over what materials people buy and throw away, but it can influence the community and industry to eliminate unnecessary waste.'

- **Operational principles**: these are more specific, to help set objectives and guide your actions in day-to-day practice. They help establish a distinct way of doing things, in part by recognising issues that traditionally may not always have been within the engineer's direct remit. In Figure 1.5, they guide the *processes* that you need to adopt at appropriate stages of each project.
- **Individual principles**: it is as individuals, acting alone or collectively, that you make decisions, and can influence a project outcome to be more sustainable. In Figure 1.5, these draw from the *values*-based aspects of sustainability, reflecting your professional and personal ethos. They cut across all four absolute principles, and reflect ethical and professional responsibility and also individual personal behaviour.

Table 2.1 Principles for sustainable infrastructure

Objectives, goals		Approaches	
Absolute principles			
A1 Environmental sustainability – within limits	A2 Socio-economic sustainability – 'development'	A3 Intergenerational stewardship	A4 Complex systems
[Edwards' list of principles*]			
[Respect for limits] [Economic restructuring]	[Fair distribution]	[Stewardship] [Intergenerational perspective]	[Interdependence] [Nature as model and teacher]
Operational principles			
O1.1 Set targets and measure against environmental limits	O2.1 Set targets and measure for socio-economic goals	O3.1 Plan long term	O4.1 Open up the problem space
			O4.2 Deal with uncertainty
O1.2 Structure business and projects sustainably	O2.2 Respect people and human rights	O3.2 Consider all life-cycle stages	O4.3 Consider integrated needs
			O4.4 Integrate working roles and disciplines
Individual principles			
I1 Learn new skills – competences for sustainable infrastructure			
I2 Challenge orthodoxy and encourage change			

* Edwards (2005)

This gives us the structure shown in Table 2.1. The top absolute level contains the four key principles from Section 1.3, and illustrates below this how they cover Edwards' seven common themes. Next down, the operational level sets out, under each column, the principles that we propose to allow you to apply the absolute ideas to infrastructure delivery practice. Lastly we suggest individual principles – in effect, competences – that you need to use, to do this.

Within these proposed operational principles fit many of the established general sustainability *principles* and *tools* which are recognised in the literature. We have not mapped these onto the table here to show where they fit, to avoid confusion. They have their own literature, which you can follow up if necessary; we have listed them, with a brief description and references, in Appendix A. In addition, those practices that provide *tools* relevant to infrastructure (such as life-cycle analysis, relevant to 'consider all life-cycle stages') are described in Chapter 13.

These key principles are expanded in more detail in the later sections of this chapter. They form the core structure upon which we will suggest sustainability practice – ways in which they can be regularly embraced and adopted in infrastructure projects – in Part II.

2.2.1 Linking the four absolute principles for infrastructure

Infrastructure provides the interfaces to allow human society to have a given quality of life (Principle A2) (see Figure 1.2(b)) within the limits of the Earth's environment (Principle A1), and so infrastructure projects are usually driven by A2, but cannot avoid impacting A1. They are linked through the global environmental limits impact equation, and the terms in that equation reflect both the A1 and A2 objectives, but also the A3 and A4 approaches.

This so-called 'IPAT' equation (Figure 2.1) links the factors that contribute to the problems we are facing – growth in consumption, overpopulation and the state of technology (Ehrlich and Holdren, 1971). At a global level, this is easy to grasp: overall environmental impact is a function of the total population, of our average level of affluence, and of the average environmental efficiency – use of resources, energy, land and waste per unit – of the technology used to provide that affluence. *T* is reduced as technology becomes more efficient, and also reflects non-technological issues such as product reuse and the organisation of production (Muldur, 2006).

The equation can also be applied at a country level, as was seen in Figure 1.1. On that graph, the horizontal position of each country's point is a measure of its 'affluence', measured in this case by human development index (HDI); how far it is up the vertical scale, or its ecological footprint (per person), is a measure of the resulting environmental 'impact' per head of 'population'. Countries that have achieved the same HDI but at a lesser eco-footprint are using a more sustainable technology approach, with an average lower environmental impact, to provide that affluence. The vertical spread of countries' eco-footprints on the graph, for the same HDI, shows that these 'technology' choices

Figure 2.1 How infrastructure links and involves all four fundamental principles

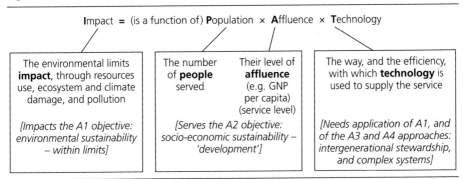

The environmental limits **impact**, through resources use, ecosystem and climate damage, and pollution	The number of **people** served	Their level of **affluence** (e.g. GNP per capita) (service level)	The way, and the efficiency, with which **technology** is used to supply the service
[Impacts the A1 objective: environmental sustainability – within limits]	*[Serves the A2 objective: socio-economic sustainability – 'development']*	*[Needs application of A1, and of the A3 and A4 approaches: intergenerational stewardship, and complex systems]*

vary widely, and are by no means just fixed by some global average state of technical invention.

- Of countries just attaining the threshold HDI of 0.800, the eco-footprint ranges from a near-sustainable 2.0 to 6.00 – a range of three times in the technology efficiency (in its broadest interpretation) applied.
- Of well-developed countries with an average HDI of 0.900, the eco-footprint ranges from 4.5 to 11.00 – a range of 2.4 times in technology efficiency.

The equation also applies to you, in every infrastructure organisation, its strategy, and the projects you build to implement it. Each country 'point' on the graph is merely the sum of every decision made by that country's government, infrastructure providers, companies and engineering practitioners. The wide range of technology efficiencies that results is partly due to each country's geography and history, but also shows how much influence your own decisions can have. The notes in the boxes below the equation in Figure 2.1 show how, in every project, you actually decide on the values to put in each term on the right-hand side of the IPAT equation.

- P – How many people (customers/users), and who, are we serving?
- A – What level of affluence (satisfied socio-economic need) will our infrastructure provide to those customers/users through our chosen level of service?
- T – What technology environmental efficiency (in the broadest sense of the word) will we use to provide that service, and how effective can we make it?

P and A determine *demand*; they answer the *socio-economic* development question (A2): What social need (level of service) are we serving, and for which group of people? T determines how demand will be *supplied*; it answers the key *socio-economic* sustainability question of cost, and the *environmental limits* sustainability questions (A1) of 'effectiveness', as defined by eco-efficiency, including carbon efficiency. Together these answers determine the impact – I, on environmental limits, at global, regional and local scales – of the strategy and the project. As the right-hand box shows, you need to now learn to greatly improve T, through both incremental and game-changing innovation, helped by a new focus on A1 – respect for limits, and better understanding through applying the approaches in A3 – intergenerational stewardship, and A4 – complex systems. Thus all four fundamental principles are linked, in every infrastructure project.

Past technology practice has delivered major socio-economic improvements, lifting many people out of poverty and enhancing quality of life – but at the price of environmental impacts which we can now see are far too high for the Earth's carrying capacity. The IPAT equation, interpreted at a global level, shows how all three terms on the right-hand side, multiplied together, determine overall impact. Population control – in terms of growth rates and total numbers – is absolutely critical; as is questioning how high a level of consumption-driven affluence societies really need (Jackson, 2011; Wilkinson and Pickett, 2010). You can influence these two through your private actions as individuals, and as citizens and consumers. (Note, however, that provision of the right infrastructure, for education, health and enhancing the position of women is the most effective way, apart from draconian government control, to reduce birth rates, and thus population growth. So infrastructure can contribute to this, too.)

In your role as infrastructure engineers, on each project your key opportunity is both to challenge the real need in the 'demand' definition (the $P \times A$ terms), and to change the technology (in its broadest sense) that you use in order to gain more socio-economic (A2) service benefit for less environmental impact (A1) and at less cost – or, in that well-known phrase, to do more with less. To achieve that, we have defined and quantified the fundamental principles, and developed the operational principles shown in Table 2.1, to drive your innovative sustainability *practice*. We develop these in the following sections.

2.3. Principle A1: Environmental sustainability – within limits

Before we can operationalise A1, we need to define better what environmental limits we mean, and how to quantify them. The measure we referred to in Figure 1.1 was 'eco-footprint', and this covers many of the aspects needed (see Appendix A). It has been applied at country and city level in practical ways; and, in a few cases, it has also been used at infrastructure strategy or project level.

To ensure a more comprehensive view, we can build on the work by the Stockholm Environmental Institute (SEI), which produced the nine 'Earth-system boundaries' used to define the outer 'environmental ceiling' shown in Figure 1.2(c). We now use these to define and quantify our A1 absolute objective.

In this definition, the concepts of planetary *thresholds* and *boundaries* are critical. They can be explained in terms similar to those of *failure stress* and *design stress* in structural engineering. Thresholds define trigger points when planetary-scale Earth systems, such as the atmosphere and its climate processes, can switch into non-linear positive feedback activity, leading to runaway change or collapse. For example, in climate science, such real points are carbon dioxide (CO_2) concentrations causing temperatures high enough to trigger the complete loss of the summer arctic ice, to complete melting of the Greenland ice sheet, the loss of the carbon sinks of the Amazon rainforest, or the release into the atmosphere of solid methane hydrates stored in frozen tundra and seabed deposits. These are analogous to the point reached at *failure stress* when an over-stressed structural component actually starts to fail and collapse.

Structural engineers avoid this danger by applying a *factor of safety*, to define a *design stress* to work to, which is safely clear of the *failure stress*, given the knowledge of the uncertainties involved. In a similar way, the SEI suggests planetary *boundaries*, as working limits safely far enough away from their *thresholds*, also taking into account uncertainty. Of course, the difference at the planetary scale is that the impacts of getting it wrong are world changing, and the uncertainties vary, but are often very large – but the principle of designing for safety is exactly the same, so engineers should be able to accept it.

Using this concept, Table 2.2 shows the nine Earth-system boundaries suggested by the SEI (Rockström *et al.*, 2009). The table summarises key facts from the SEI's paper, and

Table 2.2 Proposed planetary boundaries and infrastructure measures for principle A1
Source: Rockström et al. (2009)

Earth-system process	Scale	Key factor controlling it	Suggested boundary limit	Proposed infrastructure measure	Why measure is different from the SEI's – implications for 'practice'
Boundaries with *direct* impacts from infrastructure projects – construction and operation					
Climate change*	Global – threshold known	Atmospheric CO_2 (ppm)	350 ppm [already exceeded]	CO_2e emissions, (tonnes CO_2e)	For projects, use this calculable *output*, not the wider *outcome*
Land-use change*	Regional – threshold unknown	% of global land cover converted to crops	<15% of global ice-free land surface	Land area *used* (ha) – directly; and surrounding area *influenced* – indirectly Plus an ecosystem quality and wilderness measure assessing its biodiversity	Is an available project figure; can set targets / It is hard to assess an individual project's biodiversity impact, but we can assess relative biodiversity value of land used or provided, and seek habitat enhancement to replace losses; seek habitat enhancement, to mitigate loss
Biodiversity loss*		Extinction rate, per million species per year (E/MSY)	<10 E/MSY		
Freshwater use*		Consumptive 'blue' water use, km³/year	<4000 km³/year, globally	Net water use/year (megalitres/year)	Direct *output* measure used by project
Chemical pollution		Concentrations of persistent organic pollutants – (POPs)	To be determined	All POPs emitted/year, separately (loads and concentration)	Direct project *output* measure – minimise and comply with standards

Earth-system processes	Scale	Key factor controlling it	Suggested boundary limit	Infrastructure project approach, for – 'practice'
Boundaries with *indirect* impacts from infrastructure projects – construction and operation				
Ocean acidification	Global – threshold known	Carbonate ion concentration, average global surface ocean saturation state	Sustain 80% of pre-industrial saturation of surface ocean	Identify, measure and minimise use of chemicals and emissions affecting these boundaries, in all stages of projects – from materials, their manufacture, transport and operation
Atmospheric aerosol load	Part global, part regional – threshold unknown	Overall atmospheric particle concentration – regional basis	To be determined	When countries mandate limits or reductions, follow them
Stratospheric ozone (O_3)		Stratospheric O_3 concentration in Dobson units (DU)	< 5% reduction from pre-industrial level of 290 DU	
Global phosphorus and nitrogen cycles		Phosphorus: inflow of phosphorus to ocean – increase compared with natural weathering. Nitrogen: amount of N_2 removed from atmosphere for human use (Mt N/year)	Phosphorus: max to be <10 times the natural background. Nitrogen: limit industrial and agricultural fixation of N_2 to 35 Mt N/year	

* Processes included in ecological footprint analysis

adds our proposals on how infrastructure projects should respond to them. The boundaries fall into two groups, shown separately in the table, and with different forms of response

- those which play a significant part, and are most *directly* caused by the construction and operation of infrastructure projects – such as CO_2e emissions and land take. These need specific recognition, and often target setting
- those which play a smaller part, but still may be *indirectly* affected by the construction and operation of infrastructure projects – chemical use such as nitrogen and phosphorus, organic compounds, aerosols and particulates.

For the directly impacted boundaries, the second from right column proposes measures that you should use to set targets and measure impacts. These differ from the SEI measures because those are mainly large-scale *outcomes*, whereas infrastructure engineers need to set targets and measure in terms of project *outputs*. For the indirect boundaries, our more general proposal is that you identify, measure and minimise the use of relevant chemicals and emissions. Broadly, the first four, asterisked, boundaries in Table 2.2 are the ones included in 'ecological footprint' (see Appendix A), the measure that we used in Chapter 1.

If the global resource use (and pollution) is kept below the thresholds of the nine processes (green inner circle in Figure 2.2) the Earth will remain in the *Holocene*, the current geological era that enabled the development of humans. On the other hand, if this 'safety zone' is surpassed the world may transit into the *Anthropocene*, a geological era where humans shape the environment and its ecosystems. This may trigger negative and irreversible feedbacks, with huge uncertainties for the future development of humans.

Although incomplete, as the indicators for two of the nine subsystems still have to be quantified, Rockström *et al.*'s framework usefully identifies the global environmental challenges, and where major interventions are necessary. Figure 2.2 shows that the current situation is already outside the 'safety zone' for three of the nine subsystems – namely, climate change, biodiversity loss and the nitrogen cycle. Furthermore, it shows that the critical boundaries for ocean acidification and the global phosphorus cycle may soon be passed.

2.3.1 Operational principle O1.1
– Set targets and measure against environmental limits

This principle is applicable to your practice, through every stage of project planning and delivery described in Part II, from business strategy and infrastructure project scoping, through design and construction, into operation and end of life.

The previous section discussed global-scale environmental limits, while recognising that there may also be regional or local ones that you need to respect. All such limits will help you set sustainability objectives and targets for infrastructure projects. They differ greatly in their significance – depending on the *scale*, *duration* and *reversibility* of the impacts they reflect. A summary of the major differences in impacts is given in Table 2.3.

Figure 2.2 A safe operating space for humanity. (Adapted from Rockström *et al.* (2009))

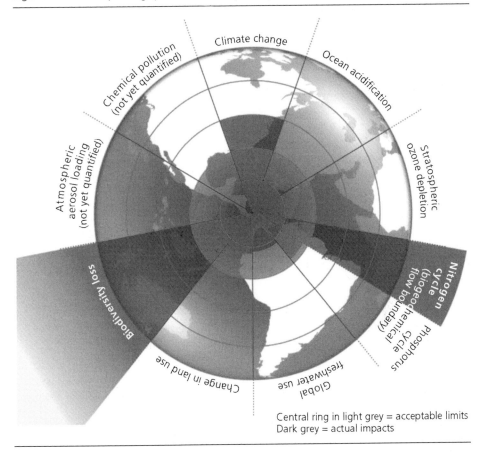

Central ring in light grey = acceptable limits
Dark grey = actual impacts

The table shows how different are each impact's combination of scale and duration; so they need to be dealt with differently.

- **Very long-term impacts** are irreversible, or would take more than a single lifetime to reverse. They reflect the planetary boundaries discussed above; and their duration exceeds, often greatly, the operational life of the project. Although this is about environmental impacts, we have included here human death and disability, to highlight the non-acceptability of such irreversible impacts. Just as respecting health and safety principles reflects that human death or disability is not tolerated as an inevitable price to pay on construction projects, so we now say that irreversible environmental impacts are unacceptable too. If they are not regarded in this way, they will kill far more people over time than site accidents. Within this category, there are:
  - **Global impacts** contribute to effects that affect the whole balance of planetary systems; for example, damage from temperature rise driven by CO_2e emissions to the atmosphere, or the loss of a globally significant ecosystem such as the

Table 2.3 Scale and duration differences in environmental (and some health) impacts*

Scale	Duration		
	Very long-term (~100 years) or *irreversible* impacts	Operational life impacts (~5–50 years)	Construction stage impacts (~1–5 years)
Global	Damage from climate change due to CO_2e emissions Loss of species, soils or ecosystems which may be key to global systems		
Regional	Loss of regionally key species, soils, ecosystems Loss of inhabited sites, with people moved Loss of 'wild land',† key 'heritage' assets, or key archaeology Radiation impacts, including nuclear waste	Land-use and soil changes Damage or pollution to ecosystems or water resources Materials quarrying and extraction impacts Loss of 'tranquillity'‡ in surrounding land, due to visual intrusion, and noise, light, dust and air pollution	
Local and community	Local deaths or disabilities caused by project in use		Noise, dust, light, traffic Air, soil and water pollution
Project site	Fatal or disability – causing accidents on site		Site accident injuries Dirt on the road

* Impacts in each box also apply to the 'scale' boxes below
† 'Wild land' is 'large areas with spectacular scenery and high wildlife value ... very little evidence of human activity ... typically include mountains, tracts of blanket bog, river margins and rugged coastlines. ... A connection with nature and a sense of wildness can be found in many places, often within or close to urban areas. These wild places are where most people have the opportunity to enjoy, value and care for nature.' (John Muir Trust, n.d.)
‡ Tranquil areas are 'places which are sufficiently far away from the visual or noise intrusion of development or traffic to be considered unspoilt by urban influences' (CPRE, 2007)

Amazon rainforest. These trends would take centuries to reverse, even if we knew how, and the damage caused in the meantime is irreversible. They are quite literally 'world changing'. The logical response is to apply the precautionary principle (see Appendix A) and set absolute preventive boundaries, not to be exceeded, which any project must not breach. ('The precautionary principle says, in effect, that because the stakes are so high, we have to weigh even the most dramatic benefits against the prospects of even more destructive consequences. ... When the whole world is at risk because of the scale of human intervention, then a new scientific approach is required that takes the whole world into consideration' (Rifkin, 2004).)

- **Regional or more local impacts** include the displacement of people needed for some dam constructions, which are effectively irreversible in their human-life implications, even if notionally the dam could be removed. Also the half-life of radiation, or the time to decommission and recover a site used for nuclear power generation, is very long term; and the loss or disabling of human life is, of course, irreversible (although in detached economic valuations, such life is replaceable by others). However, these impacts are not world changing, and in some cases (other than death or disability) the benefits of a project may justify accepting some impacts.

■ **Operational life impacts** last for 5–50 years (or more), depending on the design life of the asset. They are large, regional in scale, and either continuous over a long-term impact or separate repeating impacts. However, they are (notionally at least) reversible or replaceable at end of life, by the physical removal of the assets. They should not be regarded as inevitable. Projects that cause such impacts should also include building directly mitigating components (e.g. installing noise barriers, creating new ecosystems or forestry) to minimise the 'net' impact.

■ **Construction-stage impacts** last typically 1–5 years, matching the construction period. They are mainly local in nature, and are continuous or separate repeating impacts during that period. An example that creates serious problems is the creation of extra construction traffic. These impacts end when construction is completed, and then recovery from their effects can happen. They are, to some extent, inevitable, and must be minimised and mitigated by careful attention during design and construction.

In densely populated countries such as the UK, infrastructure projects can have a particular impact on such landscape qualities as wilderness and tranquillity, which may generate less pressing concerns in more expansive and remote areas.

Because these qualities each represent a subtle combination of separate attributes, it is difficult to quantify them exactly, although techniques such as hedonic pricing and contingent valuation (see Section 13.6) can reflect how local communities and other beneficiaries attach importance to them. Figure 2.3 shows the huge loss of 'tranquillity' in the UK between the 1960s and 2007. We all must give more attention to keeping both wild and tranquil places, because they make a vital contribution to human quality of life, are important as wildlife sanctuaries, and are wonderful to appreciate when you experience them.

Principle O1.1 requires that these limits and impacts, with their varied significance, must be taken into account in our *practice*. They should be reflected as business strategy objectives, in project scoping, and in choosing solution options in design and construction (see Part II). Other relevant concepts, measurement tools and indicators are summarised in Part IV.

Figure 2.3 Loss of 'tranquillity' in the UK

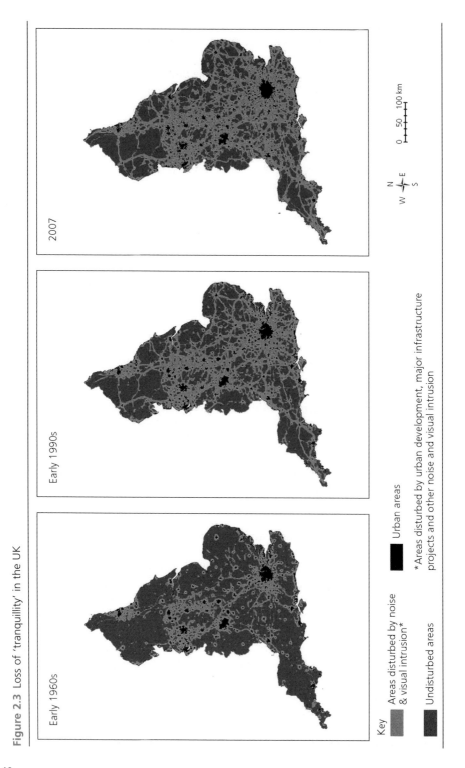

2.3.2 Operational principle O1.2
– Structure business and projects sustainably

This principle is applicable particularly to your practice in the early stages of project planning and delivery described in Part II; in setting business strategy (see Chapter 3) and scoping infrastructure projects (see Chapter 4), and sometimes also in outline design (see Chapter 7).

The standard linear model of products in most economies – buy, use, dispose – coupled with an economic system demanding growth as a frequently unquestioned mantra, leads to a fundamentally unsustainable system. It results in a tendency to overconsumption, using ever more resources, and emitting ever more pollution and CO_2e, with no regard for natural limits. Applied, perhaps unconsciously, to scoping an infrastructure project, it can lead to an overreliance on building new infrastructure to meet any need, when a more innovative solution would be to work out how to continue providing the service within existing physical capacity (e.g. soft solutions for traffic regulation). Too often there is an acceptance that some environmental damage and social cost are inevitable, to be dealt with by adding mitigation costs and, maybe, social compensation afterwards. This always, of course, adds cost to projects, and such past practice has led to the often wrong, but pervasive, assumption that 'being more sustainable always costs more money'.

Even with such a 'product' business model and project structure, you can strive for resources and energy efficiency in design, construction and operation, but you are working with one hand tied behind your back. To free up that other hand, you can structure your business and projects more sustainably in the first place, earlier in the delivery sequence. This means adopting *service* business models that naturally align commercial success with less resource and energy use, and consider project scoping within a *sustainability hierarchy* that naturally favours more sustainable solutions. These approaches build in, from the start, drivers that naturally will favour more sustainable solutions, before getting into the detail.

2.3.2.1 Use a service, not product, business model

The traditional infrastructure or utility contract with the customer/user is to sell them your 'product' – water, energy, building space, road or rail capacity – and charge a unit price for it. With this usual *product* model, to increase profit, or margin, you must sell more product, and therefore cause more impacts. This is fundamentally unsustainable, in that the commercial incentive to make more money is directly opposed to the environmental one – which is to reduce impacts and use fewer resources.

However, once strategy and project scoping boundaries have been extended to include the 'in-use' stage (see Section 2.5), you can consider adopting a *service* model. An example is given in Box 2.2. This apparently small change, which involves switching the price to the service, transforms the way in which the incentives work. The commercial incentive to make more money now goes in the same direction as the environmental one to reduce impacts and use less resources. It is a powerful example of integrating sustainability into the business model (Stahel, 2012).

Box 2.2 The 'service' business model

'Instead of selling the customer a product [*say, gas for heating*] that you hope she'll be able to use to derive the service she really wants [*a warm flat*] . . .

- provide her that service directly at the rate and in the manner in which she desires it;
- deliver it as efficiently as possible;
- share as much of the resulting savings as you must to compete, and pocket the rest.'

(Stahel, 2012)

Example: Parisian *Chauffagistes* sell warmth, not fuel

They are paid, by an annual price per square metre, for the *service* of keeping a flat's floorspace within an agreed comfort range of temperature. There are 160 firms, providing 28 000 jobs.

Competition drives the per square metre price down, and pushes innovation for sustainability because, the less fuel energy firms use, and the more efficient and long-lasting their heating equipment, the more money they make (Hawken *et al.*, 1999).

The example in Box 2.2 is in the energy sector, but service provision in transport, other utilities, buildings and even construction can all achieve sustainability gains from this approach.

2.3.2.2 Apply a sustainability hierarchy

Applying the waste hierarchy is a sustainability principle that has become familiar to many. In the hierarchy, solutions are listed in priority order, from prevention, through reuse, recycling and other recovery, before disposal options are considered as a last resort (see Box 2.1).

We have applied the same idea to formulating projects at the scoping stage, to list a sequence of levels of types of action, in which the bottom one is the least sustainable (the old business-as-usual practice), with a rising list above it, which follows the likely ranking of sustainability impacts, including all the most critical ones discussed above. The most sustainable approach is at the top. This creates a *sustainability hierarchy* of project scope options for environmental sustainability. The rule is to always start at the top, and seek to apply the highest approach in the hierarchy possible. This is summarised in Figure 2.4. Although for some projects the actual environmental impacts may not exactly rank in the hierarchy, the principle is that you should always consider possible project options in this sequence.

With respect to the bottom level (level 6), it is now just bad engineering to assume that large projects will inevitably have large, impossible to avoid impacts, and that these can be left unaddressed. The message throughout this book is that, by being more innovative at all stages of project delivery, even large projects can be delivered in line with level 5 – at least. For really unavoidable impacts, any losses should be more than made up for by the creation of additional environmental assets such as, for example, wetland areas or wildlife corridors.

Figure 2.4 A hierarchy of project options for environmental sustainability

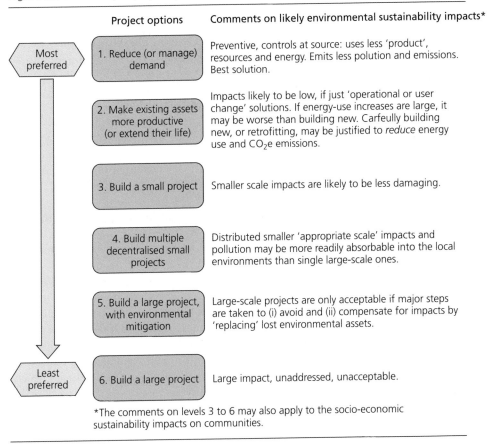

Cost will always be a key factor in scoping choices, and the whole-life cost-efficiency of the alternatives may not match this sustainability preference ranking. However it is wrong to assume this from the outset *without testing it*. Options higher in the hierarchy use less of everything, and so they may well be better at 'doing more with less'.

The sustainability preference for multiple small projects (level 4) before one large one (level 5) reflects the principle of *appropriate scale*; that is, dispersing impacts, as the notes in Figure 2.4 suggest. Such decentralisation, perhaps achieved by multiple simpler and smaller solutions, goes counter to engineers' strongly embedded assumptions that economies of scale are best. The often accepted goal is to achieve maximum efficiency through optimisation, but this can lead to significant vulnerabilities when systems fail. Conversely, smaller decentralised solutions, while perhaps providing some redundancy or spare capacity, can achieve greater overall resilience in maintained service provision. However, this mindset is difficult to change. It also perhaps goes against engineers' (usually unexpressed) preference for working on large, complex

43

(and so 'more interesting'?) projects. These can also, as a result, be difficult to implement, and may be subject to opposition from some stakeholder groups.

By applying this sustainability hierarchy early on in scoping projects, you will build in a useful trend to minimise impacts against all environmental limits, before you get into the more detailed stages of delivery.

2.4. Principle A2: Socio-economic sustainability – 'development'

In order to operationalise A2, we need to define better what socio-economic goals we mean, and how to quantify them. The social dimension is the least studied and most often overlooked pillar of sustainable development. One definition suggests that social sustainability is:

> how individuals, communities and societies live with each other and set out to achieve the objectives of development models, which they have chosen for themselves taking also into account the physical boundaries of their places and planet earth as a whole.

(Calantonio, 2009)

This also notes that, traditionally, social sustainability has dealt with basic human needs (such as housing), education and skills, equity, employment, human rights, poverty and social justice. Emerging themes in this area also include: demographic changes through both ageing and international migration; empowerment, participation and access; identity, sense of place and culture; health and safety; social mixing and cohesion; social capital; and well-being, happiness and quality of life.

The socio-economic measure used in Figure 1.1 was the HDI. This covers many of the aspects needed, and is regularly applied at country or regional level in a workable way, but has not been applied at infrastructure strategy or project level. Work in parts of India has shown strong correlations between growth in aggregate infrastructure development and rising levels of human development (Vijaymohanan, 2008). Indian slums often do not have access to basic services, namely water, sanitation, energy, roads, solid waste and rainwater management. Empirical evidence suggests that, when basic infrastructure provisions are met, slum dwellers shift their focus from lower order aspirations to higher order aspirations such as health, education, housing and land ownership. Specifically, Parikh et al. (2012) have found that energy provision enhances productivity and enables slum dwellers to shift their aspirations upwards.

The goals for socio-economic sustainability can be defined and quantified by starting from the 11 socio-economic goals that Oxfam used to define the inner 'social foundation', as shown in Figure 1.2(c). Table 2.4 reproduces these goals, and summarises key global facts reported by Oxfam, while adding our proposals on how infrastructure should respond to them. For comparison, the three asterisked social foundation goals in Table 2.4 – healthcare, education and income – are the ones measured by the HDI, the measure of A2 that we used in Chapter 1.

Table 2.4 Proposed socio-economic sustainability goals and infrastructure responses, for principle A2
Global data taken from Raworth (2012)

Social foundation goal	Extent of people's global deprivation	Percentage (at year)	Proposed infrastructure response – meeting a demand
Goals typically *directly served* as outputs of infrastructure projects			
Food security	Undernourished	13% (2006–2008)	Set direct, quantified *output* measures as *primary* targets (e.g. numbers served, service level provided) for all projects, addressing any of these issues
Water and sanitation	No access to an improved drinking water source	11% (2010)	
	No access to improved sanitation	39% (2008)	
Healthcare*	No regular access to essential medicines	30% (2004)	Where feasible, include other issues as *secondary* targets. (e.g. education might attach some healthcare and food security too)
Education*	Children not enrolled in primary school	10% (2009)	
	Illiteracy among 15–24 year olds	11% (2009)	
Energy	No access to electricity	19% (2009)	
	No access to clean cooking facilities	39% (2009)	
Mobility¶	No access to reasonable mobility**	To be determined	
Resilience†	Facing multiple dimensions of poverty and vulnerability to disasters	To be determined	Identify and set specific resilience measures for all projects
Goals which can be delivered as broader *indirect outcomes* of infrastructure projects			
Income*	Living below $1.25 (PPP) per day	21% (2005)	Consider as integrated local/regional *secondary* targets, for all projects
Jobs	E.g. Labour force not employed in decent work	To be determined	Use quantified *output* measures – numbers helped, and by how much
Wealth – including ownership¶	Unfair global distribution of wealth – proportion owned by poorest 50%‡	1% (2006)	

Table 2.4 Continued

Social foundation goal	Extent of people's global deprivation	Percentage (at year)	Proposed infrastructure response – meeting a demand
Social equity	Living on less than the median income in countries with a Gini coefficient (a measure of inequality of income or wealth) exceeding 0.35§	33% (1995–2009)	Make special efforts to achieve improvements in the three goals above
Gender equality	Employment gap between women and men in waged work (excluding agriculture)	34% (2009)	Improve this in targeting the first three secondary goals above
	Representation gap between women and men in national parliaments	77% (2011)	Not relevant in infrastructure
Voice	E.g. Living in countries perceived (in surveys) not to permit political participation or freedom of expression	To be determined	Ensure that dialogue gives communities a strong voice, always

¶ Added by authors.
* Goals included in the measurement of HDI.
** 'Poverty, inequality and social exclusion are closely tied to personal mobility and the accessibility of goods and services. Evidence of the economic role of transport in promoting better living standards and greater well-being can be seen in the effects of both overall public investment in transport infrastructure, and in the impacts of specific transport policies, projects and multi-project plans.' (International Transport Forum, 2011).
† *Resilience* is a critical characteristic of complex dynamic systems, being the capacity of a system to tolerate disturbance (Fiksel, 2003).
‡ The richest 2% of adults in the world own more than half of global household wealth according to a path-breaking study released today by the Helsinki-based World Institute for Development Economics Research of the United Nations University (UNU-WIDER). The most comprehensive study of personal wealth ever undertaken also reports that the richest 1% of adults alone owned 40% of global assets in the year 2000, and that the richest 10% of adults accounted for 85% of the world total. In contrast, the bottom half of the world adult population owned barely 1% of global wealth.' (Mindfully, 2006).
§ World Bank (2011).

The goals fall into two groups, and with different implications for a response.

- Goals that typically are already set in terms of service need, as *primary* objectives of infrastructure projects. One infrastructure related service which is missing from the Oxfam list is 'mobility', which we have added to Table 2.4.
- Goals that usually are *not* set as objectives of infrastructure projects, in current practice. Again, we have added one we think is missing from the Oxfam list – 'ownership', beyond just 'jobs' and 'income'.

The first group of socio-economic goals is generally dealt with by existing practice. You should continue what already happens; that is, use them as the driver for specific infrastructure projects aimed at improving people's access to those goals – food, water, energy, mobility, etc. The one exception to this is the last – *resilience* – which has not commonly been considered as an essential infrastructure goal. We include it in this part of Table 2.4 because building in resilience against future risk, climate change in particular, is now an essential requirement for all infrastructure.

Considering the second group of goals, infrastructure investment can support and 'enable' growth in a country's socio-economic 'quality of life' and economy. This is particularly true in developing economies, but also in mature ones. In South Africa, for example, it contributes about 50 jobs per year per $1 million of investment (Watermeyer, 2011) – during infrastructure construction, and to operate and maintain it during its lifetime:

> South Africa's state-led infrastructure drive will be crucial in realising the country's target of creating five-million new jobs by 2020.
>
> (Brand South Africa, 2012)

In spite of this, at present most infrastructure projects miss the opportunity to make an explicit contribution to these goals. So here your sustainability response needs to change existing practice, and to take the opportunities that infrastructure provision creates, to improve these other socio-economic goals, on every project.

2.4.1 Operational principle O2.1
– Set targets and measure for socio-economic goals

The principle summarised in Table 2.3 recognises two categories of socio-economic goals for infrastructure projects; direct service outputs – what the infrastructure is for – and broader, indirect outcomes, which can also be achieved through designing, constructing and operating the project. The first is applicable particularly to current standard practices in setting business strategy and scoping of projects. The opportunity for the second lies in particular in a new emphasis within procurement, and how you select and use the supply chain in design, construction and operation.

The two categories are recognised in the ISO 10845 series of standards for construction procurement, which identify two levels of 'universal procurement system' objectives (ISO, 2010):

- **primary objectives**: the procurement system shall be fair, equitable, transparent, competitive and cost effective, *e.g. 'build a stadium for the World Cup'*
- **secondary objectives**: the procurement system may…. promote objectives additional to those of the primary objective, *e.g. 'create new construction jobs and specialist contractors, and train up specialist workers for construction jobs'.*

To operationalise the primary objective requires, as usual, that you identify the people and their number to be served, and the level and quality of service to be provided by the infrastructure. There are two extra emphases for greater sustainability. The first is to ensure that you only decide strategic, and then project scope, *output* targets after understanding their real role in improving the wider *outcomes* that individuals and communities actually need. The second is that you recognise the *system* interfaces within which the infrastructure asset will exist – for example, the people displaced from their homes by a dam project (see O2.2 – Respect people and human rights), or the extra commuter traffic generated by a new building (see A4 – Complex systems) – and understand and scope targets or constraints for these impacts too. These will then drive the project detail in response.

To operationalise the secondary objectives requires new practice. As stated in the right-hand column in Table 2.4, you can use the procurement process for every infrastructure project to serve such extra secondary objectives, and in a way that is, if possible, not just an add-on but makes an integral contribution to the project's success. So your response must be to use construction opportunities to help meet wider local, regional and even national socio-economic sustainability goals:

> Definite relationships exist between employment opportunities, available skills, entrepreneurship, and the use of small-scale enterprises in the creation and maintenance of assets. The construction strategies adopted can be used to address social and economic needs and concerns and, depending upon how they are structured, to facilitate the economic empowerment of marginalised sectors of society in a focussed manner. Thus, the process of constructing assets can be just as important as the provision of the assets themselves.
>
> (Watermeyer, 1999)

So, you can use procurement through the project supply chain to enhance the first three indirect outcomes goals in Table 2.4 – income, jobs (including training) and wealth, generated and retained, in the local community. You can do this by setting such goals and choosing who you want to include in the supply chain. Furthermore, you can address the next two indirect outcomes – social and gender equity – by targeting the use of companies and people in order to improve these ratios.

Another key goal here – community wealth – is also critically connected with project siting and land take. In many areas local wealth is mainly in property and rural land ownership, not money. So projects with a large land take – dams, pipelines, transport – which displace local people are particularly damaging, and money alone often cannot compensate for loss of property and livelihood. Projects should 'deal fairly' with displaced people.

Some key components within this O2.1 principle are social needs, sustainable cost-effectiveness, and pricing social 'externalities', and these are discussed next. Part II discusses setting targets as part of *practice*; and measurement tools are described in Part IV.

2.4.1.1 Serve social needs
Social needs are key drivers for infrastructure provision, and are to be taken account of at the strategy and project scoping stages (see Part II).

Social need usually provides the fundamental driver of the scope for an infrastructure project. This is often defined in terms of *outputs*: for instance, 'We will provide water to this town of 10 000 people, at a rate of 150 litres/person/day, at WHO potable water quality, at the lowest possible cost.' However, defining the scope immediately in terms of assumed *outputs* may miss sustainability opportunities. For instance, in this case the required *outcomes*, for the individual and the community, are actually as follows.

- Nourishment and health – through water for drinking and cooking. The quantities needed here are basic and fixed, and the quality 'at the tap' must be potable.
- Attractive living conditions, inside and outside the house, and possibly growing food – through water for cleaning of all kinds, for (often) toilet flushing, garden watering and car washing. The quantities required to meet these needs can be much reduced by behavioural, pricing and technical measures for demand management; and the water quality here can be lower than potable standards.

So one key part of this principle is: before defining outputs, identify the social need at this *outcomes* level. This can lead to project scoping and solutions that serve the community in a more innovative and sustainable way, and which are also cheaper and more affordable.

More widely, both individuals and their communities have socio-economic needs that are linked. The classic examination of an individual's needs is Maslow's hierarchy of needs, with its later developments (Huitt, 2007; Maslow and Lowery, 1998). These include, working up from the most basic: physiological needs, safety/security, belongingness and love, and esteem. The physiological and security/safety needs are those most emphasised in Table 2.4. For international development, they are also the basis of most of the UN's Millennium Development Goals (UN, 2011). They are likely to be the drivers for most infrastructure projects.

Infrastructure projects will also interact with socio-economic sustainability at the community level, particularly in urban areas. Several cities have developed their own version of social sustainability definitions. Box 2.3 shows one from Vancouver, Canada (Gates and Lee, 2005). In Vancouver's case, this is followed by a detailed set of components and objectives: basic needs, such as housing; individual capacity, such as opportunity for learning and self-development; and community capacity, such as networks and organisations.

In the UK, the Highways Agency developed a single framework to address social sustainability at corporate, project and operational levels (Mitchard *et al.*, 2011). The framework comprises 21 aspects, grouped under four sections: health and well-being, social cohesion, inclusion and equality. A special issue of the journal *Engineering Sustainability* on the human dimension was published in the Proceedings of the ICE in March 2011 (164(ES1)).

Such detailed definitions, produced by local organisations and people, may be different for each particular place, but they then provide a good sustainability test for infrastructure

---

Box 2.3 Social sustainability – a definition

For a community to function and be sustainable, the basic needs of its residents must be met. A socially sustainable community must have the ability to maintain and build on its own resources and have the resiliency to prevent and/or address problems in the future.

There are two types, or levels, of resources in the community that are available to build social sustainability (and, indeed, economic and environmental sustainability) – individual or human capacity, and social or community capacity.

Individual or human capacity refers to the attributes and resources that individuals can contribute to their own well-being and to the well-being of the community as a whole. Such resources include education, skills, health, values and leadership.

Social or community capacity is defined as the relationships, networks and norms that facilitate collective action taken to improve upon quality of life and to ensure that such improvements are sustainable.

To be effective and sustainable, both these individual and community resources need to be developed and used within the context of four guiding principles – equity, social inclusion and interaction, security, and adaptability.

City of Vancouver 2005 (Gates and Lee, 2005)

---

projects in that location – by asking: 'How will the project help or hinder these specific local components of socio-economic sustainability?'

2.4.1.2 Sustainable cost-effectiveness

You will be very familiar with the drive for cost-effectiveness – gaining the most service for the least cost – as a fundamental principle in infrastructure decision-making, at all stages of project delivery. (While there are very many different definitions of *cost-efficiency*, *cost-effectiveness* has a more specific definition, as the ratio of the amount of service provided to the cost of providing it. Cost-effectiveness analysis (CEA) is a form of economic analysis that compares the relative costs and outcomes (effects) of two or more courses of action. Typically, the CEA is expressed in terms of a ratio where the denominator is a gain in, for example, health obtained from a measure and the numerator is the cost associated with the health gain.) There are many good standard instructions on how achieve cost-effectiveness (Snell, 2010); and within this, the process of *risk assessment* is another key standard methodology (Ostrom and Wilehelmsen, 2012). Cheapest cost as the overriding choice criterion first became conditioned by environmental considerations in the 1990s, through requirements such as BATNEEC (Best available Technology Not Entailing Excessive Cost) (see Appendix A).

Here, we use the term 'sustainable cost-effectiveness' to add a very specific emphasis on taking key sustainability principles into account while making cost-based infrastructure decisions. This means that, whenever doing cost analysis and comparisons, as appropriate you should consider the following.

- Take account of all stages of the project, and calculate whole-life costs in line with principle A3 (Intergenerational stewardship), and build in *resilience* (see Section 2.5).
- Take care to consider all the whole-system costs within the boundaries of the appropriate system within which the project works, and consider *integrated needs*, in line with principle A4 (Complex systems) (see Section 2.6).
- Include in any cost–benefit analysis (CBA) the price of *environmental and social externalities* (see below). However, a CBA and a notional price should not be used to cover any global or irreversible environmental *boundary* impacts. Rather, these should be met by setting an absolute target that no option can exceed (see Section 2.3).

Finally, remember to always test and challenge the myth that 'being more sustainable always costs more'. Project teams charged just with cost-saving targets may not actually deliver the most cost-effective projects, because cost-saving fatigue means that this alone no longer drives much innovation. By comparison, setting hard sustainability targets – such as large CO_2e reduction targets – may deliver more cost-effective projects too (see Box 11.1 in Chapter 11).

2.4.1.3 Pricing social and environmental 'externalities'

There is a problem with the standard principle of seeking cost-effectiveness when it fails to take into account the social or environmental damage or benefits caused by a project, because they are 'common goods,' which are 'free'. One way to correct such 'market failure' is to apply the principle that all socio-economic and environmental 'externalities' – both costs and benefits – should be given shadow prices and included in CBA.

CBA is acceptable for cost evaluations and option choices at strategy or project scoping stage, as long as environmental and socio-economic externalities are priced. It is also sometimes applicable at the outline design stage, when alternatives are evaluated; however, this must be done with a clear understanding of its assumptions, and the questions that CBA does not answer (see Section 13.6).

One caveat, when pricing such a 'good', is that it is important to examine carefully its real value in the context of the actual project. For instance, for a transport project, time saved by private car drivers on roads (who cannot do any useful work while they are driving) is usually valued as 'working time' lost. If the project is a public transport system, such as rail or luxury bus, modern Wi-Fi and power sockets can make this an attractive place to actually get more work done. So valuing the time saved in the same way is not justified. Given that the real impact of most business travel is that generally it takes time away from home and family, it would be more correct to value such time in terms of wider 'quality of life' lost.

As well as economically modelling priced externalities in CBA, the best way to take account of the cost of the unpriced damage in project choices is for you to require the actual project itself to pay real extra financial compensation for, say, property being lost (in the UK, the proposed HS2 railway line (incidentally, largely justified on the grounds of travel-time reductions) plans to pay 10% extra compensation to households closest to the line) or to actually create local jobs and wealth. You can also require the project itself to include the replacement of damaged or lost environmental assets, such as land take, and to bear the financial cost of this. This means not accepting any level 6 options in the project sustainability hierarchy (see Section 2.3), but using level 5 as the minimum acceptable option.

2.4.2 Operational principle O2.2
– Respect people and human rights

This principle is, of course, applicable throughout every stage of project planning and delivery. It embodies a wide range of very well-known core ideas, many of which have been established as ethical principles (e.g. the UN Charter of Human Rights (UN, 1985)), as well as in much global and local regulation and legislation. In relation to infrastructure projects, some of the most familiar places where these concepts are expressed may be in the codes of ethics of professional bodies (see, for example, the *ICE Code of Professional Conduct* (ICE, 2008), as one of very many, which are for every profession involved with infrastructure) and health and safety legislation. In the business area this is covered by frameworks for corporate (social) responsibility (see Section 13.8); for instance, as summarised by the Global Sullivan Principles of Social Responsibility (see Appendix A).

While these mechanisms lay out formal obligations to be met, there are also more informal responsibilities to ensure proper justice is achieved for all user groups and communities affected by a project. This means making sure everyone's voice is heard in the development of a proposal, and that equitable access to its services is provided. The adoption of effective consultation processes and a willingness to act on what they reveal is a direct corollary of this principle. These issues closely match the last three important goals in Table 2.4: enabling and supporting *voice*, *social equity* and *gender equality*.

Broad involvement should be encouraged in the ways in which projects are defined and accepted, and the engineering team should show a willingness to share knowledge and achieve mutual learning with all stakeholders. Not only can you encourage and strengthen normal practices such as stakeholder engagement, but you can try to create a genuine two-way dialogue with everybody involved. For example, fairness regarding the affordability of the infrastructure service for all stakeholder groups may be a key concern. Such considerations often go to the heart of understanding the real needs for the engineering solutions that are being proposed, and which require you to be scrupulous in terms of transparency and justification of decision-making.

While reaching consensus with stakeholders is desirable, you need to understand all the motives and arguments, particularly for those schemes where the outcomes disproportionately affect the lives of individuals. A trend analysis can reveal changes in society that lead to the need for new services (e.g. an ageing population or more people living alone), and a stakeholder analysis can systematically help you understand which groups should be consulted when formulating the solution of the problem (Muldur, 2006).

In the delivery of infrastructure projects, the human dimensions of sustainability, such as how conflicts can arise from different interpretations of the problem by different groups, and how different views may give rise to alternative plans and proposals, become core factors for you. Four common dilemmas affect this process including

- conflicts of strategies versus concerns, where negotiating positions are mistaken for genuine legitimate interests, and these differences are sometimes unacknowledged or misunderstood
- the tension between those involved in negotiation having the freedom to be creative and innovative in problem-solving, while at the same time maintaining and representing the positions of organisations and groups on whose behalf they are negotiating (and who are not directly at the negotiating table)
- differing groups' perceptions of scales and time horizons
- (mis-?)using 'professional expert' credentials to argue for a pre-preferred outcome (Laws and Loeber, 2011) in what has been called an 'inflexible design and defend attitude' to negotiation.

You can deal with these dilemmas by recognising that technical projects are also arenas for learning and problem-oriented negotiation. Some practical strategies that you can employ are: involving stakeholders early and often, being transparent about procedural commitments, and consciously creating the conditions that make learning and reframing feasible. Another key response is to understand how 'place-based' and human factors need to be considered in design if sustainable outcomes are to be achieved. (This forms part of our discussion of complex systems in Section 2.6.)

2.4.2.1 Stakeholder engagement
An important principle in ensuring social justice is to allow people a proper *voice* – to take care over how people are engaged with infrastructure projects at all stages. This is the principle of 'stakeholder engagement'.

Many infrastructure professionals have long been cautious about engaging with 'non-experts', and so the 'design and defend' approach has been well entrenched, when proposing projects to the public. Sometimes the politically correct language of stake-holder engagement has been used without any substance beneath it. A challenging criticism of this approach was put by Arnstein, which is sometimes, unfortunately, justified:

Many planners, architects, politicians, bosses, project leaders and power-holders still dress all variety of manipulations up as 'participation in the process', 'citizen consultation' and other shades of technobabble.

(Arnstein, 1969)

Partly because of this perception, over recent years trust in politicians, large companies with big projects and professionals has been eroded in many countries, and as this has happened demands for more transparency have increased. This has been characterised by saying that society's expectations in stakeholder engagement have changed, from: 'trust me' to 'tell me' to 'show me' to 'involve me' (Willems, 2001). In practice, there is

Table 2.5 Levels of stakeholder engagement

Ladder of citizen participation (Arnstein, 1969: 217)			Community engagement 'ladder of participation' (North Yorkshire County Council, 2010)	
Degrees of citizen power	8. Citizen control 7. Delegated power		Empowering	Placing decision-making in the hands of the community
	6. Partnership		Collaborating	Working in partnership with communities in each aspect of the decision, including the development of alternatives and the identification of the preferred solution
			Involving	Working directly with communities to ensure that concerns and aspirations are consistently understood and considered (e.g. partnership boards, reference groups and service users participating in policy groups)
Degrees of tokenism	5. Placation**		**'the ground rules allow have-nots to advise, but retain for the power-holders the continued right to decide'	
	4. Consultation		Consulting	Obtaining community feedback on analysis, alternatives and/or decisions (e.g. surveys, door knocking, citizens' panels and focus groups)
	3. Informing		Informing	Providing communities with balanced and objective information to assist them in understanding problems, alternatives, opportunities, solutions (e.g. websites, newsletters and press releases)
Non-participation	2. Therapy* 1. Manipulation*		*'a substitute for genuine participation ... to enable power-holders to "educate" or "cure" the participants'	

a range of levels of engagement. These are summarised Table 2.5, which reflects the progression in stakeholder engagement, from bottom to top.

Arnstein's summary of the levels is on the left-hand side of the table. Her 1969 paper is a strong critique of the general disempowerment of communities, and is worth reading as a corrective to the bland 'consultation' language that can be used to disguise that reality. To her, the only really acceptable form of engagement is 'citizen power'. The right-hand side of the table shows a modern local government version (Calantonio, 2009) of these levels, which we have matched to Arnstein's, giving each a summary description of the corresponding kind of engagement actions. The lowest acceptable level is *informing* (Arnstein's level 3), and the top one is *empowerment*, corresponding to full community control.

The engineer's traditional 'design and defend' approach is probably only, at best, *informing* (Arnstein's level 3). One reason why the engineering profession has found it difficult to move up through the levels may be because of worries that, in giving up the overriding power of 'expert' knowledge and accepting non-expert views, the 'professional' will lose control of the decision.

Empowerment by achieving full community control is unrealistic for many large projects (but, how that community is affected, for good or bad, by the project is still something that can be much improved – see Section 5.3), and Arnstein's labelling of 'retain for the power-holders the continued right to decide' as merely 'placation' can be unfair. Full stakeholder control may indeed be appropriate in some situations, such as deciding strategy, particularly for local infrastructure; but, clearly, in most projects for regional or national infrastructure, at the project scoping or feasibility stages the final decision remains with the sponsoring and funding infrastructure organisation.

Table 2.5 shows that these concerns are wrong. There are tenable levels of engagement in between the extremes. Answering the 'justice through participation' (Fenner *et al.*, 2006) questions (Figure 1.5) then becomes really practical, because the answers help define which level of engagement we are using.

- How has a fair foundation for this scheme been developed with the stakeholders?
- Who is involved in establishing a base of agreed positions (facts as well as aspirations)?
- Have genuine concerns been considered openly; is there a willingness to modify designs? Is the basis of decision-making established and known to all likely stakeholders at the outset?
- Has the extent to which participation can and will affect decisions been agreed? Who carries responsibility for explaining what cannot be altered, and why?
- What are the steps in the process for managing disagreement; with whom are these discussed?

One difficulty in asking these and other related questions is that engineers have often been trained to focus almost exclusively on the technical aspects of what in reality is a complex socio-technical system (Geels, 2004). So the technical specialist in charge may be seen by some to impose solutions, in what we have referred to earlier as a 'propose and defend' attitude. The alternative approach is for you to have a stronger, iterative dialogue with all stakeholders, to agree needs, goals and solutions (see Section 2.5). This requires you to accept that the most elegant technical solution – judged around the performance of a single variable (e.g. maximising car travel speeds so as to minimise journey times) – sometimes may be suboptimal if it does not carry acceptability from a wide range of stakeholders. The best solution, therefore, may be seen as not being the finest possible in terms of hardware design, but the one that satisfies the widest group of users and others affected by the decisions. It is sometimes the 'second-best option' that has the wider ability to meet multiple needs. Your recognising this is a step towards dealing with 'wicked problems', which are perceived from a number of different viewpoints, with real complexity in defining the project scope (see Section 2.6).

So the principle is that you should adopt from Table 2.5 the highest possible level of stakeholder engagement that is possible– at level 6, 'involving', at least. How your *practice* can respond to these ideas is discussed in Part II (see Section 5.4).

2.4.2.2   Health and safety

There has been much progress in this area in many countries (but by no means everywhere), but construction is still one of the most dangerous sectors to work in. In the UK, for example:

> There have been significant reductions in the number and rate of injury over the last 20 years or more. Nevertheless, construction remains a high risk industry. Although it accounts for only about 5% of the employees in Britain it still accounts for 22% of fatal injuries to employees and 10% of reported major injuries.
>
> (Health and Safety Executive, 2012)

The much stronger management focus on health and safety over the last 20–30 years has become an essential part of good engineering. This has involved a varied range of changes, including strong leadership by individuals and companies, and tighter government legislation and enforcement. Some examples of these actions are given in Box 2.4. In short there has been a persuasive demonstration that better Health and Safety does not conflict with more efficient construction but actually enables it; so it is not a trade-off choice.

Box 2.4 Examples of applying strong health and safety principles

1. **Leadership and discipline.** A water sector company brings in a new CEO from the coal mining sector. He has very strong principles, from experience, on health and safety as a fundamental culture. He applies it, with top-down discipline, to all design and construction for his company, as a part of all their work, and for all procurement qualifications.

2. **Procurement impact.** A water sector company, about to award a substantial framework contract, eliminates the preferred contractor just before award because of a bad accident on one of its other sites. This example of a serious new requirement for health and safety greatly concentrates the minds of those in the supply chain.

3. **Creative communication.** A health and safety trainer, talking for the first time to the workers on a large construction site, makes it personal: 'The sector average death rate is 2.5 deaths per 100 000 workers, per year. On this site, that implies that two of you will die during this project – does anyone want to volunteer?'

(Source: Author's experience)

2.5.   Principle A3: Intergenerational stewardship

This principle follows directly from Bruntland's assertion that actions taken today should not comprise 'the ability of future generations to meet their own needs' (Bruntland, 1987). To make it clearer, an earlier definition of stewardship for future generations is that:

> at the core of the idea of sustainability, then, is the concept that current
> decisions should not damage the prospects for maintaining or improving living
> standards in the future ... This implies that our economic systems should be
> managed so that we live off the dividend of our resources, maintaining and
> improving the asset base so that the generations that follow will be able to live
> equally well or better.
>
> (Repetto, 1985)

This simply requires you not to take decisions and actions now that close off options for future generations to live sustainably; or, as Tony Blair famously summarised, it is 'not cheating on the kids'. Engineers must avoid handing on a negative legacy to our children.

The track record on this is mixed. The provision of urban infrastructure has been an enormous force for good since the industrial revolution, contributing to the essential fabric of the modern world, and enjoyed by many generations whose lives, as a result, have been generally better than those of their forbears. But these developments have sometimes come at a cost, with unforeseen consequences of congestion, pollution, social dislocation and loss of biodiversity emerging as a result of the complex interactions that have evolved in the modern world. There can also be positive changes of course; the unloved quarry that becomes a valuable wildlife sanctuary, for example.

At the heart of this principle is your willingness to take an anticipatory view of the kind of future you want to create, and how the infrastructure that you develop now can continue to add benefit, and avoid damage, into the future. While the engineered services that society enjoys have often provided buffers against environmental extremes (drought, flood, food security, disease transmission), they sometimes have done so at the cost of a lock-in to expensive technical solutions that do not respond well, or cannot be adapted quickly, to the changing and uncertain circumstances faced in this century. Bequeathing assets that lack the necessary resilience to respond, for example, to a range of plausible future climate scenarios is one way in which engineers have already constrained those who come after them.

To operationalise this responsibility, you must think well into the next generation, by planning long term; and keep the next generation's options open, by considering all stages of a project's life. We discuss these aspects below. You must also try to avoid unexpected future consequences of your actions now. This involves recognising complex interdependencies, and dealing with uncertainty, which lies within the province of complex systems (see Section 2.6).

2.5.1 Operational principle O3.1
– Plan long term

A project that is based on the principle of intergenerational equity should (Morrison-Saunders and Hodgson, 2009)

- demonstrate enduring economic and wider value for future generations
- set out liability and management processes for any future negative (and positive) impacts, including on land use
- provide legal/commercial ways to hold the various 'players' accountable for their commitments, into the future.

Your strategy making and project scoping must look far enough into the future to be able to answer such requirements – so you must *plan long term*. As a 'generation' is typically considered to last, say, 25 years, that sets that period as the minimum timescale for the long term. In 2013, as we write, some long-term planning looks at mid-century (up to 2050), whilst climate change, ice cover loss and the certainty of rising sea levels poses questions on a 100-year timescale. The biggest challenge that all this gives you is, of course, related to the future and its uncertainties! Because of the difficulties this raises with regard to *forecasting (prediction)* – the traditional method used for planning – many planners have preferred to put the long term into the 'too difficult' box, and plan for only the short term. This is not enough to achieve sustainability.

Long-term planning for infrastructure involves three components, including

- imagining what the *surrounding world* will be like, within which our infrastructure must work
- deciding what existing, or changed, characteristics the *infrastructure* must have (e.g. must it be decarbonised?)
- estimating how the changes in the first will interact with the changes in the second, to *determine actual demand, use* and *impact* (good and bad).

The difficulty now is that, as you see the global challenges of climate change, population demographics and energy costs (to name just some) increase, you know that the future will not – indeed must not – be like the past. So, the traditional forecasting and future risk analysis methods used by infrastructure strategists and asset planners become inadequate. You need new approaches that can address those three components of long-term planning more explicitly. These include foresighting (including scenario planning) and visioning and backcasting (see Section 13.9).

The key principle is to plan long term, using the right tools and questions to explore where you want to go, and to identify the likely real uncertainties that may exist. Then you can structure your strategy (see Chapter 3) or project scoping (see Chapter 4) to minimise the risks.

2.5.2 Operational principle O3.2
– Consider all life-cycle stages

The total sustainability benefits and impacts, and costs, of an infrastructure project can be out of the view of a project planning team, as they occur in different and remote locations, 'downstream' in the supply chain, or over some longer time horizon after many of years of operation, or during decommissioning. They are often dominated by the long *in-use* stage; here the CO_2e emissions, for instance, are typically at least an order of magnitude larger than those of the construction stage (see Section 10.2). Work on the typical costs of owning an office building for 30 years has suggested a construction costs/maintenance costs/costs of the operation being carried out in the building (including staff costs) ratio of $1 : 5 : 200$ (Evans *et al.*, 2004). (Others have questioned these figures, noting that the three costs for each individual building are affected by a plethora of factors, yielding a wide variation in ratios (Hughes *et al.*, 2004).)

These impacts are a result of all the decisions made in the previous project delivery stages, through detailed design and construction, as well as how operators and users act – more or less sustainably. This interdependence emphasises the need to consider all life-cycle stages of the project from the start of project planning. The core principle here, therefore, is for you to assess the benefits, impacts, costs and hence viability of an infrastructure project on a 'whole-life' basis.

This principle is applicable particularly to your practice in the earlier stages of project planning and delivery – in setting business strategy, scoping infrastructure projects and choosing outline designs. It requires that you take account of all the later stages – design, construction, operation, and end of life – from the start, and so may well change the emphasis and approach that you take in each of those later stages.

To serve this principle in *practice*, the methodology of whole-life costing (see Section 13.7) is well established; and for the (usually) most important of environmental targets, whole-life carbon accounting has been developed (see Section 13.3). Other infrastructure impacts can be assessed on a whole-life basis by adding the impacts from each stage in its life; and for more complex combined environmental impacts, life-cycle assessment is available (see Section 13.2).

2.6. Principle A4: Complex systems

When considering infrastructure strategy, planning assets or scoping projects, you are working out how you want them to interface externally with their surroundings. These interactions are described by the 20th-century science of complex adaptive systems. This deals with how everything is interconnected, in unpredictable ways, and cannot be determined by means of applying our familiar Newtonian mechanics. Many infrastructure problems and unexpected consequences have arisen because of a failure to think, at these early stages in the project, in terms of the whole (interactive) system and to embrace the complexity this entails (see Section 1.3).

The key characteristic that separates complex adaptive systems from merely 'complicated' ones is that they have emergent properties. (In systems theory *emergence* is the way in which complex systems and patterns arise out of a multiplicity of relatively simple interactions. Emergence is central to the theories of integrative levels and of complex systems.) These arise, not as characteristics of the individual components, but as a result of the intricate relationships – the two-way interdependence, not just one-way interaction – between components. These might include a whole range of interactions at psychological, social, political, technical and economic levels. A good 'primer' for understanding this is the short book *Thinking in Systems*, by Donella Meadows (2008), one of the authors of the seminal Club of Rome's Report, *The Limits to Growth* (Meadows *et al.*, 1972). Complexity however is not restricted to the inter-dependencies between physical systems. Distinctions have been made by Fratini *et al.* (2012) concerning functional complexity and relational complexity. The former is related to the physical dimensions of the urban space and to the range of functions assigned to technical objects (e.g. infrastructures). The latter is related to humans and in particular to the different views and perspectives of the actors and organisations involved in the decision making process. It is often this relational complexity, and specifically the lack of integration between agencies, which can hinder the adoption of new sustainable practices.

So, in infrastructure, there are links between engineering systems, such as transportation, buildings or water supply networks, and *social systems* – their *users* – as manifested by urbanisation, communication and public health, and also *environmental systems* – their *natural or urban surroundings* – in the aquatic, atmospheric or terrestrial environment (see Figure 2.5(b)). There are flows of materials, wealth, energy, labour, waste and information between these systems. A useful tool for 'mapping' and understanding these interactions is 'systems dynamics' (see Section 13.11). The complexity, dynamics and non-linear nature of these interdependent systems suggest that the notion of 'sustainability' as a steady-state equilibrium is not realistic (Fiksel, 2006). Climatic, technological or geopolitical changes will disrupt cycles of energy and material flows. The achievement of sustainability will, therefore, require the development of resilient and adaptive technological and societal systems that imitate the dynamic attributes of natural systems – the idea of using nature as model and teacher, or *biomimicry* (see Appendix A). Some steps you can take to ensure sustainability when working within complex systems are given in Box 2.5.

The change in approach that enables this requires

- opening up the problem space, allowing redefinition of the problem. This is one of the four components needed to operationalise the principle of dealing with complex systems. The others are
- dealing with uncertainty

Box 2.5 Operating in complex systems requires this approach

- Addressing multiple scales over time and space.
- Capturing through systems dynamics points of leverage and control.
- Representing problems to an appropriate level of complexity.
- Capturing stakeholder perspectives in various domains.
- Understanding system resilience relative to foreseen and unforeseen stressors.
- Managing variability and uncertainty.

(Adapted from US EPA (2007))

- considering integrated needs
- integrating working roles and disciplines for project delivery.

An example of how thinking about complexity can change the definition of an engineering problem can be seen in the management of pollution emissions. This has been done for many years using end-of-pipe treatments and controls, which focused on pollutants from chimneys (to atmosphere) or sewers (to surface watercourses). More recently, new 'upstream' approaches have been developed, such as waste minimisation and pollution prevention. As additional environmental stressors became recognised, the evaluation and choice of pollution control required greater understanding of the overall context of the problem (US EPA, 2007). This is an example of 'opening up the problem space', which is discussed next.

2.6.1 Operational principle O4.1
– Open up the problem space

The best way to explain this principle is to use an example, as summarised in Figure 2.5. Consider a water company, which needs to provide water to more customers and at the same time is required to provide water for drinking at a higher quality. The traditional approach is shown in Figure 2.5(a). The problem is assumed to lie within the assets owned by the company, so the only available solutions are to seek more water sources, to increase abstraction and to add more water treatment. The increased use of resources and energy is unlikely to make this a sustainable solution.

Figure 2.5 Opening up the 'problem space' for solutions

Arena of action	Water Co.
Infrastructure > (Add/change it?)	Constructed and operated assets and energy

(a) The narrow – focused, **assets space** –
only one 'box' for solutions

Arena of action	User	Water Co.	Urban & Natural
Standards > (Change them?)	Health, taste, convenience	Design and operations manuals	For environment, air, climate
Infrastructure > (Add/change it?)	Household pipes, equipment, energy, use	Constructed and operated assets and energy	Urban design and energy use, 'nature' in catchments
Demand/inputs > (Control at source?)	Sewage, waste, water	Operating methods	Run-off, drought, pollution

(b) The wider, deeper **'system' space** –
many more boxes for solutions

Figure 2.5(b) shows what happens when you take into account the wider *system* surrounding the water company's assets, to enlarge the 'problem space'. The extra row below the assets takes into account the source or demand for services; the row above them shows that what the assets have to provide depends on the standard that has been set for the service. Both of these offer possible change opportunities. The extra columns cover the systems that 'sandwich' the water company's infrastructure assets: to the left, the users, and to the right, the natural and urban environment. The 'users' column includes the plumbing in users' houses and the appliances that use water; the 'environment' column includes the rural catchment from which water is drawn and discharged back to.

With these additions, there are now nine 'boxes' of 'problem space' in which to look for solutions, not one. This enables you to consider several lower resource/energy use, more sustainable options. For instance

- improve 'raw' water quality into the treatment plant, to reduce treatment needs, by working with farmers to minimise use of pesticides, and nitrogen and phosphorus in fertiliser, in the catchment (bottom right box)
- increase only the quality of the water actually used for drinking/cooking, not at the treatment plant, by adding in-line polishing treatment at the customer taps used for drinking (top left box)
- avoid the need to abstract any more water, by managing demand down, using metering, progressive block tariffs and assistance with low-water-use devices in the home (lower left box).

If such a problem-space matrix was drawn for a road transport system, 'more working at home', to reduce building new commuter capacity, would be in the lower left box; road pricing to optimise use of existing roads would be in the lower centre box; setting lower mandatory speed limits to optimise traffic flows would be in the top centre box – and so on. All of these water and transport options are already in use in various locations, so it is not that the solutions have not been thought of or cannot be implemented. The principle is that you can always apply this thinking, in the strategy (see Chapter 3) and project scoping (see Chapter 4) stages, to ensure that such wider options are always investigated, and not missed. This works particularly well when looking for solutions higher up in the 'sustainability hierarchy'.

Your engaging with options in these extra 'boxes' means planning with less direct control, and with more outside issues affecting outcomes. Coupled with the need to plan long term, this means you have to learn to deal with more uncertainty.

2.6.2 Operational principle O4.2
– Deal with uncertainty

With complex systems comes uncertainty, an inherent property arising not only from a lack of data, but also from the nature of the interrelationships themselves. In these circumstances (exemplified well by predictions of global warming in which global climate models have to pile uncertainty onto uncertainty), you have to make decisions before conclusive scientific evidence is available. The potential consequences of making the 'wrong' decision can be huge, and may be based on values that are not universally accepted by all parties.

Problems occur where the knowledge base is characterised by large uncertainties, multiple causality, knowledge gaps and imperfect understanding. In these circumstances more research may be required, but this is likely to lead to even more unforeseen complexities. Your judgement is required as to whether information is really objective, valid and reliable, or is subject to conflicts of interests. So, two factors increasingly combine to make your decision difficult: high decision stakes are found in the personal interests of individuals who are disproportionately affected by the outcomes of engineering decisions and schemes; and increasing uncertainty leads to risks that cannot be fully quantified.

With low uncertainty and low decision stakes, there is a technically 'right' answer, and your decisions lie within traditional expertise and conventional problem-solving methodologies. When you assume this, you tend to operate as an advocate for a single preferred (sometimes predetermined) solution. But increasing uncertainty, coupled with possible outcomes of decisions that may make people act and argue irrationally, generates a range of conflicting views on the available options. This creates an unpredictable situation, in which you may better use your technical knowledge to act as an 'honest broker' (Azapagic and Perdan, 2011). Your role then changes to one where you use your skills and expertise in a fair and neutral manner to arbitrate between a range of technical options, not to advocate just one.

One way of dealing with uncertainty is to use modelling and decision-making approaches that support dynamic, adaptive management rather than static optimisation. In these, the implications of adopting different technical options need to be understood in terms of how they contribute to system resilience. If more alternative solutions are considered, and these are assessed against a wider set of choice criteria, you may need tools for multi-objective decision-making (see Section 13.10), involving trade-offs between conflicting goals, as well as ways of dealing with missing or uncertain information.

Uncertainty comes in a variety of forms (Brugnach *et al.*, 2008) including

- **unpredictability** arising from complex systems that are constantly adapting to new situations, often through non-linear and chaotic behaviour
- **incomplete knowledge** is perhaps due to a lack of information or data, or to the unreliability of that data, or to the lack of a complete theoretical understanding of the problem, or even sheer ignorance
- **uncertainty that arises from multiple knowledge frames**, where a problem is defined differently by different stakeholders.

So, for example, a situation of water shortage might be seen either as a problem of insufficient water supply or as excessive water consumption. If the first, the solution to be pursued is likely to be a technical one, whereas, if the second, the solution may be seen to involve a change in the behaviour of consumers. Sometimes the conflicting views of how a system should be managed may all be plausible and legitimate, and arise from where to place the boundaries of the system or the different interpretations about the urgency of the problem, or even different values and beliefs (this reflects looking in different 'boxes' in Figure 2.5(b)). Combining these can lead to a range of uncertainty questions, as shown

Table 2.6 Types of uncertainty*

	Unpredictability	Incomplete knowledge	Multiple knowledge frames
Natural systems	How will climate change affect weather extremes?	What are the historic water levels in the river?	Is the main problem deterioration of river water quality *or* reduction in ecosystem services?
Technical systems	What will be the impacts of an inter-basin water transfer scheme?	How much water can be saved with new irrigation technologies?	Should flood walls be built *or* flood plains created?
Social systems	How strong will stakeholders' reaction be to the next flood?	What are the economic impacts of a flood for different stakeholders?	Should water scarcity be dealt with through tariff structures *or* efficiency savings?

* Modified from Brugnach *et al.* (2008)

in Table 2.6. The principle is that you identify which type of uncertainty you are dealing with, and then adopt an appropriate strategy to deal with it.

2.6.3 Operational principle O4.3 – Consider integrated needs

This principle is applicable particularly to your practice in the early stages of project planning and delivery, in setting business strategy and scoping infrastructure projects. You need to take account of the many interdependencies in the built environment and in urban settings. Some are within the urban system, and some are between it and the fluctuations and changes in surrounding natural systems.

Infrastructure engineers consciously shape the built environment. An example of good practice is creating infrastructure that simultaneously minimises ecosystem damage, is energy and resource efficient, and contributes to healthy, vibrant and cohesive human habitats (Fenner *et al.*, 2006). Examples are to adopt integrative approaches linking transportation, mixed-use development and brown-field regeneration, or to avoid developments in flood plains, and to encourage biodiversity in town and city landscapes through the creation of blue-green corridors which integrate water management functions with green infrastructure.

Often this requires the coming together of agencies and institutions at the city level to work together in a coordinated and integrated response, for multiple benefits. Thus the introduction of sustainable urban drainage systems (Hoyer *et al.*, 2011) can also provide additional green space and habitat, encouraging biodiversity, help sequester carbon and reduce the urban heat-island effect, as well as add to the visual aesthetics of an urban area, providing recreation and amenity opportunities, and even serving traffic-calming measures. But these benefits do not accrue in isolation, and need

Box 2.6 San Francisco's Better Street Plan

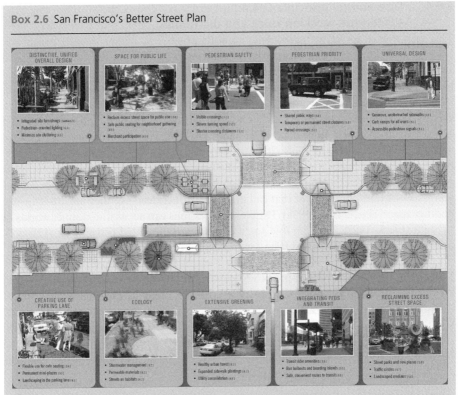

Image reproduced courtesy of San Francisco Planning Department

In San Francisco an integrated approach has been taken to wastewater treatment, stormwater management, street greening and community revitalisation. Eight coordinated plans have been developed that link stormwater management and street design with urban forestry and water quality, green space corridors and traffic calming. The goal is a set of unified guidelines that coordinates redevelopment of streets while integrating ecological function into a more pedestrian friendly, less car-dominated city (San Francisco Planning Department, 2010).

careful integrated planning and forethought by many bodies, with disparate responsibilities. Good examples of how such integration can work are San Francisco's Better Street Plan (San Francisco Planning Department, 2010) (Box 2.6) and Portland's Grey to Green Initiative (City of Portland, 2008).

Another key response is to understand how you can consider 'place-based' and human factors in design. Holistic design will seek to deliver several functions through one project (Owen *et al.*, 2011). By broadening the design inputs to include more social and economic factors specific to the project location, you can use different, more participative design processes that provide outcomes which respond to a wider range of concerns and provide a greater contribution to sustainability.

2.6.4   Operational principle O.4.4
– Integrate working roles and disciplines

This principle is applicable particularly to your practice in outline design, design and construction. The scale and complexity of modern projects, with the wider issues and skills we have raised above, cannot be addressed effectively by any one professional group or discipline; they need wider groups. A shared approach to the problem does not necessarily emerge simply by putting a range of specialists together in a group. A technical specialism, or different groups in the supply chain, can be stereotyped or misconceived, and hence mistrusted, by those who have not had direct contact with them, or because of the lack of a shared language. This situation can also arise because engineers are highly competent individuals who are guided by their own different personal and professional norms, and who prefer to remain in control of their own work. So, professional boundaries and issues of territory can exist that can make individuals feel 'like they are crossing into another space' (Pirrie *et al.*, 1998). Different professions can act as if in distinct worlds of thought (Fleck, 1997). This can lead to different expectations and perceptions of success, and these 'silos' can make innovation for sustainability hard.

Long term, this separation will need to be dealt with through the education process, but on current projects you will need to take steps to avoid it, by employing a multi-disciplinary 'project team'. 'Teams' differ from 'groups' in that their members should see themselves as interdependent (rather than independent), having mutual goals, shared leadership and, hopefully, collective rewards. This will encourage the team to have the widest of perspectives, to listen to other team members when they disagree, and to communicate well, to generate a strong creative process.

One of the best ways to achieve such teams is for you to use modern partnered and programme management approaches to procurement (see Section 6.4). This creates teams that satisfy the definition above, work together and respect each other. They learn to generate shared knowledge for the project, across their traditional 'silo' boundaries, both between different disciplines, and between client, consultant and contractor (who are all in the team). The evidence shows that such teams are more creative. These processes are interwoven throughout a project, as shown in Figure 2.6.

However you do it, innovation for sustainable infrastructure will be easier if you can create such teams, and nurture these approaches within them. As such teams develop, their members' positive enjoyment of learning to share knowledge across these boundaries can overcome any initial negative assumptions and constraints. This encourages each individual, some of them also fuelled by enthusiasm for working on more sustainable solutions, to realise that they can learn their own new practice, and help enable the changes that sustainable infrastructure needs.

2.7.   Individual principles
– for action for sustainability

In Chapter 1, we advocated that we must now incorporate sustainable development as a core of good engineering, and a fundamental part of safe service performance. It follows that any individual principles for action for sustainability must lie within our overall

Figure 2.6 Interrelationships between multidisciplinary knowledge creation processes (Fong, 2003)

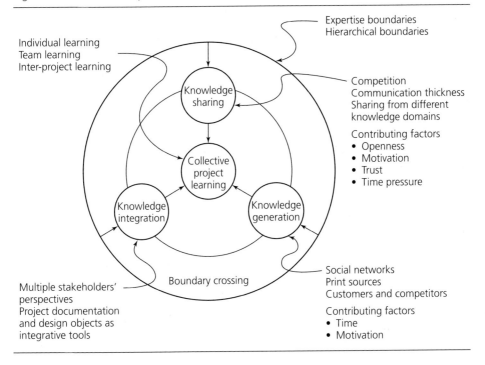

professional ethics. A good statement of these, which is really applicable to all technical disciplines, has been made by the Royal Academy of Engineering, UK. Its four fundamental principles are: (1) accuracy and rigour; (2) honesty and integrity; (3) respect for life, law and the public good; and (4) responsible leadership – listening and informing. Within the third principle, the statement covering (but not using the word) 'sustainability' says:

> Professional Engineers should give due weight to all relevant law, facts and published guidance, and the wider public interest. They should: minimise and justify any adverse effect on society or on the natural environment for their own and succeeding generations.
> (Royal Academy of Engineering/Engineering Council, n.d.)

Usefully, and in line with this book's intent of turning principles into practice, it has also published *Engineering Ethics in Practice: A Guide for Engineers*, which contains detailed case studies, built around real-life dilemmas (Royal Academy of Engineering, n.d.).

We believe that the way to achieve the Royal Academy's principles is for you to use 'better questions to be asked at every stage of project delivery' (Fenner *et al.*, 2006), and specifically to ask yourself, and answer, the question: 'Am I acting sustainably?' By doing this you can develop good sustainability practice in your everyday decision-making. Finally, because moving faster to sustainability requires game-changing innovation, to help you take part you also need to apply two key individual principles.

2.7.1 Principle I.1 – Learn new skills; competences for sustainable infrastructure

All these ideas in Chapter 2 really amount to one overriding need: simply to be able to think more broadly about problems and solutions, drawing on support from other disciplines for help. This ability to 'widen our horizons' is shown in Figure 1.5; it asks for sustainability issues beyond the traditional engineering scope in order to be able to frame and form part of the solution.

To achieve this requires you as an individual to exercise some new skills. Some suggest (Wiek *et al.*, 2011) that your first step is to gradually widen your key competence beyond that of 'problem solver' (which remains vital in itself) to that of 'innovator', 'change agent' or 'transition manager'. To achieve that goal you need to recognise and develop the following capabilities, which reflect all the principles we have been discussing.

- First, think of problems at the **systems level**, described as: 'the ability to collectively analyze complex systems across different domains (society, environment, economy, etc.) and across different scales (local to global), thereby considering cascading effects, inertia, and feedback loops'.
- Second, become able to form an **anticipatory view** of the future. This involves thinking in the short and long term, to test the likely consequences of decisions against a range of possible circumstances.
- Then, encourage **value-focused thinking**, which tests project objectives and outcomes against concepts such as justice, equity, social–ecological integrity and ethics, as well as traditional measures of technical and economic performance.
- Further, **think strategically** about issues, and how transitions can be made to more sustainable solutions. An example here on global warming is to appreciate the extent to which future problems can be reduced, or simply accepted and adapted for.
- Finally, develop your skills at the **interpersonal level**, including communication, negotiation, collaboration and leadership. This has been described as being able to 'facilitate diversity across cultures, social groups, communities, and individuals'.

These new skills are mainly about having the knowledge to affect the 'what' of the change needed, in a project's content, planning and delivery. You will develop these skills as you apply the principles described in this chapter to help secure a sustainable future through your own *practice*, which we go on to discuss in Part II. The last, interpersonal, skill can also help you contribute to the 'how' of the change – how you can help persuade everyone else to change their practice too.

2.7.2 Individual principle I.2 – Challenge orthodoxy and encourage change

Even if you are enthusiastic about trying out all the new practice in Part II, your opportunities to do so will be constrained by the attitudes of everyone around you, and of the 'orthodoxy' – the current business-as-usual practice – within which you work. This is an interacting 'cloud' of barriers to change – regulations, codes and manuals, contracts, incentives, measurement systems and mindsets – that has been called the 'socio-technical lock-in' (Lowcarbonworks, n.d.).

It is not surprising that the current professions, clients and business models associated with infrastructure still require considerable transformation to allow Part II's sustainability practice to be applied widely. And that is the whole justification for this book. Changing the status quo is a separate new skill in itself, of being the 'change agent' or 'transition manager' that we referred to above; being able to 'challenge orthodoxy' and 'encourage change'. You probably have more of this skill than you realise. Part III is devoted to helping you to develop it.

Key principles: summary

Four fundamental principles can guide practice for delivering sustainable infrastructure. Each has 'operational' subprinciples.

A1. Environmental sustainability – within limits
O1.1 Set targets and measure against environmental limits
O1.2 Structure business and projects sustainably

A2. Socio-economic sustainability – 'development'
O2.1 Set targets and measure for socio-economic goals
O2.2 Respect people and human rights

A3. Intergenerational stewardship
O3.1 Plan long term
O3.2 Consider all life-cycle stages

A4. Complex systems
O4.1 Open up the problem space
O4.2 Deal with uncertainty
O4.3 Consider integrated needs
O4.4 Integrate working roles and disciplines

And apply two individual principles:

I1.   Learn new skills – competences for sustainable infrastructure
I2.   Challenge orthodoxy, encourage change

Suggested Further Reading is listed after References.

REFERENCES

Arnstein SR (1969) A ladder of citizen participation. *AIP Journal* **35(4)**: 216–224. Available at: http://lithgow-schmidt.dk/sherry-arnstein/ladder-of-citizen-participation.html (accessed 26 July 2013).

Azapagic A and Perdan S (2011) *Sustainable Development in Practice – Case Studies for Engineers and Scientists*, 2nd edn. John Wiley, New York.

Brand South Africa (2012) Infrastructure drive 'key to job creation'. Available at: http://www.southafrica.info/business/economy/development/infrastructure-160812.htm (accessed 26 July 2013).

Brugnach M, Dewulf A, Pahl-Wostl C and Taillieu T (2008) Toward a relational concept of uncertainty: about knowing too little, knowing too differently, and accepting not to know. *Ecology and Society* **13(02)**: 30.

Bruntland GH (ed.) (1987) *Our Common Future: The World Commission on Environment and Development*. Oxford University Press, Oxford.

Calantonio A (2009) Social sustainability: linking research to policy and practice. Presented at Sustainable Development – A Challenge for European Research, Brussels.

City of Portland (2008) Grey to Green Initiative. Available at: http://www.portlandonline.com/bes/index.cfm?c = 47203 (accessed 26 July 2013).

City of Sydney (2012) *Interim Waste Strategy: Managing the City of Sydney's Resources for a Sustainable Future*. Available at: http://www.cityofsydney.nsw.gov.au/__data/assets/pdf_file/0019/122914/InterimWasteStrategy.pdf (accessed 26 July 2013).

CPRE (2007) Developing an Intrusion Map of England. Campaign to Protect Rural England, London. Available at http://www.cpre.org.uk/resources/countryside/tranquil-places/item/1790-developing-an-intrusion-map-of-england (accessed 26 July 2013).

Dresner S (2008) *The Principles of Sustainability*, 2nd edn. Earthscan, London.

Edwards AR (2005) *The Sustainability Revolution Portrait of Paradigm Shift*. New Society Publishers, Gabriola Island, BC.

Ehrlich P and Holdren J (1971) Impact of population growth: complacency concerning this component of man's predicament is justified and counterproductive. *Science* **171(1)**: 211–217.

Evans R, Haryott R, Haste N and Jones A (2004) The long-term costs of owning and using buildings'. In *Designing Better Buildings: Quality and Value in the Built Environment* (Macmillan S (ed.)). Taylor & Francis, London, pp. 42–50.

Fenner RA, Ainger C, Cruickshank HJ and Guthrie P (2006) Widening horizons for civil engineers – addressing the complexity of sustainable development. *Proceedings of the ICE – Engineering Sustainability* **159(ES4)**: 145–154.

Fiksel J (2003) Designing resilient, sustainable systems. *Environmental Science and Technology* **37(23)**: 5330–5339.

Fiksel J (2006) Sustainability and resilience: towards a systems approach. *Sustainability: Science, Practice, & Policy* **2(2)**: 14–21.

Fleck J (1997) Contingent knowledge and technology development. *Technology, Analysis and Strategic Management* **9(4)**: 383–398.

Fong PSW (2003) Knowledge creation in multidisciplinary project teams: an empirical study of the processes and their dynamic interrelationships. *International Journal of Project Management* **21(7)**: 479–486.

Fratini CF, Geldof GD, Kluck J and Mikkelsen PS (2012) Three Points Approach (3PA) for urban flood risk management: a tool to support climate change adaptation through transdisciplinarity and multifunctionality. *Urban Water Journal* **9(5)**: 317–331.

Gates R and Lee M (2005) Policy Report Social Development. Report to City of Vancouver from Director of Social Planning.

Geels F (2004) *Technological Transitions and Systems Innovations: A Co-evolutionary and Socio-technical Analysis*. Edward Elgar, Cheltenham.

Hakkinen T (2009) The ISO framework for defining sustainable buildings. Presented to UNEP Sustainable Buildings & Climate Initiative (SBCI) Sustainable Buildings & Climate Index (SBCIndex) Global Guide for Building Performance, Paris, Session 2: Pathways to Setting the Bar Building Energy Performance. Annex 2c. Available at: http://www.unep.org/sbci/pdfs/Paris-ISOframework_briefing.pdf (accessed 26 July 2013).

Hawken P, Lovins A and Lovins H (1999) *Natural Capitalism – The Next Industrial Revolution*. Earthscan, London.

Health and Safety Executive (2012) *Construction – Work Related Injuries and Ill Health*. Available at: http://www.hse.gov.uk/statistics/industry/construction/construction.pdf (accessed 26 July 2013).

Hoyer J, Dickhaut W, Weber B and Kronawitter L (2011) *Water Sensitive Urban Design: Sustainable Stormwater Management in the Cities of the Future: Principles and Inspirations for Sustainable Stormwater Management in the City of the Future*. JOVIS Verlag, Berlin.

Hughes W, Ancell D, Gruneberg G and Hirst L (2004) Exposing the myth of the 1:5:200 ratio relating initial cost, maintenance and staffing costs of office buildings. *Proceedings of the 20th Annual ARCOM Conference, 1–3 September 2004, Edinburgh, UK* (Khoswowshahi F (ed.)), Vol. I.: ARCOM, Reading, pp. 373–382.

Huitt W (2007) Maslow's hierarchy of needs. *Educational Psychology Interactive*. Valdosta State University, Valdosta, GA. Available at: http://www.edpsycinteractive.org/topics/regsys/maslow.html (accessed 26 July 2013).

ICE (2008) *ICE Code of Professional Conduct*. Institution of Civil Engineers, London. Available at: http://www.ice.org.uk/getattachment/1ebe1f7e-7b36-43a2-a4eb-520901cc01cc/Code-of-professional-conduct-for-members.aspx (accessed 26 July 2013).

International Transport Forum (2011) Economic Perspectives on Transport and Equality. Discussion Paper No. 2011-09. Available at: http://www.internationaltransportforum.org/jtrc/DiscussionPapers/DP201109.pdf (accessed 26 July 2013).

ISO (2008) ISO 15392:2008 Sustainability in building construction – General principles. International Standards Organisation, Geneva. Available at: http://www.rpd-mohesr.com/uploads/custompages/ISO_15392_2008%28E%29-Character_PDF_documentm.pdf (accessed 26 July 2013).

ISO (2010) ISO 10845-1:2010 Construction procurement – Part 1: Processes, methods and procedures. International Standards Organisation, Geneva. Available at: http://www.iso.org/iso/catalogue_detail?csnumber=46190 (accessed 26 July 2013).

Jackson T (2011) *Prosperity without Growth: Economics for a Finite Planet*. Routledge, London.

John Muir Trust (n.d.) Wild Land Campaign. Available at: https://www.jmt.org/wild-land-campaign.asp (accessed 26 July 2013).

Laws D and Loeber A (2011) Sustainable development and professional practice: dilemmas of action and strategies for coping. *Proceedings of the ICE – Engineering Sustainability* **164(ES1)**: 25–33.

Leopold A (1949) *A Sand County Almanac* Republished in 2006: Ballantine Books, New York.

Lowcarbonworks (n.d.) *Insider Networks. Human Dimensions of Low Carbon Technology*. Centre for Action Research in Professional Practice, University of Bath. Available at: http://www.bath.ac.uk/management/news_events/pdf/lowcarbon_insider_voices.pdf (accessed 26 July 2013).

Maslow A and Lowery R (eds) (1998) *Toward a Psychology of Being*, 3rd edn. Wiley & Sons, New York.

Meadows DH (2008) *Thinking in Systems*. Chelsea Green Publishing, White River Junction, VT.

Meadows DH, Meadows D, Randers J and Behrens III WW (1972) *The Limits to Growth*. Universe Books, New York.

Mindfully (2006) The World Distribution of Household Wealth. Available at: http://www.mindfully.org/WTO/2006/Household-Wealth-Gap5dec06.htm (accessed 26 July 2013).

Mitchard N, Frost L, Harris J, Baldrey S and Ko J (2011) Assessing the impact of road schemes on people and communities. *Proceedings of the ICE – Engineering Sustainability* **164(ES3)**: 185–196.

Morrison-Saunders A and Hodgson N (2009) Applying sustainability principles in practice: guidance for assessing individual proposals. Presented at IAIA09 Impact Assessment and Human Well-Being, 29th Annual Conference of the International Association for Impact Assessment, Accra.

Muldur K (2006) *Sustainable Development for Engineers A Handbook and Resource Guide*. Greenleaf Publishing, Sheffield.

North Yorkshire County Council (2010) What is the community engagement 'ladder of participation'? Available at: http://www.northyorks.gov.uk/index.aspx?articleid=9308 (accessed 26 July 2013).

Ostrom LT and Wilehelmsen C (2012) *Risk Assessment: Tools, Techniques, and Their Applications*. Wiley, New York.

Owen A, Mitchell G and Clarke M (2011) Not just any old place: people, places and sustainability. *Proceedings of the ICE – Engineering Sustainability* **164(ES1)**: 5–11.

Parikh P, Chaturvedi S and George G (2012) Empowering change: the effects of energy provision on individual aspirations in slum communities. *Energy Policy* **50**: 477–486.

Pirrie A, Wilson V, Elsegood J, Hall J, Hamilton S, Harden R, *et al.* (1998) *Evaluating Multidisciplinary Education in Health Care*. SCRE, Edinburgh.

Raworth K (2012) *A Safe and Just Space for Humanity: Can We Live Within the Doughnut?* Oxfam, Oxford.

Repetto R (ed.) (1985) *The Global Possible: Resources, Development, and the New Century*. Yale University Press, New Haven, CT.

Rifkin J (2004) A precautionary tale. *The Guardian*, 12 May.

Rockström J, Steffen W, Noone K, *et al.* (2009) Planetary boundaries: exploring the safe operating space for humanity. *Ecology and Society* **14(2)**: article 32. Available at: http://www.ecologyandsociety.org/vol14/iss2/art32 (accessed 26 July 2013).

Royal Academy of Engineering (n.d.a) *Engineering Ethics in Practice: A Guide for Engineers*. Available at: http://www.raeng.org.uk/societygov/engineeringethics/pdf/engineering_ethics_in_practice_short.pdf (accessed 26 July 2013).

Royal Academy of Engineering/Engineering Council (n.d.b) Statement of Ethical Principles. Available at: http://www.raeng.org.uk/societygov/engineeringethics/pdf/Statement_of_Ethical_Principles.pdf (accessed 26 July 2013).

San Francisco Planning Department (2010) San Francisco's Better Street Plan. Available at: http://www.sf-planning.org/ftp/BetterStreets/proposals.htm#Final_Plan (accessed 26 July 2013).

Snell M (2010) *Cost Benefit Analysis: A Practical Guide*, 2nd edn. Institution of Civil Engineers, London.

Stahel W (2012) The business angle of a circular economy – higher competitiveness, high resource security and material efficiency. *EMF*, 15 May.

UN (United Nations) (1985) *Charter of the United Nations and Statute of the International Court of Justice United Nations*. UN Department of Public Information, San Francisco, CA. Available at: http://www.un.org/en/documents/charter/preamble.shtml (accessed 26 July 2013).

UN (1992) Rio Declaration on Environment and Development. UN Doc./CONF.151/5/ Rev.1. Available at: http://www.un.org/documents/ga/conf151/aconf15126-1annex1.htm (accessed 26 July 2013).

UN (2011) The Millennium Development Goals Report 2011. Available at: http://www.un.org/millenniumgoals/pdf/%282011_E%29%20MDG%20Report%202011_Book%20LR.pdf (accessed 26 July 2013).

US EPA (Environmental Protection Agency) (2007) *Sustainability Research Strategy*. Office of Research and Development, US Environmental Protection Agency, Washington, DC.

Vijaymohanan PN (2008) *Infrastructure, Growth and Human Development in Kerala*. Munich Personal RePEc archive Paper No. 7017. Available at: http://mpra.ub.uni-muenchen.de/7017 (accessed 26 July 2013).

Watermeyer RB (1999) Socio-economic responsibilities: the challenge facing structural engineers. *The Structural Engineer* **77(17)**: 22.

Watermeyer R (2011) The critical role of consulting firms in the acceleration of infrastructure delivery and the improvement of the quality of life. In *New Perspectives on Construction in Developing Countries* (Ofori G (ed.)). Spon, London.

Wiek A, Withycombe L and Redman CL (2011) Key competencies in sustainability – a reference framework for academic program development. *Sustainability Science* doi 10.1007/s11625-011-0132-6.

Wilkinson R and Pickett K (2010) *The Spirit Level: Why Equality is Better for Everyone*. Penguin, London.

Willems R (2001) Executive Vice President, Shell Chemicals: 3rd Middle East Refining & Petrochemicals Event. Bahrain, personal communication, 31 October.

World Bank (2011) *Measuring Inequality*. Available at: http://web.worldbank.org/WBSITE/EXTERNAL/TOPICS/EXTPOVERTY/EXTPA/0 contentMDK:20238991~menuPK:492138~pagePK:148956~piPK:216618~theSitePK:430367,00.html (accessed 26 July 2013).

FURTHER READING

Abraham M (2005) *Sustainability Science and Engineering*, Vol. 1: *Defining Principles*. Elsevier Science, Oxford.

Dresner S (2008) *The Principles of Sustainability*, 2nd edn. Earthscan, London.

Edwards AR (2005) *The Sustainability Revolution Portrait of Paradigm Shift*. New Society, Gabriola Island, BC.

Ehrlich P and Holdren J (1971) Impact of population growth: complacency concerning this component of man's predicament is justified and counterproductive. *Science* **171(1)**: 211–217.

Gagnon B, Leduc R and Savard L (2009) Sustainable development in engineering: a review of principles and definition of a conceptual framework. *Environmental Engineering Science* **26(10)**: 1459–1472.

Meadows DH (2008) *Thinking in Systems – A Primer*. Chelsea Green Publishing, White River Junction, VT.

Moriarty G (2008) *The Engineering Project, Its Nature, Ethics and Promise*. Pennsylvania State University Press, University Park, PA.

Muldur K (2006) *Sustainable Development for Engineers: A Handbook and Resource Guide*. Greenleaf Publishing, Sheffield.

Part II

Practice

This book grew out of the authors' paper 'Widening horizons for engineers' (Fenner *et al.*, 2006), which suggested a set of useful questions to ask to drive greater sustainability. Many chapters in this part start with a quotation; where these are not externally attributed, they are taken or adapted from that paper.

Throughout the stages of this part we suggest actions that will help apply the principles described in Part I, Chapter 2. The chapter structure matches the typical stages of project delivery, as shown in Figure PII. This will allow you to see where you are working in the

Figure PII Opportunities for more sustainable outcomes at typical stages of infrastructure project delivery

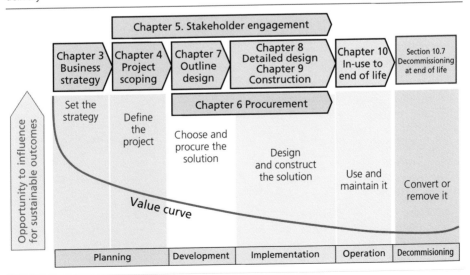

sequence, to help you ask the right questions at the right time, and to take action at the right opportunity.

Different infrastructure sectors and countries have subdivided and labelled these stages in their own ways. The figure uses generic descriptions, which we think will allow you to identify where you are in the process.

REFERENCES

Fenner RA, Ainger CA, Cruickshank HJ and Guthrie P (2006) Widening horizons for engineers: addressing the complexity of sustainable development. *Proceedings of the ICE – Engineering Sustainability* **159(ES4)**: 145–154.

Sustainable Infrastructure: Principles into Practice
ISBN 978-0-7277-5754-8

ICE Publishing: All rights reserved
http://dx.doi.org/10.1680/sipp.57548.077

Chapter 3
Business strategy

3.1. What this stage involves

All the really important mistakes are made on the first day.

(Hawken *et al.*, 1999)

Developing business strategy is the first step in planning, at the very 'front end' of the infrastructure process. This is where owners define what service they must deliver to the public by reference to their purpose, business model and hence strategy. The business strategy unit, the asset management and planning section and, if in existence and operating separately, the sustainability group, will review their purpose, vision, mission, objectives and strategy against current reality and future trends, as well as their customer/user 'market' (which may be real or virtual, if the organisation is a public body). Such a review is usually based on some kind of 'current and future risk' model and assessment. This leads to a decision on what the strategy is, and what has to be done to implement it.

This process determines the customer service model, usually for a fixed planned period into the future, and so assesses the expected demand for the infrastructure service being provided. In turn, this information is used to determine the asset investment needs, as the service supply required to meet those demands. Within this, the organisation must decide

- the serviceability of its existing assets – and hence, the operation and maintenance plans and capital replacement needs (see Chapter 10)
- the gaps in service provision, which may require the development of new projects.

The depth of thinking, and assessment of any need for change in the way this is done, may be constrained by government policies and current regulatory frameworks (as most utility and infrastructure owners are regulated in some way), or by the length of franchise periods. This interaction with external regulation may be a key factor in incentivising or constraining innovation for sustainability.

This process will lead the organisation towards adopting certain objectives, and also set the long- to middle-term timeframe of their planning. For instance, in the UK water sector, the regulator Ofwat requires detailed business plans for 5-year periods, and a strategic direction statement looking 25 years ahead. Such regulation may place extra constraints on developing the innovative approaches needed to cope with climate change and to achieve sustainability for this sector.

Because this initial stage determines everything that follows, you need to apply here all of the fundamental principles discussed in Part I. Environmental (A1) and socio-economic (A2) sustainability should be brought into strategy, while the broader perspectives provided within intergenerational stewardship (A3) and complex systems (A4) help define how to approach and do the planning work. Putting these principles into practice in the business strategy stage will help you create the *potential* for the most sustainable outcomes.

3.2. Sustainability into strategy – and what can engineers do?

What ambitious goals and targets are set, which stimulate creativity and allow innovation?

Sustainability decisions lie at the core of an infrastructure business strategy, whether or not this is consciously acknowledged. This is because making strategy mirrors the famous global environmental impact IPAT equation (see Section 2.2).

All infrastructure organisations are familiar with the primary socio-economic needs of their sector (A2), because these are inherent in deciding demand, and the cost of supply. However, many have not always considered the environmental impacts (A1) of their investment decisions. The IPAT equation shows that this is always fixed by the other determinants, but often it is not apparent at this early stage, because it emerges from a series of decisions made later in the project delivery process. So, if you do not explicitly set environmental objectives at the same time as your socio-economic ones, you actually have made a hidden choice – that the environment matters less. The 'nested' diagram shown in Figure 1.2(b) shows why this is precisely the wrong way round. The Earth's global environmental limits are the first and absolute constraint on what we can safely do. So adopting strategic environmental objectives, set against limits, is one way of changing priorities to reflect this constraint.

Most infrastructure organisations now produce explicit broad commitments on sustainability, which often appear in the 'tag lines' associated with organisations' logos. For example: 'Building a better world' (MWH); 'We shape a better world' (Arup); 'To create, enhance and sustain the world's built, natural and social environments' (AECOM). It is a good and honest start to declare such an intent, but it is harder to quantify such sustainability goals, and to truly embed them in strategy, objectives and asset planning. So practice and performance in infrastructure organisations varies widely, and your opportunities to intervene for greater sustainability, and how you approach them, will depend on the position that your organisation has reached.

3.2.1 What can engineers do?
Whatever the starting point, your opportunities to put sustainability principles into practice lie within the choices made during the strategy stage. There are five key actions that you can take, based on the operational principles set out in Chapter 2.

- Set sustainability objectives and targets, and choose boundaries (O1.1 – Set targets and measure against environmental limits; O2.1 – Set targets and measure for socio-economic goals; O3.2 – Consider all life-cycle stages; O4.1 – Open up the problem space).
- Plan strategically for sustainability – timescale and uncertainty (O3.1 – Plan long term; O4.2 – Deal with uncertainty).
- Choose the business model – sell the 'service' rather than the 'product' (O1.2 – Structure business and projects sustainably).
- Join up thinking – to implement strategy (O4.4 – Integrate working roles and disciplines).
- Consult with key stakeholders, particularly customers, regulators and funders, so that the strategy will be understood, supported and have legitimacy (O2.2 – Respect people and human rights).

We describe the first four of these below; the last one is covered in Chapter 5.

3.3. Set sustainability objectives and targets, and choose boundaries

How commonplace is it to take action, before legislation and regulation require change?

Most infrastructure organisations will work within some aspect of government *policy* or planning. Within this they will set themselves some *purposes* (possibly formalised in a form of public service agreement (PSA), or through the terms of a sector regulator's remit), and from this a series of *strategic objectives* will follow. Each objective should then inform a hierarchy of performance measures and targets that cascade down through the stages of project delivery.

An example for a roads organisation is shown in Figure 3.1. The policy is at the top, the strategic objective is below it, and the next level down defines one key performance

Figure 3.1 A strategic objectives hierarchy: roads. (Adapted from MWH New Zealand sources)

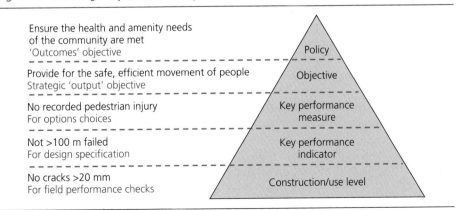

measure which covers one aspect of that objective. This translates that aspect into a measurable target, which you would use in project scoping (see Chapter 4) and would be one criterion for a solutions choice (see Chapter 7). Only one aspect is shown; several others would be required, to reflect the overall 'safe, efficient movement of people' output objective.

3.3.1 Set targets and measure against environmental limits (principle O1.1)

To safeguard planetary environmental resources (see Section 2.3) you need to include strategic objectives that respect global or regional environmental limits. The critical impacts include the following.

- **Global warming control**: set limits to reduce carbon dioxide equivalent (CO_2e) emissions (often referred to as the 'carbon footprint' – see Section 13.3) – both those 'embodied' in any construction and those produced directly from asset operations.
- Set limits to **land and resource use and biodiversity impacts**: both direct and influenced.
- Limit **freshwater use**: with respect to local or regional resources.
- Minimise **chemical pollution**: by limiting the total persistent organic pollutants (POPs) emitted.

It may be convenient to combine some of these within broader strategic objectives; for instance as a total 'ecological footprint' of the organisation, or, for buildings, 'all investments will be to BREAM Excellent standard' This is a good step to take, but the most critical objectives – carbon emissions, and land use and biodiversity – need their own specific limits. Strategic goals to limit general resources use – energy, water and materials – may also be useful. All objectives need to be measured by 'total use' and 'use per user/customer', as well as in terms of 'unit' efficiencies. (This is explained in Section 7.5.2.)

You also need to consider the *indirect* impacts (see Table 2.2) of ocean acidification, atmospheric aerosol load, stratospheric ozone, and the global nitrogen and phosphorus cycles. Your strategy should commit to identifying, measuring and minimising any use of chemicals and emissions affecting these impacts. When countries mandate limits or reductions, follow them. However, for any infrastructure involving large impacts on nitrogen or phosphorus (e.g. chemicals, agriculture, treatment processes), you should set specific limits, because these cycles already greatly exceed (nitrogen), or are close to (phosphorus), their global boundaries (see Figure 2.2).

3.3.2 Set targets and measure for socio-economic goals (principle O2.1)

To reflect the need to serve wider socio-economic goals (O2.1) you need to reflect these goals in the strategy. Of the socio-economic sustainability goals listed in Table 2.4, the relevant ones from the upper list – food, water, energy, mobility, i.e. what the infrastructure is for – are likely to already drive strategy. The exception to this is the last one in the list – building in *resilience* against future risk, (particularly the uncertainties arising from global warming) – which should now appear as an essential strategic objective.

You also need to consider strategic objectives for the *indirect outcomes* in the lower part of Table 2.4 – social outcomes such as social and gender equity, and having a voice; and economic outcomes such as creating wealth, jobs and income. They might include

- focusing on specific groups of users/customers in providing the service (e.g. movement, for disabled people; or poverty reduction for communities)
- targeting improvements in affordability (e.g. minimising the number of customers who are suffering energy and/or water poverty; or affordable access to transport systems)
- setting targets for providing secondary benefits, in jobs, enterprise development and ownership, through ensuring investment in projects and operations for local/regional communities.

Finally, your overall strategy should reflect an acceptance of the absolute 'one-planet' limit on resources use and emissions, by aiming to provide a successful service to customers/users without building in an automatic preference for *growth*.

3.3.3 Set limiting targets for critical objectives

Once you have set objectives, some must be translated into specific *targets*, rather than just a commitment to minimise, maximise or 'consider' them. All these new objectives can provide new sustainability criteria by which alternative project solutions should be ranked (see Chapter 7), but specific targets are to set goals to be met by all strategies, then translated into project scoping (see Chapter 4). Such clear targets help clarify your policy and objectives, particularly for limiting global warming, and they stimulate innovation. So, with respect to average global temperature rise:

> While we must treat the 2°C threshold as a hard limit, our strategies for delivering vibrant businesses and a growing economy within this should be as flexible and innovative as we can make them.
>
> (Brown, 2012)

One feature of many infrastructure sectors (particularly when they are closely regulated) is that they tend to set specific targets only for their main 'what the infrastructure is for' objectives. In the UK, for instance, water quality and pollution targets for water utilities are set by EU directives, and then translated into UK regulatory demands. CO_2e emissions reductions have been agreed politically at EU level, and the UK Government has set legally binding and timed CO_2e reduction targets, but it dictates these only to the energy sector, and leaves others to decide for themselves, under various forms of financial inducement and taxation. Because of their large impacts, all infrastructure organisations should set strategic targets, at least for reducing CO_2e emissions.

In practice, you may need to invoke a combination of regulatory, legal, market and voluntary drivers to argue a successful business case for organisations to set wider sustainability objectives and targets. Using the UK CO_2e emissions example again, general government guidance to organisations on setting CO_2e reduction targets says:

Recommendation 12: Set a reduction target and choose the approach to use. Once you have measured and calculated your total GHG emissions, setting an emission reduction target is the logical next step.

(DECC/Defra, 2009)

The guidance then invokes cost, sector leadership, and branding or market share reasons for setting targets. Many business cases may rest mainly on this mixture of arguments, but you will need to customise it to fit your sector (see Chapter 11). The move by Drax Power Station to maximise biomass fuel use in a previously all coal-fired station shows a strategy shift in response to sustainability drivers (Box 3.1).

Box 3.1 Drax Power Station, UK starts to burn biomass

In July 2012, Drax confirmed that it plans to transform itself into a predominantly biomass-fuelled generator. Initially, Drax plans to convert three of its six generating units to run on sustainable biomass; the first unit will be converted in the second quarter of 2013, and the second a year thereafter.

Drax Power Station in North Yorkshire is already the largest, cleanest and most efficient coal-fired power station in the UK. Our biomass plans will not only strengthen our environmental leadership position, but further enhance our reputation to stay at the forefront of developments to establish effective alternative fuel technologies for electricity generation in the UK.

(Drax, 2012)

3.3.4 Extend the boundaries for your objectives

Another key intervention that you can make to spur innovation for sustainability is to widen the *boundaries* that your organisation recognises it has *influence* over and takes some responsibility for. This recognises the complex system within which the infrastructure sits, described by the operational principle O4.1 – Open up the problem space (see Section 2.6).

The traditional well-established approach is to take account, as in cost accounting, of all those things that can be directly controlled and paid for within the organisation's own infrastructure assets. This is a sound foundation, but infrastructure organisations have several interfaces with their assets, across which they exert much influence. You can cross these, to take into account the wider system space surrounding the infrastructure assets. They include

- with your supply chain, which builds projects and supplies operations – with large sustainability impacts, on energy, resources and global warming CO_2e emissions
- with your customers/users, and their behaviour – reflecting benefits and impacts of the service during its in-use operation
- with the surrounding communities and environment, both urban and rural, which may gain benefit or suffer impacts in various ways.

Using the UK CO_2e example again, the UK Government guidance on setting boundaries is:

Recommendation 3: Measure or calculate emissions that fall into your scopes 1 and 2. **Discretionary:** Measure or calculate your **significant** scope 3 emissions in addition to your scopes 1 and 2.

(DECC/Defra, 2009)

The guidance defines 'significant' first as a factor of scale – 'What are the largest indirect emissions-causing activities with which your organization is connected?' – and then in terms of importance to the organisation's business, importance to stakeholders and potential for reductions. On these criteria, supply chain emissions, and infrastructure 'in use', will clearly qualify as significant.

Any practice of defining strategy to deliver positive sustainability performance must also apply operational principle O3.2 – Consider all life-cycle stages. This further confirms that it is essential that the boundary within which strategy is measured includes 'in-use' performance. In Section 10.2, Table 10.1 shows the many ways in which interactions between asset designers, operators, user behaviour and policy during the in-use stage give opportunities for greater, or less, sustainability gains. For instance, over the lifetime of a road, the vehicles using it will produce much higher CO_2e emissions than those associated with the road construction itself. These vehicles can be influenced by a combination of active traffic management, car-ownership arrangements and licensing and taxation policy. Such interactions occur in each sector, and it is necessary to try to identify and influence such opportunities as a result of planning strategy.

This requirement to plan across wider boundaries, and cover all life-cycle stages, can introduce uncertainty and discomfort in comparing alternative strategies, when some are more controlled and certain than others. Timescale and uncertainty in strategic planning practice are discussed next; and the issues are explored further in Chapter 4.

3.4. Plan strategically for sustainability – timescale and uncertainty

How are methods such as scenario planning used to explore a range of futures?

The traditional timescales for strategic planning are relatively short, whereas the implications of some sustainability trends, such as climate change, are not very apparent in the short term, but are much larger than expected in the long term. So principle A3 – Intergenerational stewardship (see Section 2.5) requires that for sustainability we must put into practice the operational principle O3.1 – Plan long term. In Section 2.5, we suggested that this means at least 25 years ahead, or in many cases to 2050. Loss of ice cover due to global warming and the resulting certainty of rising sea levels poses questions for 100 years ahead. By adopting such long-term perspectives you can help avoid unexpected consequences, build in resilience, and avoid possible 'lock-in' to expensive technical solutions that do not respond well, or cannot be adapted, to the changing circumstances you will face.

The biggest *practice* challenge that all this gives you is, of course, how to know the future, and its uncertainties! Because of the difficulties this gives for *forecasting (prediction)* – the traditional method used for planning – many planners have preferred to put the long term into the 'too difficult' box, and plan only short term. This is not enough. Your long-term planning for sustainability involves three components (see Section 2.5).

- Imagining what *the surrounding world*, within which the infrastructure must work, will be like.
- Deciding what different characteristics *the sustainable infrastructure* must have – must it be decarbonised, for example?
- Estimating how the changes in the first will interact with the changes in the second, to *determine actual demand*, *use* and *impact* (good and bad).

Your difficulty now is that, as the challenges of global warming, population demographics and energy costs increase (to name just some), the future will not – indeed must not – be like the past. So, the traditional methods of forecasting and limited analysis of future risks used by infrastructure strategists and asset planners become inadequate. You need new approaches; which can address those three components more explicitly. Two useful ones – foresighting (including scenario planning) and backcasting are described in Section 13.9.

Taking such a long-term forward look involves also recognising complex interdependencies, and the need to deal with uncertainty. The best approaches are discussed in the next sections.

3.4.1 Apply a 'visioning approach' for sustainability

As you are seeking greater sustainability, the second component above – deciding what characteristics the infrastructure must have – must involve making a break from past practice. You want to look far enough ahead so as not to be constrained by current barriers and mindsets: 'we can never do that'. For this reason, applying a 'sustainability funnel' approach, asking 'Where do we want to be by (say) 2050?', is useful. Figure 3.2 shows an application, in this case focusing on the challenge of global warming.

All the long-term risks and opportunities arising from both adapting to and controlling global warming are shown by the pressures on the narrowing sides of the funnel. You should first attempt to quantify those changes as well as you can, to define where the 2050 end of the tunnel is. Looking out to 2050 forces asset planners to raise questions that are really useful to ask, but may be very difficult to answer. As Figure 3.2 shows, your choice is then between adopting a 'strategic innovation' approach, by aiming investment at the long-term sustainable objective (in 2050), or aiming at a shorter term, through a more 'reactive compliance' response. The result might show that the 2050 end of the funnel is sufficiently well defined to be able to take 'low regrets' or 'no regrets' decisions to invest in large, long design-life assets in spite of some uncertainty.

Figure 3.2 Strategic visioning approach: choosing a strategic response to global warming (Adapted from The Natural Step (2013))

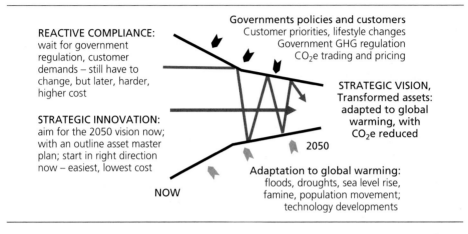

REACTIVE COMPLIANCE:
wait for government
regulation, customer
demands – still have to
change, but later, harder,
higher cost

STRATEGIC INNOVATION:
aim for the 2050 vision now;
with an outline asset master
plan; start in right direction
now – easiest, lowest cost

NOW

Governments policies and customers
Customer priorities, lifestyle changes
Government GHG regulation
CO_2e trading and pricing

STRATEGIC VISION,
Transformed assets:
adapted to global
warming, with
CO_2e reduced

2050

Adaptation to global warming:
floods, droughts, sea level rise,
famine, population movement;
technology developments

An example of doing this in the water sector is the Australian ACT Government's decision on the Murrumbigdee to Googong pipeline (ACETW, 2009). The government carefully looked 20 years ahead, to decide on difficult choices between strategic options involving very long-term dam and pipeline assets.

But long-term planning does not necessarily mean building long design-life infrastructure. This might create a technical lock-in in the future, which can be difficult to escape from. If the uncertainty is high, you can adopt 'adaptive management', building in resilience, flexibility and adaptability, and perhaps deciding on a deliberately shorter design life (see Section 4.5) for the infrastructure. (The UK Climate Impacts Programme (n.d.) gives useful approaches for dealing with uncertainty in climate change risks. Viable choices in the light of future uncertainty and climate-change risks include options that are: no-regrets, low-regrets, win–win and flexible/adaptive management. See also Appendix A.)

It may be hard for your infrastructure strategists and asset planners to change and adopt these methods, rather than undertake forecasting, because they must acknowledge that *we cannot control the future*. One result of this is that the vision may remain just that – only a 40 years ahead aspiration – while your asset planners go back to planning the next 5 years on the old 'predict from the past' basis. This short-term path is unlikely to provide a trajectory in the 'vision' direction. This is where the follow-up 'backcasting' is essential. It forces your planners to envisage specific steps and changes, backwards from the vision, to connect back to the present day, free from current limitations and constraints. Then, you will be able to build these changes into the shorter term detailed planning and project scoping.

Such practices can help you to take account of more uncertainty, over longer timescales, than you have done in the past; but new approaches to dealing with uncertainty are still needed.

3.4.2   Decide strategy in spite of complex uncertainty

Operational principle O4.2 is 'Deal with uncertainty'. In one sense, engineers do this every day; the established practice of *risk assessment* (Actuarial Profession/Institution of Civil Engineers (ICE), 2005) is dedicated to that end. This is usually good at dealing with the most obvious type of uncertainty, which arises from **incomplete knowledge** (Table 2.6). This occurs typically through a lack of data, or its unreliability, or even the lack of a complete theoretical understanding of an issue. You can generally handle this by using an extension of existing practice. This can include the following well-used approaches.

- Collect *better data*; carry out *more research*; apply *range estimation*, with confidence limits; use *simulation models* to do sensitivity analysis; use *expert opinions*.

You can also use these newer ones

- allow *uncertainty propagation* in models; do *scenario analysis* (see Section 13.9); adopt the *precautionary principle* (see Appendix A).

However, there are also two other kinds of uncertainty, arising from complexity (see Table 2.6). Pure **unpredictability** arises from complex systems constantly adapting to new situations, often through non-linear and chaotic behaviour. Ways to deal with this include the following (Brugnach *et al.*, 2008).

- **An assets-based approach**: *exert physical control* (e.g. build a dam to control irregular river flows). This is a large infrastructure option, with large sunk costs and a lack of flexibility; and failure risks extensive damage. A similar option could be to *create additional capacity* in the asset, with some of the same downsides. Such an approach could also include: *identify multiple future scenarios* and develop robust solutions under each; diversify measures so that at least one measure will be effective if others fail.

Or, alternatively.

- **Adaptive-management-based approaches**: these include *deal with the consequences* of the uncontrollable phenomenon and not the phenomenon itself (e.g. move coastal communities rather than provide sea defence); *use multiple strategies* to control negative effects in the consequences chain; *use temporary adaptation strategies*, i.e. measures that are feasible within the timeframe of an unfolding event (e.g. Thames barrier); *improvise*, based on good monitoring, communication and coordination in crisis situations.

A useful guide to modern strategic business risk and handling uncertainty is given by the Institute and Faculty of Actuaries and the ICE in the UK. They recognise the need to analyse 'systemic risk':

> It is helpful to think of a business as a complex system of people and assets, which operates within the even more complex system of the outside world. ... It

is therefore worth devoting time, effort and resources to the development of deeper thinking about the purpose of the business, how it interacts with its environment and the wider world, and how it can be adapted to cope better with a wide range of uncertain future possibilities.

Institute and Faculty of Actuaries/Institution of Civil Engineers (2011)

One way you can introduce the objectives and opportunities of sustainability is to put them into familiar 'risk' language, and add them into the organisation's strategic planning approach. The systemic risk approach recognises the interactions between risks, which may be missed in traditional risk management (MacAskill and Guthrie, 2013).

Finally, **multiple knowledge frames** may be held among the different stakeholders, causing each of them to define a problem differently. Ways to deal with this are covered in Chapter 5.

3.5. Choose the business model – sell the 'service' rather than the 'product'

Where and with whom do the benefits of the scheme lie, and who wins and who loses, could the scheme be adapted to be win–win?

To operationalise O1.2 – Structure business and projects sustainably, in Section 2.3 we introduced the principle of using a service model, not a product model, to provide an infrastructure service. This switches from pricing the product to pricing the service. It transforms the effect of the commercial incentive to make more money, by aligning it with the environmental incentive to use fewer resources and reduce impacts. This is a fundamentally more sustainable business model.

You can adopt such a 'service' model wherever possible in practice. Sector examples might be the following.

- In transport, road vehicles are leased, not bought. The lease includes the cost of fuel, servicing and insurance, and the terms incentivise the driver/user to minimise mileage and wear and tear and to maximise fuel efficiency. Contracts between car-leasing firms and traffic management/road toll organisations might even record and reward active good driving behaviour.
- Water and energy utilities adopt metering and tariffs structures that are a disincentive to using more product, and actively intervene with customers to help them reduce usage, perhaps with 'pay as you save' terms. A good example is the business model provided by Business Stream (2013) in Scotland to its industrial and commercial water customers. It combines strong customer service with assistance to use less water.
- In buildings, facilities management contracts are set in the same way as the French *Chauffagistes* (see Box 2.2). This can be extended to include not only heating and ventilation, but also air-conditioning, lighting, cleaning and all other building services.

The detail of payment arrangements may differ from case to case, but the fundamental common feature is that the customer pays less to maintain a satisfactory service, and the infrastructure service provider makes a good commercial return by using/selling less material, and less energy-intensive 'product'. In nearly all such examples, conservation behaviour by the user can assist the efficiency and reduce the cost of the service by the provider. So there is added attraction in relating service prices to performance targets, with 'gain/pain share' arrangements in which both parties save money if they collaborate to further reduce the cost and use of resources. An example of applying this approach to infrastructure supply chain procurement is given in Chapter 6.

3.6.   Join up thinking – to implement strategy

What safeguards ensure the performance of the scheme is taken into account over *all* its stages?

It can be difficult for your organisation to 'cascade down' a new sustainability strategy so that the actual infrastructure on the ground delivers it. It takes a long time to change asset performance averages, because assets have long lives and the turnover is slow. You may, understandably, be cautious and risk averse in adopting new approaches, because infrastructure often impacts people's health and well-being and the environment, and you are under public scrutiny. Introducing new objectives may require a significant change in mindset, which requires persuasion.

So to implement sustainability effectively you need to help your organisation think more intelligently and creatively about these complicated problems, establishing clear links between the strategy targets and actual implementation, both across all parts of the organisation, and down through each stage of project delivery. This amounts to adopting a practice of *collective project learning*, by *generating*, *sharing* and *integrating* the new knowledge needed across all stages of project delivery and between multi-disciplinary project teams (see Figure 2.6; Pritchard and Chesterman, 2013). This implements principle O.4.4 – Integrate working roles and disciplines. It requires consistently applied objectives, and supply chain procurement and contracts that incentivise the strategic outcomes the organisation wants to achieve. Your final requirement is to foster good connections with the sustainability champions at each stage. This is summarised below, and addressed again in the later sections of the book.

3.6.1   Apply sustainability objectives consistently, throughout project delivery

The strategic objective must frame and direct the decisions to be taken at each stage. For example, if the strategy sets a target to reduce CO_2e emissions, then this must reappear at each stage, to drive decisions at an increasing level of detail, as shown in Figure 3.3. These different stages are often undertaken by separate parts of the organisation, so it is particularly important that clear instruction is given to, say, the asset planners for project scoping (see Chapter 4), and the capital investment managers for procurement (see Chapter 6).

Figure 3.3 Cascading objectives through project delivery stages

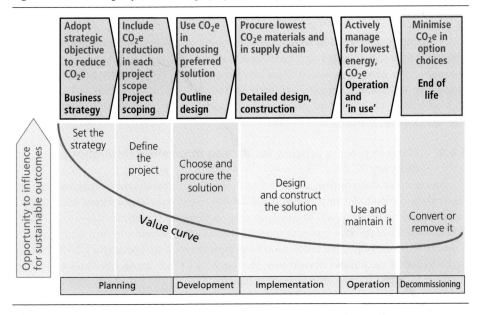

This does not just apply to environmental objectives; you can use it equally to target a service at a disadvantaged social group, or to require procurement to deliver secondary socio-economic outcomes in terms of jobs or training.

3.6.2   Use 'joined up' forms of contract and performance measurement

When you have set the project scope and have also paid attention to the sustainability objective in procurement, then make contract terms and the way in which performance is measured match this. You can incentivise and reward behaviour by the owner and project supply chain to collaborate in designing and building the asset for 'maximum whole-life sustainability'. This could include cost, CO_2e emissions, ecosystems impact and socio-economic objectives (see Chapter 6). You can apply the same requirement to incentivising operators and users, through contracts and terms of use, to the same objectives (see Chapter 10).

3.6.3   Arrange effective handover between successive stages in project delivery

Each stage often involves different departments of the asset owner and different supply chain players. It is therefore important that you have systems that encourage people to collaborate across each delivery-stage boundary, to enable the seamless transfer of the objectives for more detailed interpretation in the next stage. This is particularly important from outline design through to detailed design, and from construction to operation. Principle O4.4 – Integrate working roles and disciplines – applies here. Examples include involving users/operators in outline design, and engaging constructors to interact with designers on 'constructability' during design. During the later stages of construction, involve the operators to collaborate with the constructors, in planning and training for equipment handover, commissioning and operation.

These examples may seem to be just standard good practice, as indeed they are. Nevertheless, innovative projects are particularly dependent on good hand-on processes, because of the need to *generate*, *share* and *integrate* the new knowledge to pass on. If innovative sustainable projects are seen to fail, the innovation itself is likely to be blamed, rather than just bad practice. So, innovation for sustainability requires your strong attention to best practice in delivery.

3.6.4 Have in place a process for driving innovation throughout delivery

For the reasons described in Section 3.6, risk-averse infrastructure organisations find innovation difficult. To implement sustainability, help your organisation to develop a *strategic innovation process*. Do not limit this, as thinking sometimes does, to just technology, but embrace also business models, methodologies, relationships and mindsets. You can learn to help drive innovation by adopting individual principle I.2 – Challenge orthodoxy and encourage change (see Section 2.7). To see approaches and techniques that can help you do this effectively, see Part III.

Business strategy: summary

- Set sustainability objectives and targets within strategy, and choose wider boundaries; both against environmental limits and for socio-economic goals. Set absolute limit targets for critical objectives.
- Plan strategically for sustainability: use a 'visioning' approach to examine longer timescales, and learn to decide strategy in spite of complex uncertainty.
- Choose a business model that sells the 'service' rather than the 'product', aligning commercial success with sustainability.
- Develop an 'adaptive' climate change strategy.
- Implement your strategy throughout all stages of project delivery; apply sustainability objectives consistently, align your forms of contract and performance measurement, and arrange effective handover between stages.
- Adopt a process that drives innovation throughout delivery.

REFERENCES

ACETW (2009) Progress Report and Recommendations to ACT Government, December 2008. Water Security/Major Projects, presented to 5th Annual Water Summit.

Actuarial Profession/ICE (2005) *Risk Analysis and Management for Projects*, 2nd edn. ICE, London.

Brown B (2012) Hard limits – flexible strategies. In *The Future in Practice: The State of Sustainability Leadership 2012*. Cambridge Programme for Sustainability Leadership, Cambridge. Available at: http://www.cpsl.cam.ac.uk/SearchResults.aspx?searchStr=hard%20limits (accessed 29 July 2013).

Brugnach M, Dewulf A, Pahl-Wostl C and Taillieu T (2008) Toward a relational concept of uncertainty: about knowing too little, knowing too differently, and accepting not to know. *Ecology and Society* **13(2)**: 30.

Business Stream (2013) Available at: http://www.business-stream.co.uk (accessed 29 July 2013).

DECC/Defra (Department of Energy and Climate Change/Department of Food, Environment and Rural Affairs) (2009) *Guidance on How to Measure and Report Your Greenhouse Gas Emissions*. Available at: http://www.defra.gov.uk/publications/files/pb13309-ghg-guidance-0909011.pdf (accessed 29 July 2013).

Drax (2012) *About Drax's Biomass Plans*. Available at: http://www.draxgroup.plc.uk/biomass/cofiring_plans (accessed 29 July 2013).

Hawken P, Lovins A and Lovins H (1999) *Natural Capitalism – The Next Industrial Revolution*. Earthscan, London.

Institute and Faculty of Actuaries/ICE (2011) *Handling Uncertainty – The Key to Truly Effective Enterprise Risk Management*. ICE, London.

MacAskill K and Guthrie P (2013) Risk-based approaches to sustainability in civil engineering. *Proceedings of the ICE – Engineering Sustainability* **166(4)**: 181–190.

Pritchard S and Chesterman D (2013) Leading Complex Projects: Learning through Action Research; the Ashridge Journal 360, Summer 2013. Available at: http://www.ashridge.org.uk/website/content.nsf/w360/360 + 2013 + Summer + Leading + complex + projects: + Learning + through + Action + Research + ?opendocument (accessed 9 September 2013).

The Natural Step (2013) *The Funnel – Society is Being Squeezed*. Available at: http://www.naturalstep.org/en/the-funnel (accessed 22 July 2013).

UK Climate Impacts Programme (n.d.) *Identifying Adaptation Options*. Available at: http://www.ukcip.org.uk/wordpress/wp-content/PDFs/ID_Adapt_options.pdf (accessed 29 July 2013).

Sustainable Infrastructure: Principles into Practice
ISBN 978-0-7277-5754-8

Chapter 4
Project scoping

4.1. What this stage involves

So tell me what you want, what you really, really want!

The Spice Girls

This stage sounds simple. Asset planners take the strategic analysis (see Chapter 3) and turn it into a list of projects to fill the gap. This may include entirely new projects, but also 'capital replacement' projects to replace worn-out or inefficient assets and maintain serviceability. Increasingly that must include a wider scope, minimising energy use and carbon dioxide equivalent (CO_2e) emissions (see Section 10.6). For each project they then work out a project scope, which should be a performance specification that defines the required project *output*. When all projects are added together they combine to produce an *outcome* that satisfies the strategic service need. The scopes will also include some constraints on projects, such as location or scale or technology to be used.

The process may also be called 'project appraisal', or 'pre-feasibility'.

There is complexity in this process, because it is actually encountering principles A3 – Intergenerational stewardship, and A4 – Complex systems, and making sometimes unconscious assumptions about the *timescale* and the *complex adaptive system* within which the project will operate. This includes how operator or customer/user behaviour will actually respond to each new asset, and about how it interacts with the other components of its surrounding systems. Getting these assumptions wrong can lead to *unexpected consequences* (see Section 2.6). At worst, they may cancel out any service gain from the new asset.

4.2. Sustainability in project scoping – and what can engineers do?

What agencies and other organisations are involved in adopting a coordinated approach to infrastructure provision?

The traditional approach in project scoping has often been relatively narrow, with organisations looking only within their own assets – what they pay for and control directly. The focus is on maintaining *service efficiency* by examining the *risks*. Infrastructure

Figure 4.1 Widening the project scoping space

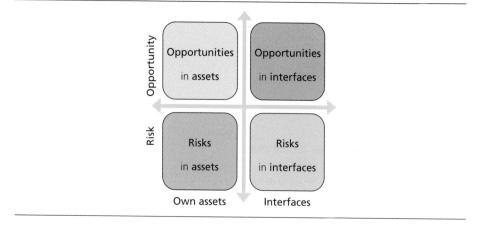

sectors have their own sophisticated and well-embedded methodologies for asset risk assessment and project scoping. These have been effective, but from a sustainability viewpoint many suffer from the limitations of narrowness, perhaps not recognising the real complexity of the context in which the project will exist. In the matrix shown in Figure 4.1, this pays attention only to the bottom left-hand box – risks in assets.

To scope more sustainable projects, you need to adopt *systems thinking* methodology (see Section 13.11), and apply principle O4.1 – Open up the problem space (see Section 2.6). How can you translate this into practice? Various infrastructure sectors have their own approaches, including wider scoring tools such as CEEQUAL (see Section 13.5), but one generic way of doing it is for you to add the idea of looking at *opportunities* and across *interfaces*, not just at *risks* in *assets*. This introduces four areas for possible action, instead of only one, as shown in Figure 4.1, which is a simpler version of Figure 2.4 that is more applicable to project scoping.

By adding the *interfaces* you can look at where the project will interact with its system. This can be the socio-economic aspects of user and customer needs and behaviour, local communities and the supply chain; and also the environmental and technology aspects of the rural natural environment, urban planning and design, and other interacting projects. Adding in the *opportunity* viewpoint gives you a wider, more holistic perspective, which is more likely to spark innovative ideas and produce multiple benefits. We give examples below.

4.2.1 What can engineers do?

Hopefully, the boundaries of your organisation's strategic risk and gap analysis will already have been extended to include its area of influence, and the in-use stages of its services (see Section 3.3). If so, this will directly feed into the *interfaces* areas, when looking for project scopes to respond to the risk and gap analysis. Even if this approach has not been raised at the strategy stage, project scoping provides you an opportunity to apply the ideas.

Whatever the starting point of your infrastructure sector or organisation, you have four main opportunities to put sustainability principles into practice in project scoping. The following are the key actions that you can take, based on the operational principles set out in Chapter 2.

- Ensure asset planning is anchored in strategic sustainability objectives (O1.1 – Set targets and measure against environmental limits; O2.1 – Set targets and measure for socio-economic goals).
- Include projects across the interfaces, and take opportunities (O4.1 – Open up the problem space; O4.3 – Consider integrated needs).
- Structure your project scope sustainably (O1.2 – Structure business and projects sustainably; O3.1 – Plan long term; O4.2 – Deal with uncertainty).
- Ensure that the project scope defines required performance, not a pre-judged solution.

The above actions are discussed in the following sections. Another thing you can do is to read Chapter 10, which deals with the in-use to end-of-life stages of project delivery, to help give context for the third point above. This will pick up early on the issues that will be critical to success at those stages, and frame the scope to take them into account.

Each of these ideas may challenge existing methodologies, so such widening of scope can be troubling for your asset planners and engineering teams. It takes you into areas outside your direct control, where uncertainty of response is greater and outputs are harder to measure. It involves interactions with stakeholders and society outside your comforting arena of technical predictability (see Chapter 5). In spite of this, it is exciting, worth the effort and essential. It is at this early delivery stage that you can create the largest scope for sustainable solutions.

4.3. Ensure asset planning is anchored in strategic sustainability objectives

Which long-term aims are considered as important drivers to responding to today's immediate problems?

It can be a challenge to ensure that longer term innovative strategic objectives which you developed by top-down visioning (see Section 3.4) are actually carried into the pragmatic, bottom-up, shorter term asset planning process which determines project scopes. The process typically begins by asking: 'What do we have now?' A risk analysis is then done and an incremental change approach applied. This may prevent sufficient weight being given to the strategic objectives, with them being regarded as 'nice to have but not essential', and the process may become locked into the 'reactive compliance' mindset that the strategy was trying to avoid (see Figure 3.2).

The result may be a project scope that underestimates the extent of future change, building assets that have to be changed to a different purpose, or abandoned, before the end of their design lives. These assets will be unsustainable, and wasteful in cost and affordability terms, so you need to avoid this outcome.

4.3.1 Develop a future sustainable assets 'destination', to guide asset planning

One way for you to anchor visionary strategic objectives into project scoping is to carry out a long-term (say 2050) master plan exercise – with the key assumption that *you have no assets now*. If you were starting again, with no constraints from existing assets, what would you build? This provides a good anchored way, intelligible to asset planners, to explore how different your future needs will be, compared with what you have now. Backcasting (see Section 13.9) can then help you work out how to get from the present to that different future. Such output can generate for you an outline, but asset-based, framework against which to test the direction of more conventional predictions, and give you short-term redirection. You can then ask of each possible scope: 'Is this heading in the master plan direction?' and 'Can we see how it can help us get there?' If the answer is 'no', then you need to rethink, with more innovation.

Of course in making such a plan you will have to deal with future uncertainty, such as: 'What will have been the effects of global warming?', 'Where will customers be, what needs will they have?' and 'What technologies will be available?' There are good arguments for doing this, is spite of the uncertainty. You can use better climate science as it becomes available for prediction of global warming, and tools such as scenario planning (see Section 13.9) and system dynamics (see Section 13.11) can help you test the responsiveness of possible actions. Predicting as far as 2050, at least, is necessary to appreciate the large changes that adaptation to global warming will itself require. Perhaps most importantly, identifying the extent of the uncertainty will avoid overcertainty in your organisation's planning (see Section 3.4), and build more resilience into the ensuing project scoping.

4.3.2 Translate sustainability objectives into project performance specifications

This refers to the need for interconnected thinking (see Section 3.6). You need to translate your extra environmental and socio-economic objectives (see Section 3.3) into project scoping, so they can cascade down through project delivery. For example, referring back to Figure 3.3, you must carry down the strategic target to 'reduce CO_2e emissions to X' as a specific CO_2e output target in each project scope. Similarly, for each project you can use the opportunity to include local community or regional jobs, wealth and business creation targets (see Section 3.3). This was also one conclusion of a report by the Omega Centre (2010) on introducing sustainability into project scope and appraisal, which considered several of the tools and practices we discuss in this book, and identifies multi-criteria analysis as a key tool for appraisal including sustainability objectives (see Section 13.10). The report recommended:

Step 9: Ensure that the environmental, social and institutional goals adopted in the design and appraisal processes are firmly embedded in arranging the funding, procuring the contract team for construction and setting up the operation.

(Omega Centre, 2010)

If you have had no opportunity to set such objectives at the strategy stage, then project scoping provides another chance for you to raise the missed issues, using the practice discussed in Chapter 3. Your infrastructure organisation may have made some high-level sustainability statements as part of its public relations programme, and you may be able to quote and use these as a 'hook' for raising these issues at the scoping stage.

Finally, infrastructure organisations are continually under pressure to extend, and to reduce the costs of, services – to do more with less. 'Cost reduction compared with last time' is frequently a performance target in project scoping. Engineers are used to this requirement to deliver cost-efficiency, but overfamiliarity can take you into a zone of cost-reduction fatigue. One utility has found that setting new challenging sustainability targets, such as a '50% reduction in embodied CO_2e' (see Box 11.7) has both succeeded and delivered more cost reductions too. So translating such sustainability objectives and targets into project performance targets can re-motivate your project teams and reinvigorate innovation.

4.4. Include projects across the interfaces, and take opportunities

Who has responsibility for seeking integrative solutions (e.g. between hard (build) and soft (non-build) measures)?

There are three types of infrastructure project scope arising from the different quadrants in Figure 4.1.

- The traditional 'supply to meet new demand' project, by extending the provision or quality of the *product* and building new *assets*. This is in 'risks in assets', and is likely to be the least sustainable.
- A variety of 'manage (or reduce) demand' solutions, achieved by inducing changes across the interface with the user/customer, or by influencing upstream behaviour, so as to continue to supply the *service* they need but without new assets. This may invoke the service business model (see Section 3.5). These solutions are within 'risks and opportunities in interfaces', and are likely to be more sustainable because they use less resources and cause less emissions and pollution.
- A project directly serving sustainability, generated by looking for hidden opportunities within your own assets. This is in 'opportunities in assets'.

We examine the second and third types below.

4.4.1 Scope projects across the interface with users/customers

Referring back to one of the three examples from Section 3.5, this approach might mean that a water demand-management project could involve, rather than building new supply capacity

- upgrading 'smart' metering and data collection (technical assets, but within the water company's boundary)
- adopting a tariff structure that rewards using less water (customer interface)
- selling or providing low water use appliances for customers (this could target poorer customers) to help them reduce water use, possibly with 'pay as you save' terms (intervening in household assets, and more customer interface).

Such projects are more complex, because they require more parties of different backgrounds to collaborate, but they can have multiple benefits, matching the principle O4.3 – Consider integrated needs. For instance, this project could not only satisfy customer demand for the water service (social), but also reduce demand on water resources and reduce CO_2e emissions (environmental), and make water costs more affordable for poor customers. This would reduce water poverty while still providing sufficient income to the water utility (economic).

A difficulty in persuading your asset planners to scope such 'manage-demand' projects is that they involve some loss of control, and extra uncertainty and complexity. Deliberately designing assets to assume a flat or lower future demand, assuming success is achieved by other means in changing behaviour (rather than extrapolating upwards the past trend), is a real act of faith for an asset planner. Planners fear that if the behaviour change is unsuccessful, service to users will suffer, and poor asset planning will be unfairly blamed. So your adopting such project scopes requires strong statements of intent, and leadership, education and support for employees, in this new mental territory. One example of this leadership is Southern Water's 100% water metering programme, for demand reduction, in the UK, which is discussed in Box 4.1.

Box 4.1 Southern Water tackles water scarcity and water poverty – together

---

This started as a technical programme – install meters, to reduce demand – but it became a user-focused programme, giving customers detailed help to save water and money.

Leadership needed to persuade the board and asset planners, consult users, undertake a cost–benefit analysis, and make a cost-effective case to the regulator (this is the first such UK programme). In addition, the organisation had to work with the local community and schools, and gain customer trust through a partnership with a non-governmental organisation (Groundwork). All this enabled a culture change to be made in the Southern Water customer team.

Southern Water accepted that metering can raise bills for poor, large families. Together with Groundwork, Southern Water is tackling this issue directly, by working with families to save water and money. Since the metering programme began, 60% of customers with meters have seen a reduction in their bills, on average a reduction of £11 a month. Of course, reducing usage also helps customers lower their energy bills, on average by £200 a year.

(Fielding-Cooke, 2012)

You may be able to reduce the scope of a new project by first examining what the user 'really, really wants', in order to try to reduce demand. This at the heart of the Rocky Mountain Institute (RMI) Factor 10 Engineering approach (Franta, 2006; Rocky Mountain Institute, n.d.) to whole system design

- downstream before upstream
- demand before supply
- application before equipment
- people before hardware
- passive before active
- quality before quantity
- and lastly, capture multiple benefits from single expenditures.

Their case studies are good demonstrations of these ideas, and are worth reviewing.

4.4.2   Involve local communities and supply chains

There are opportunities for local communities to be involved with projects, as users, employees and operators, supply chain providers or owners (see Section 5.3). The best time for you to start considering these opportunities is at the scoping stage. Many infrastructure systems involve distributed facilities that will require local data collection, alarm systems and regular but simple maintenance. The traditional economy-of-scale approach assumes that only a centralised approach can be relied on, with ITC-based SCADA systems and centrally based maintenance staff. These are brittle systems that can be expensive, and susceptible to communications loss (see, for example, the critique of Australian national urban policy by Hill (2011)). A more resilient alternative might be to rely on decentralised, trained, part-time, local staff. Adopting this approach requires a change of mindset as to what is most sustainable for the future. A good example of road maintenance in South Africa, serving multiple needs, is described in Box 10.2.

4.4.3   Consider possible project consequences

The examples given above involve accepting that there can be a two-way interaction between new infrastructure operation and customer/user behaviour. A new building will generate more or less traffic, depending on its interaction with urban plans and design, its car-parking facilities and its relation to public transport.

Such interactions can generate systemic 'emergent' changes, even when these are not intended. Demand for an *assets* project for extra water treatment can actually be caused by farmers' excess pesticide and fertiliser use on land in the catchment area, which is damaging water quality. Such a situation led Wessex Water, UK, to adopt a project scope *across the interface* with local farmers – the company helped farmers use less pesticide and fertiliser. This more nearly met the principle of 'polluter pays', and provided an integrated project with multiple sustainability benefits. It greatly reduced resource use, CO_2e emissions and pollution, compared with 'build new treatment' solutions, while also reducing the water company's costs by a factor of about 6 (Wessex Water, 2011), and assists farmers' efficiency and income.

So, an essential part of project scoping is to examine the *system* that each project sits within, in order to anticipate likely interactions. You can then open up and explore the potential scope, to minimise unwanted effects and take advantage of beneficial ones.

4.4.4 Seek new opportunities in your assets

In one sense, the example above also illustrates the 'seek new opportunities' approach. The farmers who participated in the scheme discovered that their land assets, managed more carefully, could deliver more benefits, beyond growing food: achieving better water quality for the water company, more biodiversity for wildlife, and more income. Potentially, such land-based approaches can reduce flood risk too.

A technology-based example is shown in Box 4.2. The treatment plant delivers multiple benefits: pollution control and renewable energy for the city, plus a reduction in net cost and CO_2e emissions, and a lower effluent discharge temperature, which is better for wildlife in the harbour.

4.5.   Structure your project scope sustainably

To what extent is any natural capital, lost as an integral part of the scheme, sought to be replaced and replenished?

As you decide your project scope, you can maximise its potential for sustainability by considering three issues, which we discuss below.

- Apply a hierarchy for sustainability.
- Consider the design life.
- Use a 'sustainable cost-effectiveness' approach for cost comparisons.

4.5.1 Apply a hierarchy for environmental sustainability

The alternative project scopes that we have just discussed are examples of applying the principle of 'use less/control at source'. These are powerful ways for you to increase sustainability, because they *directly reduce* the 'P' and 'A' terms – the *demand* – in the IPAT equation, which determines environmental impact (see Section 2.2). They are at the top of the 'sustainability hierarchy' of project options for environmental sustainability within principle O1.2 – Structure business and projects sustainably. Figure 2.4 showed this hierarchy, with six levels from 1 (preferred) down to 6 (no longer acceptable), and explained why.

Your project scoping should always test the hierarchy. Start at the top, and seek to apply the highest approach in the sequence possible. Even if the scope, innovatively considered, still demands a large new project, it must be at level 5, with environmental damage fully compensated for; level 6 is no longer acceptable. You should make up any unavoidable losses by creating at least replacement environmental assets (e.g. wetland areas or wildlife corridors), and require the project to bear its real financial cost, as in the successful example shown in Box 4.3. Just the *costing* of this addition

---

Box 4.2 Sewage effluent provides district heating/cooling for Helsinki, Finland

---

The water company realised that the sewage effluent from its treatment works (serving 800 000 people) was also a large energy resource. The company added heat pumps to extract low-grade heat from the effluent, to serve the city's district heating and cooling systems (Fred, 2008).

The heating power achieved is 90 MW and the cooling power is 60 MW. The investment cost was €30–35 million.

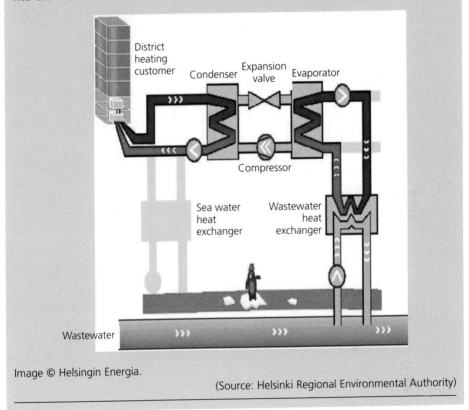

Image © Helsingin Energia.

(Source: Helsinki Regional Environmental Authority)

---

into all options may move a preferred project scope back to using less land, or to the use of 'brownfield' land.

The sustainability preference for multiple small projects (level 4 in the sustainable hierarchy) over one large one reflects the idea of using *appropriate scale*, in which impacts are dispersed and thus are more likely to lie within the absorption capacity of the local environment. A 'multiple projects' choice could also mean choosing an extended timescale, and phasing your response to meet a need in a series of smaller steps. This may be a good response for dealing with slowly changing, uncertain

101

Box 4.3 Creating new environments – Batheaston Meadows in Avon, UK, created as part of the A4 bypass

To mitigate the loss of 'natural habitat', projects should create at least as much 'replacement' green land as they destroy.

SAID Report (ICE, 1996)

Image © ICE Publishing

We cannot 'engineer' the detailed biodiversity but we can provide the potential; for nature to use.

(Charles Ainger, NCE Debate, 6 December 2001)

targets, such as adaptation to global warming. When it is feasible (which is not always), preferring this phased 'adaptive management' approach (see Appendix A) avoids over-reaction, and allows you to apply improvements in data, and understanding in the later phases. This is a useful decision-making principle that the UK Climates Impact Programme (2013) refers to as making 'no regrets/low regrets' choices.

4.5.2 Consider the right design life

The other way in which timescale is critical to infrastructure projects is in your choice of the *design life*. Most sectors have established practices, often not much discussed, which get re-adopted for detailed design. Basic design lives vary widely, being directly controlled by the properties of the materials you specify They range from 3–5 years for electronics/ITC in SCADA equipment, through 10–20 years for steel and other metals in MEICA, pumping and process equipment or vehicles, and extend to perhaps 20–40 years for road cross-sections, and to 40 to over 100 years for timber, plastic, steel, concrete and brick in civil engineering, buildings and structures. The old assumption was that the use or need for the component would not change before the end of its 'natural', material-dependent, design life.

One striking difference between sectors is in the materials and design life adopted for tanks used for storage or treatment processes. The energy and petrochemical sectors use steel and plastic, which have short design lives, while the water industry has often used concrete, which has a much longer life. This partly reflects the different historical rates of change in each sector (as well as different attitudes to the right balance between capital and maintenance cost). Such choices and assumptions are challenged by the increasing rates of change you face, with pressures arising from principle O3.1 – Plan long term – being particularly driven by considerations of global warming. Box 8.2 gives an example with a deliberate choice of a different design life, then choosing materials to match – not the other way round. This describes a water filter project which adopted a shorter 15-year life, using a refurbished concrete solution, which matched the available remaining life of the other parts of the plant on the site; after this time a new build for the whole plant is likely to be required. If the conventional solution using new concrete had been adopted, the last 25 years of material design life would have been wasted.

Sustainability pressures demand that you plan longer term. Combined with increased future risk and uncertainty, this requires that you apply principle O4.2 – Deal with uncertainty. The greater awareness of uncertainty that you gain using sustainability visioning processes such as the 'funnel' (see Figure 3.2) will likely increase the need for shorter design lives in some cases. Any project that has to be abandoned before the design life of the materials ends will have wasted money, and therefore will not have been a financially sustainable solution. Instead your sustainable response seems likely to trend to two extremes, and thus a conscious choice between them is required for each component.

- Using short design lives, for 'adaptive management' – using reusable or recyclable parts or units, and more use of steel, plastic and timber.
- Using a long life design, but building in flexibility in use, allowing for whole system re-specification, additions or reuse for other purposes.

Where existing infrastructure practice falls between these two stools, you need to reconsider it. The project scoping stage gives you an important opportunity to examine and set any overriding design-life requirement as part of the performance specification. It is frequently the point after which the project is handed over to an external supply chain for design, which can then make materials choices accordingly.

4.5.3 Apply 'sustainable cost-effectiveness' in comparing scope options

Setting a project scope often involves comparing several possible scope options, and deciding between them. This is often referred to as 'pre-feasibility studies'. The decision-making processes involved are explored in more detail in Chapter 7, but because cost–benefit analysis (CBA) is more often applied at this earlier stage, we refer to it here.

In adopting a sustainable cost-effectiveness approach you should be taking sustainability principles into account while making cost-based infrastructure decisions. This means applying practice in line with the principles identified in Section 2.4.

- Include all stages of the project and compare on the basis of whole-life costs (see Section 13.7).
- Include all the whole-system costs within the boundaries of the appropriate system within which the project works, and consider integrated needs.
- Include in any CBA the costs and benefits of environmental and social externalities (see Section 13.6). However, do not use a CBA and a notional price to cover any global or irreversible environmental *boundary* impacts. These should be met by setting an absolute target that no option can exceed (see Section 2.3).

4.6. Ensure that project scope defines required performance, not a pre-judged solution

How has technical advocacy for predetermined solutions been avoided?

As shown in the water sector story in Box 1.5, sometimes the brief defined by project scoping pre-judges the solution. In that story, 'add extra treatment' was specified, rather than defining the required performance output. This was, really: 'manage the cryptosporidium risk down to an acceptable level'. The cost and CO_2e savings were made by not building new treatment facilities. This better solution both crosses the asset interface (see Section 4.4), and comprises 'control at source' – level 1 in the sustainability hierarchy (see Figure 2.4). Having a predefined solution can severely constrain your opportunity to innovate for sustainability.

Scoping that directs a particular solution may often be unintended. The situation can arise because of a typical engineer's training, falling back on previous experience and conventional designs used successfully in the past. Many engineers first learn their trade in the later stages of project delivery, in design or construction, and then some move 'upstream' into project planning. As they then scope projects, it is easy to have in mind the way they would have solved the problem, but this is conceived against criteria that are no longer appropriate. Whatever the reason, you must try to keep the performance scope as wide as possible, with no preconceived solutions in mind. Whenever you come across solution-narrowed project scoping, you should challenge it, right at the start of the outline design stage (see Section 7.3).

So, at the end of the planning stages you have completed the business strategy and project scoping. This means that you can now move on to the development stage, which is covered Chapters 6 and 7.

Project scoping: summary

To anchor asset planning in your strategic sustainability objectives, develop a future sustainable 'destination' to guide it, and translate sustainability objectives into project performance specifications.

Widen project scopes across the interface with users/customers, involve local communities and supply chains, and consider wider possible project consequences.

Seek new opportunities in your assets, not just risks.

For a sustainable project scope, apply a scoping hierarchy for environmental sustainability, consider the right design life and apply 'sustainable cost-effectiveness' for cost evaluation.

To maximise innovation, ensure that the project scope defines the required performance, not a pre-judged solution.

REFERENCES

Fielding-Cooke J (2012) Southern Water – the future, next exit. Presented at Water and Innovation – Learning from Innovators, CIWEM, London.

Franta G (2006) A Factor 10 Solution: integrated design in architecture. In *Factor 10 Engineering for Sustainable Cities*. IABSE Henderson Colloquium, Cambridge. Available at: http://www.istructe.org/iabse/files/henderson06/Paper_05.pdf (accessed 29 July 2013).

Fred T (2008) Large-scale heat transfer from wastewater to city heating and cooling systems. *IWA WWC Water and Energy Workshop*, Vienna.

Hill D (2011) Same old new world cities. *Architecture Australia* **100(2)**. Available at: http://architectureau.com/articles/same-old-new-world-cities (accessed 29 July 2013).

ICE (Institution of Civil Engineers) (1996) *Sustainability and Acceptability in Infrastructure Development: A Response to the Secretary of State's Challenge*. SAID Report. Thomas Telford, London.

Omega Centre (2010) *Incorporating Principles of Sustainable Development within the Design and Delivery of Major Projects: An International Study with Particular Reference to Major Infrastructure Projects*. Report prepared for Institution of Civil Engineers and the Actuarial Profession. Bartlett School of Planning, UCL, London. Available at: http://www.ice.org.uk/getattachment/e0308561–9b79–4b10–9297-d490cb8c4a9e/Incorporating-Principles-of-Sustainable-Developmen.aspx (accessed 29 July 2013).

Rocky Mountain Institute (n.d.) 10xE: Factor Ten Engineering. Available at: http://www.rmi.org/10xE (accessed 29 July 2013).

UK Climate Impacts Programme (2013) *Principles of Good Adaptation*. Available at: http://www.ukcip.org.uk/essentials/adaptation/good-adaptation (accessed 29 July 2013).

Wessex Water (2011) *Catchment Management. Managing Water, Managing Land*. Wessex Water, Bath.

Sustainable Infrastructure: Principles into Practice
ISBN 978-0-7277-5754-8

ICE Publishing: All rights reserved
http://dx.doi.org/10.1680/sipp.57548.107

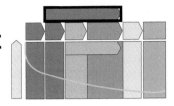

Chapter 5
Stakeholder engagement

5.1.   What this stage involves

> Your own pigs don't stink.
> A German farmer talking about community owned wind turbines (Gipe, 2012)

Stakeholders are simply all those parties or actors affected by a project or the delivery of an infrastructure service. For all organisations, these will include the following participants.

- Users or customers who receive the service.
- Employees or operators who use the assets to provide the service.
- Investors or funders who supply the finance.

Unlike many other branches of engineering, one feature of infrastructure projects (e.g. a power station, or a railway line) is that, in order to serve the wider community, they nearly all need some centralised or linear engineering structures. These must be built in, or through, a small number of affected local communities. These minority groups 'lose' by receiving negative local impacts, so that the majority of us can 'gain'. How a *fair* balance is achieved in this inevitable clash of interests is at the heart of good stakeholder engagement. So, for utilities and other infrastructure owners and operators, there are extra key stakeholders

- local affected communities
- other interest groups, who may or may not also be users/customers.

In addition, most utility/infrastructure owners are regulated in some way. This will always be by general regulation on projects and construction through normal planning procedures, but also by social and environmental regulators. Moreover, in monopoly or near-monopoly sectors, there may be sector economic regulators whose role is to 'simulate competition'. These form another, absolutely critical, stakeholder, who may well have powers to impose objectives and procedures. So, we must also add to the stakeholder list

- regulators of all kinds.

Within the discussion of A4 – Complex systems, in Section 2.6, we identified one type of uncertainty as **multiple knowledge frames**; this acknowledges that different stakeholders

107

define a common problem differently. The range of ways in which you can deal with this can vary as follows (Brugnach *et al.*, 2008).

- The traditional 'design and defend' approach – **persuasive communication** – which involves convincing others of your viewpoint (often from the power narrowly vested in your professional prestige).
- Stakeholder dialogue approaches, including: **open dialogue** across all stakeholders; possibly with a **negotiation approach** to achieve mutual gains, to seek an agreement that makes sense from multiple perspectives. You can aim for **synergistic win–win outcomes** and to avoid **win–lose positions**.
- You are aiming to avoid conflict – either **cold conflict**, where parties avoid each other, or **hot conflict**, where parties try to impose their will by force or power.

Examples of direct conflict can be found in the anti-road-building protest of the 1990s at the Newbury bypass, the M3 at Twyford Down and also the second Manchester airport runway.

Good stakeholder engagement includes all your forms of communication and relationship-forming with these groups, using the second option above rather than the other two, in order to best serve community needs and to learn from them, and to influence and comply with regulators' requirements. In addition, this engagement can also enhance the sustainability and efficiency of the service, and its delivery.

Stakeholder engagement will be needed throughout project delivery, but with different groups and in different ways in each stage. You will need different forms of engagement for each.

5.2. Sustainability in stakeholder engagement – and what can engineers do?

How has a fair foundation for the scheme been established with stakeholders?

The general art of stakeholder engagement is not the subject of this book – the interested reader can find good accounts in Chinyio and Olomolaiye (2009), Sommer *et al.* (2004) and Walker *et al.* (2008). We focus on the *commitment to doing the process* and the *quality* of the process, to put into practice operational principle O2.2 – Respect people and human rights. If you do it well, stakeholder engagement can also enhance the sustainability of the outcome. This makes it a win–win for all.

With 'regular' stakeholders, your issue is to engage them in the sustainability objectives and strategy, and use their knowledge and interest to widen the ideas raised for more sustainable outcomes, as discussed in other sections.

- **Users or customers** – jointly research their real service needs, including the potential for applying the 'service' model and behaviour change (see Section 3.5), and engage with them on the chosen solutions (see Section 7.6).

- **Employees or operators** – use their ideas, and engage them in decision-making and sustainable operation (see Sections 4.4, 6.3, 7.6 and 10.5).
- **Investors or funders** – engage them in the definition and achievement of sustainable financial success.

Current stakeholder engagement practice varies widely: from the traditional practice of 'the engineer is the expert' who will 'propose and defend', to various forms of participatory planning (the first two methods of dealing with multiple knowledge frames, discussed above). In Section 2.4, we referred to Arnstein's trenchant criticism of worst practice, which is sometimes, unfortunately, justified, and is worth repeating here:

> Many planners, architects, politicians, bosses, project leaders and power-holders still dress all variety of manipulations up as 'participation in the process', 'citizen consultation' and other shades of technobabble.
>
> (Arnstein, 1969)

Projects often do use, and show the limitations of, the traditional approach when dealing with the local community regarding complex socio-technical systems. However, practice is changing and developing rapidly, and there are many examples of good methodologies and case studies (Govert and Stahre, 2006).

5.2.1   What can engineers do?
Whatever your organisation's starting point, the key principles to consider are O2.1 – Set targets and measure for socio-economic goals, and O2.2 – Respect people and human rights. Your opportunities to put these into practice lie in making a series of choices about engagement. Your first step is to make these choices explicit; they are often made by default, or omitted, without being recognised, because they are buried within traditional practice.

So, choose the right time within delivery to take these actions, to apply the operational principles in Chapter 2

- give local communities roles in projects (O2.1 – Set targets and measure for socio-economic goals)
- choose community representatives, and methods of engagement (O2.2 – Respect people and human rights). Work effectively for sustainability, with regulators.

We discuss these in the following sections.

5.3.   Give local communities roles in projects

Have genuine concerns been considered with an openness and willingness to adapt?

Local communities affected by infrastructure projects will always suffer some local impacts, so that others can benefit. So your key question here is: 'Does the affected local community get as much compensation, jobs or wealth out of it as they could have?' This reflects the need to address socio-economic goals (see Section 2.4).

An effective 'traditional' approach would aim for good local consultation and public relations, with minimum local disruption and 'nuisance'. A sustainable approach takes this as a minimum starting point, and looks to see what wider *roles* you might give members of the local community. Could you compensate them better, or give them project roles such as designers, constructors, subcontractors, suppliers, employees or operators, or even owners? We look at each of these in turn.

5.3.1 Provide community compensation

However good the consultation and care that is taken, the locations needed for many major projects can do inescapable damage to those in local communities whose land and houses are lost, or who are forced to move. Large dam projects are particular examples of this. Other impacts on communities include the degrading of local landscapes and spoiling of familiar views, and the permanent introduction of extra noise, traffic or pollution. Compensation practice varies greatly, from not recognising local rights at all, to at least paying for compulsory purchase.

It is your responsibility to look out for those who suffer losses as a result of the project. A sustainable approach recognises that *extra compensation* is necessary for the affected community, to reward them for accepting individual suffering arising from local impacts. So, to ensure that the benefits of the project are widened as far as possible, consider the following.

- Make an honest assessment of impacts on local people, being realistic about risks, benefits and costs, using assessment methodologies that are transparent and represent best practice (e.g. the Equator Principles (2013), for major resources or pipeline projects (see Box 11.5 and Appendix A), or the World Commission on Dams' guidelines for dam construction (International Rivers, n.d.; World Commission on Dams, 2000).
- Consider also providing any missing community infrastructure needs, as part of the mitigation within the project.
- Evaluate the impact on future generations and how they may be affected long term.
- Put an early focus on direct extra financial compensation for local community members. For property loss, or disruption, pay a *premium* over the market value, as the UK's HS2 high-speed rail plans to do. (For example, the suggestion in the SAID Report (ICE, 1996) to pay over the market price for housing purchase; and Manchester Airport Extension, UK, spent £15 million on environmental mitigation and compensation (Manchester Airport, 2007).)

5.3.2 Use community members as employees, or as part of the supply chain

Short-term compensation is a basic right, but you can give local communities a larger stake. Your infrastructure project investment and operation can make major contributions to local (and national) employment and skills, and local wealth creation and retention over a long period of time via the creation of jobs (see Table 2.4). Reflecting this, ISO 10845 on construction procurement (Watermeyer, 2011) recognises that procurement strategy may legitimately have a strong 'secondary purpose': to contribute to socio-economic development in the country and community (see Section 2.4).

Priority use and support of local firms and people is often a feature of infrastructure procurement in countries that have recognised inequality issues, and implement positive development policies.

- In the USA, tenders for urban municipalities' projects can require that part of the design and construction work is done by (or subcontracted to) minority groups. An example is given in Box 5.1, and New York State has a body supporting such development (Empire State Development, n.d.).
- In South Africa, a number of pioneering infrastructure projects have invested successfully in large-scale local employment, skills training and contractor development (see Section 6.3).

Box 5.1 Requirement to use and train local firms – San Francisco, USA

Many municipal works tenders in the USA include clauses requiring that the supply chain contributes to federal and local 'affirmative action' legislation. This supports racial and gender equity.

They do this by specifying that a minimum proportion of subcontract work goes to minority-owned businesses (MBEs) and women-owned businesses (WBE). This example is from a San Francisco consulting engineer work tender (San Francisco PUC, December 2003):

Consulting work package	MBE goal (% share)	WBE goal (% share)
A – $2.5 million	9	6
B – $1.75 million	10	4
C – $1.75 million	8	4

Some countries and regions (such as the EU) have regulated in exactly the opposite way (i.e. against being allowed to favour local resources) in order to achieve an 'efficient' market. Nevertheless, the socio-economic development principle means seeking to keep wealth within the local communities affected, and you should consider how far you can do this, within the law.

The best examples of these ideas are when their application also contributes skills and motivation for the smooth running and success of the project itself. For instance, on a pipeline project, the managers knew that avoiding local cultural sites and artefacts would be critical, and they did not even know where they all were. So they directly employed local heritage leaders as advisors and inspectors, gaining their expert knowledge, and making it unlikely that they would disagree with decisions (author's personal communication).

Similarly, in Phase 1 of a public housing refurbishment project, inspectors were employed by the contractor to do final detailed quality assurance acceptance checks. However, the tenants occupying each house (who of course suffered much disruption from the contract) frequently disagreed with their acceptances. This resulted in a long 'snagging list' of complaints, further work and poor relationships. Therefore, in Phase 2 the contractor employed some of the tenants themselves as quality assurance inspectors. The process thus gained from the tenants' expert knowledge and benefitted from their authority with fellow tenants to gain acceptance of the decisions (author's personal communication).

5.3.3   Offer communities co-ownership of projects

Finally, in addition to the roles discussed so far, in some cases it may be possible to offer the community co-ownership, and the ability to share in the wealth generated by the project over the long term. This contributes to wealth (see Table 2.4).

For example, inland wind farms can be a contentious issue. The frequently adopted investment model has been that a remotely owned energy generator buys land, gains planning permission (after a fight with the community), and constructs a wind farm. One or two local farmers are highly rewarded, because their ability to sell their land for wind generation gives them a large unearned profit, but the community gets none of the renewable energy generated. Contrast this with the community ownership model, which has been used extensively in Denmark (Wind-Works, n.d.). One reason why there has been a large penetration of renewable wind energy in Denmark is because the installations are about 70% community owned, thus avoiding NIMBY-ism.

This partly reflects a long tradition of this ownership model in Denmark, combined with the relatively small scale of each project, but neither of these attributes is essential. There are an increasing number of inland wind farms in the UK in which the developer is a remote commercial generation utility, but a local community enterprise has been set up to take an ownership share, and profits long term from the development (see e.g. Energy4All, 2013). Clearly, such models cannot be applied to all types of infrastructure, but there is no reason why they might not be applied to local energy generation, transport, buildings or some utilities.

As you seek sustainability in projects, consider these possibilities – and do so before you start community consultations. Imagine the difference in community attitude that you could generate by the prospect of co-ownership.

5.4.   Choose community representatives, and methods of engagement

How have the interests of those not well represented, or not represented at all, been recognised and embraced?

In many countries, most of the community engagement at the project delivery stage is likely to involve two elements.

- A final formal 'yes' or 'no' to the project (in the UK referred to as 'planning permission') by national, regional or local government.

And, before this, the following.

- A wide range of informal or formal dialogues with community members and interest groups, to maximise the chance that the formal permission will be granted. (These choices and processes may also be applied to wider ongoing stakeholder engagement.) An example is South Lakeland District Council's stakeholder engagement strategy, which uses the same 'ladder of participation' as North Yorkshire County Council, shown in Table 2.5 (see Box 5.2).

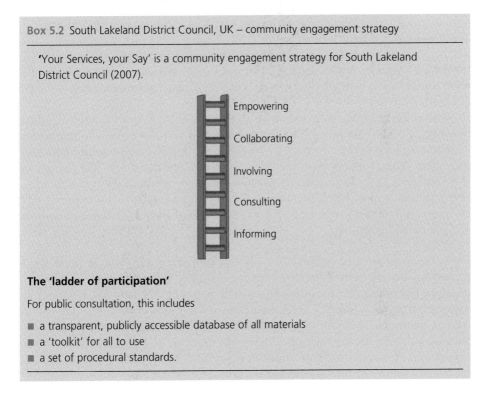

Box 5.2 South Lakeland District Council, UK – community engagement strategy

'Your Services, your Say' is a community engagement strategy for South Lakeland District Council (2007).

Empowering

Collaborating

Involving

Consulting

Informing

The 'ladder of participation'

For public consultation, this includes

- a transparent, publicly accessible database of all materials
- a 'toolkit' for all to use
- a set of procedural standards.

How this is done, particularly the second element, provides the opportunity to put into practice principle O2.2 – Respect people and human rights.

5.4.1 Do formal 'planning permission' effectively

How is the relationship with other professional and special interest groups (e.g. planners) managed?

In the first case, the formal process of decision-making will be laid down by planning regulations and related to government structures. The community will be represented by some form of local planning committee, often with elected members. They will not have *carte blanche* to accept or reject a proposal, as rules about what can be built and where will usually be constrained by regional and local plans, and additional detailed regulations. Some of these can be very extensive (see e.g. The London Plan (Greater London Authority, 2011)).

This formal planning process will require project sponsors to provide all the information needed, and will ensure that all and any voices and opinions are heard. The decision then gives, in most places, democratic control over the development.

In many countries such systems of formal approval are subject to continual struggles and change. This is often between central and local control, particularly for large projects of national importance (e.g. nuclear power stations). Areas of disagreement can centre around exactly what 'sustainability' means, and how it can be expressed in projects. The definitions we provide here may help. In deciding how much disruption it is fair for a community to suffer, you may find it appropriate to consider the real need for the project. A major project serving a national or regional public need would be due more consideration than a project serving a private commercial interest only.

In some cases, planning rules can drive more sustainable solutions, while in others they can constrain innovation. One first contribution you can make is to thoroughly understand all the formal processes, and make the required submissions in compliance with them. Then you can push the boundaries, interpretations and rule changes of these processes as far as possible, to maximise the project's sustainability contribution. One common way in which planning rules serve sustainability is by requiring that environmental impact assessments (the names vary, see Section 13.1) are carried out and submitted for scrutiny.

For contentious projects, you may face a further formal test; for example, in the UK a project may be referred to some form of planning inquiry. If this occurs the project is, in principle, less sustainable, because the decision is then made remotely, not locally, by an independent planning 'expert' – an extreme form of remote top-down power. The inspector's recommendation is then subject to a final decision by a central government minister. Such inquiries can be very legalistic, lengthy and delaying, as well as expensive. So you should aim to avoid this step if possible – by engaging in a constructive dialogue with the community before the formal processes start.

5.4.2   Engage with the community – but who represents it?

In choosing who to consult with, you can support the three social goals listed in Table 2.4: social equity, gender equality and a strong 'voice'. You need to consider any disadvantaged community members. A natural place for you to start when identifying stakeholders to represent the local community is the local democratically elected bodies such as town, village or parish councils. However, as Box 5.3 shows, this may not be enough. Infrastructure projects often raise impacts and benefits that are well outside such bodies' normal remit or expertise, and such local bodies themselves may not give a fair voice to everyone in the community. So, it is best to advertise widely, with an open-ended call for any interested parties or people to register their interest, so that the widest number of the members of the public can respond with their concerns.

Box 5.3 Chalfont St. Giles, UK – mobile phone mast

A mobile phone company negotiated a deal for an annual payment, with the (elected) parish council. Then the community found out about the proposed phone mast, and some were strongly opposed to the perceived risk from electromagnetic radiation:

'Our village under threat – this is how it will look if we don't act now!'

Going to the 'democratic' representatives may not be enough: wider engagement is needed.

(Source: Author's personal experience)

This approach has sometimes been seen as risky by project sponsors, on the grounds that it may encourage special interest groups that are likely to resist the project, as well as any individual locals with extreme views who 'oppose everything'. This is a somewhat stereotypical view, but the risk does exist. However, fairness argues that you must accept the risk; projects cannot be hidden from local interest, and it is risky to try, given the extensive use of social media. Also, local communities will know of individuals with extreme views, and may be the best people to manage such responses to the project.

Even when you have made such an open inquiry to the community, the response may still not represent 'justice' and good practice. 'Justice through participation', as shown in Figure 1.5 of the widening horizons framework (Fenner et al., 2006), raises difficult questions, such as the following.

- How can you take into account local cultural, religious, ethnic and gender issues?
- How do you act if local power structures adversely affect peoples' ability to be heard? (For instance, local culture may give no voice to women, so they may not respond.)
- How can you then recognise and include the interests of those not well represented or not represented at all (including future generations)?

There are no perfect answers to these dilemmas, but you can ask the questions explicitly and do your best to be 'fair'. You should also try to understand and to take into account views that are important but are not being represented.

5.4.3 Be ambitious, but practical, about community engagement

Many infrastructure professionals have long been cautious about engaging with 'non-experts', and so the old 'design and defend' approach has been well entrenched. Sometimes the politically correct language of 'stakeholder engagement' has been used without any substance beneath it, and Arnstein's criticism (see Section 5.2) can be justified. Table 2.5 shows 'levels of stakeholder participation', ranging from just *informing* at the bottom, through *consulting*, *involving* and *collaborating*, to *empowering* at the top. The full table is worth examining; Arnstein's harsher descriptions of the bottom levels – *therapy, manipulation* and *tokenism* – are indeed represented by worst practice.

You can adopt practical levels of engagement between the extremes. To respect people and deal fairly, aim for as high a level of engagement as possible, preferably at the *involving*, *collaborating* or *empowering* level. To help you, the following lessons have been learnt from experience with communities. (See Section 7.6.4.)

- Get the community to choose its representatives, and to manage any local extremist views (often they can this better than external organisations or professionals).
- Trust them to be able to cope with the issues; do not assume that the engineer or proposer understands the complexity and that the public does not.
- Educate them on the realities of the choices; do not allow false choices, as this represents power without responsibility.
- Involve them at the right times in the process. This may involve you seeking local knowledge and ideas early, and then doing work in the project team before bringing final options and choices back to them.
- Have a transparent audit trail.

In the end, what you can actually offer communities may be nearer to what Arnstein disparagingly called *placation*:

the ground rules allow have-nots to advise, but retain for the powerholders the continued right to decide.

(Arnstein, 1969)

To develop trust in the engagement process it is important that you to answer honestly from the start the question 'Who will decide?'. If you do this, sustainable stakeholder engagement can become practical, not just an ideal.

5.5.   Work effectively for sustainability, with regulators

What assumptions are made regarding increasing levels of regulatory control?

Your infrastructure organisation, whether public or private, will be subject to *environmental* regulation, through some form of environmental agency that sets standards which have to be met. Similarly, *social* regulation is required by a range of bodies that set health and safety standards, and define levels of service. In addition, any organisations that act in monopoly or near-monopoly sectors will have some form of *economic* regulator, which agrees customer prices, sometimes by simulating competition by comparing relative performance. Such regulators are really making sustainability decisions, because their job is to achieve that triple bottom line balance between the environmental and social standards that can be achieved, and economic affordability for customers and users.

The regulators have to set rules, and can adopt quite rigid methodologies, with a 'demonstrate compliance' mindset. This may constrain the rate of innovation required to respond to the challenges of sustainability. Sometimes this involves setting one narrow objective, without consideration of newer sustainability ones with which the regulations have not yet caught up. For example, the ever rising regional health risk and environmental standards set by the EU may be reaching the stage of diminishing returns; they have been pursued without recognising their negative impact on new concerns such as carbon dioxide equivalent (CO_2e) emissions (see Figure 1.3).

More subtly, in some countries the social and environmental objectives for investment are set so remotely by specialist regulators that both the strategy and project scoping stages are imposed on your infrastructure organisation from the outside. So your community is unable to make its own judgements about the right sustainability balance, and can lose the appetite and skills for innovation; rather, it waits to be told what to do. In contrast, in New Zealand, municipalities make their own regional plans, in which they have much more say in and ownership of the agreed range of sustainability objectives. (For a recent review of its (relative) success, see Parliamentary Commission for the Environment (n.d.).)

Economic regulators have to define regulatory boundaries around the organisation's assets, within which they have to approve 'business plans' in order to agree on price rises for customers. This boundary definition tends to make the organisation define its strategic boundaries in the same limited fashion, and ignore the sustainability opportunities at its interfaces with the 'product in-use' phase (see Sections 3.3, 3.5 and 4.4).

Also, regulators can require a very 'tight' level of proof that expenditure is justified. This can make it difficult to justify the strategic longer term investments and mindsets that provide your capacity to innovate. This includes investment in research and development of better data and modelling capabilities, and physical experimentation and pilot work.

Some economic regulators measure a set of efficiency and asset serviceability indicators across the companies they regulate, to establish comparative competition. Only performance against these indicators determines allowable pricing, so these alone drive company behaviour. Therefore, if new sustainability key performance indicators, such as CO_2e emissions, are not recognised in this set, it can be difficult to get them taken as seriously. Such pressure of comparative competition on private utilities can be strong enough to inhibit the knowledge sharing that would help rapidly spread new best practice.

However, there are some encouraging trends. For example, for privatised water companies in England and Wales the economic regulator Ofwat is pulling back to a more hands-off, outcomes-based regulation, and will include CO_2e emissions as an outcomes indicator:

[*we will*] ... shift from heavy reliance on annual regulatory returns from the companies ... make greater use of targeted information requests or focused, in-depth examinations of high-risk areas or companies.

(Ofwat, 2010)

Similarly, the UK's Environment Agency is recognising the value of a 'use-less-water' model:

the challenge is for regulatory incentives that fundamentally shift water businesses to both 'sell less, not more', and 'build less, not more'.

(Trevor Bishop, *Utility Week*)

The Environment Agency also recognises that a partnership approach with stakeholders is not incompatible with retaining a strong regulatory discipline. An example of this lies in more sustainable integrated catchment management:

Enabling local collaboration and partnership: The Catchment Based Approach to deliver Integrated Catchment Management.

(Environment Agency, 2011)

Finally, in Scotland the economic regulator, the Water Industry Commission for Scotland, is allocating specific funding for innovation to deal with climate change. The Commission's final determination on Scottish Water's business plan for 2010–2015 provides an additional £15 million of funding for Scottish Water, to take into account its duties under the Climate Change (Scotland) Act 2009, and some millions of pounds more for a range of studies, and first new investments in both mitigation and adaptation measures (Water Industry Commission for Scotland, 2009). The Commission expects Scottish Water to use this funding to stimulate collaborative innovation, between itself, the environmental regulator, local government and landowners, by making this a condition of using the funding.

In the past, the overall effect of regulatory constraints has been to inhibit radical innovation for sustainability. If your organisation is regulated in this way, or is in such a supply chain, then you need to educate and persuade the regulators and your clients to drive for more sustainable infrastructure. Encouragingly, there are increasing numbers of examples of change to demonstrate this process.

Stakeholder engagement: summary

Engage with users and local community stakeholders throughout the project delivery process, so that your strategy and projects will have legitimacy.

Local community roles can include, in rising order of preference: provide compensation; employ community members, or use them as part of the supply chain; offer the community co-ownership of projects.

Be ambitious, but practical, about community engagement. Get formal permissions effectively; empower the community, do not just 'inform' them; be careful about who represents the community; and use their local knowledge and common sense.

Work early with regulators, to open up space for innovation.

REFERENCES

Arnstein SR (1969) A ladder of citizen participation. *AIP Journal* **35(4)**: 216–224. Available at: http://lithgow-schmidt.dk/sherry-arnstein/ladder-of-citizen-participation.html (accessed 26 July 2013).

Bishop T (2009) More carrot and less stick drives better behaviour by water companies. *Utility Week*, 19th March 2009.

Brugnach M, Dewulf A, Pahl-Wostl C and Taillieu T (2008) Toward a relational concept of uncertainty: about knowing too little, knowing too differently, and accepting not to know. *Ecology and Society* **13(02)**: 30.

Chinyio E and Olomolaiye P (eds) (2009) *Construction Stakeholder Management*. Wiley-Blackwell, Chichester.

Empire State Development (n.d.) *Development of Minority and Women's Business Development*. Available at: http://www.esd.ny.gov/MWBE.html (accessed 30 July 2013).

Energy4All (2013) Available at: http://www.energy4all.co.uk (accessed 30 July 2013).

Environment Agency (2011) *Extending the Catchment Based Approach – How the Environment Agency will Support and Assist Others*. Available at: http://www.environment-agency.gov.uk/static/documents/Research/Guidance_for_hosts_v2.pdf (accessed 30 July 2013).

Equator Principles (2013) The Equator Principles III. Available at; http://www.equator-principles.com/index.php/ep3 (accessed 30 July 2013).

Fenner RA, Ainger C, Guthrie P and Cruickshank HJ (2006) Widening horizons for civil engineers – addressing the complexity of sustainable development. *Proceedings of the ICE – Engineering Sustainability* **159(ES4)**: 145–154.

Gipe P (2012) Wind Energy on Land has an Essential Role to Play in Germany's Energiewende. Available at: http://www.wind-works.org/cms/index.php?id=43&tx_ttnews[tt_news]=1924&cHash=58f8c2cb21a8350a1dcdf356e8225bd6 (accessed 30 July 2013).

Govert G and Stahre P (2006) On the road to a new stormwater planning approach: from Model A to Model B. *Water Practice and Technology* **1(1)**: doi: 10.2166/wpt.2006.005.

Greater London Authority (2011) The London Plan. Available at: http://www.london.gov.uk/priorities/planning/londonplan (accessed 30 July 2013).

ICE (Institution of Civil Engineers) (1996) *Sustainability and Acceptability in Infrastructure Development: A Response to the Secretary of State's Challenge*. SAID Report. Thomas Telford, London.

International Rivers (n.d.) *The World Commission on Dams*. Available at: http://www.internationalrivers.org/campaigns/the-world-commission-on-dams (accessed 30 July 2013).

Manchester Airport (2007) *Environment Plan – Part of the Manchester Airport Master Plan to 2030*. Available at: http://www.manchesterairport.co.uk/manweb.nsf/alldocs/8100FB8EF658808C80257364002D85FA/$File/Environment+Plan.pdf (accessed 30 July 2013).

Ofwat (2010) *Getting It Right for Customers – How Can We Make Monopoly Water and Sewerage Companies More Accountable?* Ofwat, Birmingham.

Parliamentary Commission for the Environment (n.d.) The cities and their people: New Zealand's urban environment. Available at: http://www.pce.parliament.nz/publications/all-publications/the-cities-and-their-people-new-zealand-s-urban-environment (accessed 30 July 2013).

Sommer F, Bootland J, Hunt M, Khurana A, Reid S and Wilson S (2004) *ENGAGE – How to Deliver Socially Responsible Construction – A Client's Guide.* CIRIA C627. CIRIA, London.

South Lakeland District Council (2007) *Community Engagement Strategy.* Available at: http://www.southlakeland.gov.uk/community/community-engagement.aspx (accessed 30 July 2013).

Walker DHT, Bourne L and Rowlinson S (2008) Stakeholders and the supply chain. In *Procurement Systems: A Cross-industry Project Management Perspective* (Walker DHT and Rowlinson S (eds)). Taylor & Francis, London.

Water Industry Commission for Scotland (2009) *The Strategic Review of Charges 2010–2015. The Final Determination.* Available at: http://www.watercommission.co.uk/UserFiles/Documents/Final%20Determination%20document.pdf (accessed 30 July 2013).

Watermeyer R (2011) *Briefing on ISO 10845: Construction Procurement.* Project leader: ISO 10845 and convenor of ISO TC 59 WG. Wiley-Blackwell, London.

Wind-Works (n.d.) Available at: http://www.wind-works.org/articles/community-windthethirdway.html (accessed 30 July 2013).

World Commission on Dams (2000) Dams and Development. A New Framework for Decision Making. Earthscan, London.

Sustainable Infrastructure: Principles into Practice
ISBN 978-0-7277-5754-8

ICE Publishing: All rights reserved
http://dx.doi.org/10.1680/sipp.57548.121

Chapter 6
Procurement

6.1.   What this stage involves

> If you are procuring pencils, the client–supplier relationship ends when you sign
> the contract. In procuring a project, that's when the relationships start.
>> (Charles Ainger, standard introduction in facilitating
>> 'partnering' workshops, 1997–1999)

Procurement is a key part of the project development stage. At the start of this stage,
the owner's asset management or planning department, having defined the problem to
be solved (the scope of the project(s)) hands over to the capital investment department, to achieve project delivery. The stage concludes with an agreed programme, and
a detailed procurement strategy for suppliers and other contractors. In modern
procurement practice, the department may not just develop a procurement strategy,
but will actually start to involve the supply chain – designers, key suppliers,
programme and project managers, constructors – to help develop the project solution.
(In 'traditional' approaches (see Section 6.2), constructors may not be chosen at this
early stage, leaving the selection until the detailed design has been completed (see
Chapter 8). Nevertheless, the procurement sustainability issues discussed here still
arise, but they may be more difficult to respond to with this later timing.) If you are a
supply-chain employee, this may be the first formal opportunity for you to influence
infrastructure sustainability.

In procurement, the asset owner is largely in control of its own choices. However, these
may still be constrained by sector or organisational standards of procurement practice,
which must be followed. However, your sustainability opportunities go beyond merely
meeting 'compliance' requirements. Several of the sustainability principles discussed in
Part I still need to be applied, but now in a more detailed way. So, bring the targets
and choice criteria from A2 – Socio-economic sustainability – into procurement.
Similarly, use the insights and approaches within A3 – Intergenerational stewardship,
and A4 – Complex systems, in your approach, and make choices.

Delivering infrastructure projects has always been a complex and, perhaps to outsiders, a
somewhat amazing process, in which just words and diagrams that describe the project –
the 'brief' resulting from project scoping – are transformed into a set of real physical
assets on or under the ground, with all the interacting parts relating properly to one
another, and ready to go into operation.

This is *internally* very complicated, requiring us to combine a range of interacting, multi-disciplinary, technical and management roles played by the sponsor, programme and project managers, designers, constructors, and materials and equipment suppliers. The process is also risky *externally*. It starts with a sometimes incomplete knowledge of the detailed surroundings, such as geotechnical and existing assets information, utilities, and other current and historical infrastructure, and requires management of a host of affected people, permissions and permits.

The actors required for project delivery will either be employed directly by the asset owner and sponsor – the client – or be 'procured' from external organisations that specialise in the required skills – the 'supply chain'. There has been a strong trend towards greater use of external private companies. To incentivise these players, contracts are put in place that aim to drive the achievement of a balance between (at least) the three objectives of lowest cost, shortest time and right quality.

Learning to succeed in this complex process has led practitioners to develop a wide range of separate sophisticated management skills, methodologies and manuals: of projects, design, risk, change, construction, safety and others. For a long time, dealing with the critical connection between all of these – the contracts – depended on traditional contract forms, which in turn relied on the professional role and authority of the 'Engineer'. However, since the 1990s, when many contractual relationships failed to deliver projects well, and deteriorated into adversarial delays and claims, the new professional skill of 'procurement' has been developing:

> Procurement is the process which creates, manages and fulfils contracts.
> Procurement commences once a need for goods, services, engineering and
> construction works or disposals has been identified and it ends when the goods
> are received, the services or engineering and construction works are completed or
> the asset is disposed of.
>
> (Watermeyer, 2011)

This is focused on making it all work well together, through *procurement strategy*.

6.2. Sustainability in procurement – and what can engineers do?

What protocols exist to actively manage the supply chain?

Procurement, as a professional response, has led to several trends in project delivery, not all pulling in the same direction. First, previously separate roles are now often combined into one. In particular, the role combines detailed design with construction, in order to increase the integration of these roles and minimise interface risk. A second trend is the transferring of more project risk to the outsourced design and construction team, and adopting increasingly sophisticated tendering procedures, which may include 'reverse auctions' to extract lowest possible unit prices at the beginning of the project.

However, reviews such as the Egan Report (Egan, 1998) and the work of Martin Barnes in the UK, have led to the development of new forms of contract such as the NEC3 series:

> the single most important factor for being able to deliver the [London 2012] Olympic Game venues on time was the New Engineering Contract which was used on all construction projects.
>
> (Martin Barnes; see Vaskimo, 2012)

These publications recognise the key project manager role, and the need to better clarify and reflect the roles of and interactions between all the supply chain members. In addition, the adoption of 'partnering' and target-cost contracts reflects the contribution that good relationships and incentives can make. Such arrangements put projects together into programmes, and set the separate roles within one overall 'virtual' or JV 'programme delivery contractor' organisation. This is often accompanied by 'early contractor selection', in order to get this team appointed in time to take part in the choice of the outline design solution during feasibility studies (see above).

The extra flavours that sustainability brings to this dynamic mix are as follows.

- More *complexity*, requiring you to deliver on extra objectives. These include better ecosystem protection and lower CO_2e emissions (principle O1.1 – Set targets and measure against environmental limits), and more socio-economic benefits (principle O2.1 – Structure business and projects sustainably) – all under the continued pressure for cost reduction and outcome certainty.
- Increased delivery efficiency to ensure *economic* sustainability, including user affordability, to provide more service for less money.
- More *innovation* throughout delivery, because these extra objectives cannot be reconciled with the other pressures without *cleverer solutions*.
- Stronger emphasis on *human responsibilities*, reflecting the social sustainability perspective (principle O2.2 – Respect people and human rights), which recognises you are not just 'mechanical' practitioners in the process, but have opportunities to bring your passion, relationships and purpose to dealing with these challenges.

The best of modern procurement practice is fully capable of embracing these possibilities, provided you use it. Delivering more innovative, sustainable projects successfully *requires* that you do this (see Section 3.6). If such innovative projects fail, the innovation itself is likely to be blamed, rather than just bad delivery practice. So, innovation for sustainability requires the use of the best modern procurement practice.

6.2.1 What can engineers do?

Infrastructure organisations around the world differ greatly with regard to how far they have got in this transition to modern procurement practice, and how formally they plan for it within their procurement strategy. In the international development aid sector, most infrastructure organisations tend to be required by their aid funders to use the old 'engineer-led' contract forms, and project roles are separated into separate contracts, between roles and between stages. There are great opportunities there to deliver development aid infrastructure more effectively. On the other hand, in some developed countries (the UK, Australia, the USA) efficiency pressures have led both privatised and public organisations to lead in using some or all of the above procurement trends. To some extent, the project delivery approach will match national experience and culture, and will be more hierarchical or self-managed accordingly. What matters is that a clear set of objectives, including innovation and expected behaviour, is set, and procurement strategy and contract terms are chosen to match them,

Some clients still believe that only the rather mechanistic and legalistic, 'transfer all risk, with extreme competition to get the price down' approach will deliver effectiveness. This can deliver short-term savings, but is very bad at incentivising real innovation. The better approach is to develop the relationships, and provide incentives for collaborative working. This can much better enable long-term innovation, if given time by combining the creativity of the whole supply chain to achieve a sustainability focus. There is much to learn on this from the London 2012 Olympics facilities, which were successfully delivered on time and below budget, using partnered programme management and NEC3 contracts (ICE, 2012).

So, where your organisation is now will set your starting point in using procurement to help put sustainability principles into practice. The opportunities at this stage lie in three key factors about procurement. These are listed here, with a cross-reference to the principle they represent (see Chapter 2).

- Use procurement strategy to provide extra socio-economic benefits (O2.1 – Set targets and measure for socio-economic goals).
- Use modern collaborative working relationships, enabled by the contract (O2.2 – Respect people and human rights).
- Integrate it all within a procurement strategy – for development (O4.4 – Integrate working roles and disciplines).

These are discussed in the following sections. A further procurement opportunity lies in sourcing all materials and equipment in line with principle O1.1 – Set targets and measure against environmental limits. This is discussed in Chapter 9.

6.3.   Use procurement strategy to provide extra socio-economic benefits

To what extent do you contribute to social cohesion and inclusion, and human well-being and welfare?

For good sustainability practice here, apply principle O2.1 – Set targets and measure for socio-economic goals – to your procurement. Beyond observing basic human rights with regard to the treatment of workers, in your procurement strategy you can add local/ regional *secondary* targets (see Section 2.4) to match the first three indirect outcomes goals in Table 2.4 – income, jobs (including training) and wealth (generated and retained) in the local community. Furthermore, you can address the other indirect outcomes of social and gender equality and community resilience, by choosing companies and people so as to improve these ratios. Use quantified *output* measures (e.g. numbers helped, and by how much) for this, and be as specific as possible.

In Section 5.3, we gave an example of this approach in the USA, where community members were included as employees or as part of the supply chain. Here we use projects from South Africa. South Africa is a deeply unequal developing country, but has a sophisticated professional engineering community, which is leading the way in attempts to use infrastructure investment and modern procurement practice to assist in development. (We are indebted to Ron Watermeyer, a pioneer of this process, for the South African examples used in this and other sections.)

6.3.1 Set out a procurement strategy to include extra socio-economic benefits

The practical opportunities that you can use in procurement and contracts include (Watermeyer, 1997)

- choosing materials and construction methods that are available from, and known in, the local supply chain
- preferring local contractors and suppliers in procurement, to create local jobs, develop contractors from within the community, and retain, as far as is possible, the funds expended on the project within the community
- including training and experience requirements in the project contract, to develop technical, administrative, commercial and managerial skills in the community, particularly to develop local small specialist contractors.

The results of this are best when the socio-economic benefits are not considered solely as 'add-ons', but also solve real technical and resources problems (see Section 5.3). Two projects from South Africa epitomise this; they are summarised in Boxes 6.1 and 6.2 (and a 'road maintenance' example is given in Box 10.2).

These two projects illustrate the potential for innovation in developing countries. However, this approach can also be used in developed countries – you can adopt a reasonable sustainability objective to provide local jobs and use local suppliers and contractors, to keep wealth in the local community as much as possible, and to help increase acceptance of the project by the community. This approach also serves environmental sustainability, as the use of local materials and people incurs less travel and CO_2e emissions than those sourced from further afield. You can use such arguments to make a case to specify the use of local resources, against the general legislative pressure for open economic markets and no local protectionism.

6.3.2 Combine projects into longer programmes, to allow local supply chains to respond

Modern procurement strategies can help you too. For instance, putting several local projects into a longer programme can give enough time and commercial certainty to

Box 6.1 Mabhiba Stadium, Durban – developed resources 'on the job'

The 3 billion Rand (approx. £260 million) Mabhiba Stadium in Durban, Kwa Zulu Natal was part of the major football World Cup stadium construction programme for 2010.

South Africa is very short of skilled design and construction resources, and this was made worse by the peak of their national World Cup programme, so it was clear that contractors would not have enough 'skilled trades' workers to complete in time. So, the *procurement strategy* acknowledged this (rather than the usual approach, which would make prequalification to tender *dependent* on proving you already had enough workers), and required a constructive response to this factor as part of the tender award criteria. The project then solved the lack of resources problem 'on-the- job'.

The tender specification included requirements for trades training; the tender award was based on a 90 : 10 evaluation split – price/'preference'. 'Preference' covered plans for training local labour. This ensured many small local and specialist companies were used; 90 000 hours of training were done within the contract, and wages were supplemented to allow for training.

(See ICE, 2010b; KwaZulu-Natal Department of Transport, n.d.; MOST Clearing House Best Practices, n.d.) http://www.unesco.org/most/africa16.html)

(Reproduced courtesy of Ron Watermeyer)

enable local suppliers to develop and sell their capabilities, and to reduce the environmental impact of the programme. An example of this is the UK Environment Agency's Broadland Flood Alleviation Project (BFAP), a 20-year public–private partnership (PPP) project for providing flood defence improvements:

Box 6.2 eThekwini water mains renewal – reduced poverty and developed local contractors

The eThekwini Water Mains Replacement programme in Kwa Zulu Natal was a 3-year, 1.6 billion Rand (approx. £138 million) AC water mains replacement programme. The utility maintains some 13 000 km of water mains in the semi-rural area around Durban. About 2500 km of the mains were asbestos cement pipes, at the end of their useful life; they burst frequently, and needed to be replaced.

Road traffic was heavy, and many of eThekwini's 38 municipalities had little or no information about their pipes; the same applied to other services located in road edges. So a combination of high-tech and low-tech solutions was adopted, in appropriate roles – horizontal drilling under large road crossings, and hand digging for road-edge work enabled flexibility in finding services and avoiding damage.

Image © Tom Wilcock

The hand-digging approach gave an opportunity for large-scale employment of underemployed local manual workers, and the entrepreneurial growth of small local contractors. Large contractors were in charge, and showed that their management skills can be used cost-effectively to manage small subcontractor hand labour work.

Peak mains replacement rates of 80 km/month were achieved, and about 1750 km of mains were replaced in 3 years. This reduced pipe bursts and water loss, halved maintenance crews, and for the first time pipe position data were recorded for all replaced mains.

Poverty reduction: At any one time 3800 underemployed workers were employed to excavate trenches; workers were rotated every 4 months to share out the work, so in all about 12 000 obtained some work during the project. The wages for these workers comprised 21% of project expenditure.

Develop resources: All subcontractors increased their tendering capability by an order of magnitude during the project, to grow their business.

(Source: Ron Watermeyer, 2010. Reproduced courtesy of Ron Watermeyer)

Bryan Banham, one of our plant suppliers, took the opportunity to develop his [local] business on the back of the BFAP. Although we do not offer any 'tie ups' with plant companies, Bryan realised that if he provided good machines and operators with a helpful local service then there should be plenty of work available for him. This was only made possible due to the long term nature of the project.

(Environment Agency, 2011)

So, investigate the potential of your design and construction projects to assist in delivering socio-economic benefits, consider how this might contribute to the success of the projects, and use your knowledge and creativity to help optimise this outcome.

6.4   Use modern collaborative working relationships, enabled by the contract

What steps are taken to seek long-term relationships between clients and the supply chain?

Since the 1990s, infrastructure sectors in several countries have dramatically changed the way in which clients engage their supply chains, to deliver both large projects and large programmes composed of smaller projects. One example is the UK water sector from 1990 to 2010 (Ainger, 2010). In 1989, the sector needed to double the rate of investment from £2 billion to £4 billion per year, and existing procurement and contract arrangements were not delivering. The contracts were bedevilled by extra costs, delays, poor quality, a poor health and safety record, and damagingly adversarial relationships in the supply chain, with many claims and some litigation. Some pioneer clients took new ideas from the UK's Egan Report (Egan, 1998) (see Section 6.2), and BP's 'project alliancing' (Sakal, 2005), and started to experiment and then develop these approaches, in stages. They used their regional market power to insist on radical new models for their supply chains, adopting major innovation in contracts, incentives and behaviour. Over the period 1990–2010, some key benefits were

- clients met all the standards of environmental, water quality and service improvements
- over 20 years of investment at a rate of about £4 billion/year; cumulative capital cost savings over 15 years were of the order of £4 billion
- over the same period, quality, time and health and safety performance improved greatly, and operational costs also reduced – these were *not* a 'trade-off' against lower capital cost.

The first change made was to integrate design and construction into lump-sum contracts in order to deliver projects *more efficiently*. Then the search for substantial *cost savings* led to the adoption of 'partnering' (or 'alliancing') of single teams, this being incentivised by target-cost contracts, with 'gain/pain' shares for each role in the team. Finally, the search for *risk transfer* and *management cost reductions* combined projects into longer 5–10 year programmes, some involving hundreds of individual projects, and with values of over £2 billion. All these actions were paralleled by setting up multi-project framework agreements with key suppliers.

These non-technical innovations have changed the culture and commercial behaviour of the UK water sector. There remains a large degree of variety in the detailed contract and subcontract forms used, and not all clients adopted these changes; but the new approaches and cultures have been absorbed by the supply chain to the extent that they have become widely applied. Many have adopted a 'virtual team' partnered approach, as summarised in Box 6.3. This drives collaborative joint working, which enables a culture change from adversarial to trusting. Removing interfaces allows the setting up of joint self-accountability and quality assurance, and helps allocate financial risk fairly. There is joint planning and management of all stakeholder roles between client, project manager, designer, constructor, planner, environmental and suppliers. Finally, relationships extending for 5–10 years or longer give time enough for partners to learn together, innovate and benefit. These behaviour changes have still been underpinned by recourse to the contract as a last resort, with tighter legal contract terms.

Box 6.3 Partnering – one client's definition

The creation of a team who: work together maximising each other's skills and contribution, to deliver to the client a cost-effective solution, in a spirit of teamwork, with shared objectives.

The client wants the project on time, within budget cost, at the right quality, useable and flexible, with good public relations – and no surprises!

The contractor/designer wants a fair overhead and profit, good cash flow, a known workload, clear and managed risk, a good reputation and more work.

In partnering, each looks after the *other's* objectives.

How to do it: 'Single joint team' culture and cost/value targets, shared goals, self-accountability, under commercial 'make more profit' target-cost incentives, and developing trust:

Trust grows if you start by expecting it; it fails if you start by distrusting each another.

(Source: Ainger, 2010)

Three interacting changes have been enabled by these new procurement strategies.

- Mindsets have changed: joint team working has forced the removal of 'silo thinking' and barriers between the roles, requiring a more open and innovative attitude by all the people involved – and this has freed up their engineering thinking too.
- This new creativity in the supply chain was applied in the 'solutions' stage, not just at the 'design and construct' stage, enabled by means of brainstorming

129

methods such as value management, to deliver lower cost and/or higher value project solutions.

- The change in culture and behaviour was incentivised commercially through target-cost, gain/pain contracts allowing players to profit from lower capital cost solutions – so the commercial terms 'worked with' and rewarded the new behaviours, thus driving innovation. Examples of the cost and CO_2e savings achieved are given in Boxes 8.3 to 8.5.

These approaches have great potential to enable you to deal with your new sustainability objectives, as shown in the following sections.

6.4.1   Use the 'service' business model

The introduction of programme management with incentives through target pricing was primarily driven by the search for lower cost and more efficient project delivery, not by the desire for sustainability. However, it is a good example of applying principle O1.2 – Structure business and projects sustainably – because it in fact changes the model for the supply chain from a product-based one to a service-based one (see Section 2.3).

In the old model, the work that designers or contractors were asked to price was a single project, defined by the pre-chosen solution: 'supply me this project' – i.e. as a product. The only way that profit could be increased was as a margin on doing more work through a larger project, thus increasing the resources used. In the multiple projects programme approach, you define the work as achieving (or, usually, beating) a list of the client's performance-defined target outputs – you take the risk (cost, time and quality) on achieving the client's targets. Your supply chain has inputs into choosing the solution (see Section 6.4), as well as the design and construction, and have a gain/pain share of the cumulative programme target. Members of the supply chain have multiple opportunities to achieve these outputs, while beating the target, and so may make more profit while using fewer resources. They can, for instance

- design and build fewer assets, by trying solutions higher up the sustainability hierarchy (see Figure 2.4), including 'level 1' 'no-build' solutions wherever possible
- treat outputs in groups, in order to minimise costs and resources used
- test an innovative new solution on an early project, and then apply it to later ones
- try out a risky new technology from a small new company on one project, 'self-insuring' the risk against the overall programme cost (see Section 9.3.3)
- improve their processes over time, and streamline on-costs.

None of these options is likely to be implemented, or incentivised, in the 'product' model. In the target-incentivised programme model, designers are happy to design less and contractors are happy to build less. This is a profound attitude change, and a more sustainable business model.

6.4.2   Add sustainability objectives into commercial contract cost targets

Once target costs are used in contracts, you can monetise other targets, including sustainability ones, and this brings serious commercial pressure to meet them. This has mostly been done, with critical project *completion time* targets, through pricing delay or rewarding early completion. One example is the Sydney Northside Storage Tunnel 'alliance' contract to clean up Sydney Harbour before the 2000 Olympic Games. The contract contained a very high delay penalty in the target cost – the Olympics start date was fixed! However, it also priced in social disruption and environmental spoil disposal impacts (Sydney Water, n.d.), thereby practising some aspects of operational principles O2.1 – Set targets and measure for socio-economic goals – and O1.1 – Set targets and measure against environmental limits. One obvious sustainability extension of the O1.1 principle is for you to measure embodied CO_2e, and to price and include it into all cost targets.

Sustainability requires taking a whole-life-cost approach to minimise all resources use and costs across the life of the infrastructure, applying principle O3.2 – Consider all life-cycle stages. In most target-cost projects and programmes, the solution choice (see Chapter 7) is indeed based the lowest whole-life-cost net present value, but the cost target (and also construction stage MEICA purchasing, see Section 9.4) is still usually based only on capital cost. Some clients are beginning to include operating costs in incentive targets. This approach can be difficult to test in reality, but is worth persevering with. One example is the Maroochydore sewage treatment plant alliance in the Sunshine Coast, Australia, which included key operational costs in the contract targets (author's, personal communication). Once this has been done, you can measure operational CO_2e from the energy, chemicals and transport inputs, and can price and include it. The NEC3 contract suite includes optional provisions for incentive payments against KPIs, and for damages payments against low performance (NEC, 2009).

6.4.3 Sometimes, use public–private partnership (PPP) projects – with care!

These projects include private finance, and integrate construction, operating and maintenance costs for a 15–25 year project life into an annual charge, so they represent a real whole-life-cost-based service business model (see Box 7.3 for an example of the operating-stage costs influencing design). The pricing can, in principle, be extended to include a whole range of sustainability targets, as suggested above. This structure is attractive, at least in theory.

However, there are some real concerns associated with the model: including excessive cost of finance and risk transfer, failure to transfer last-resort risk, and the inflexibility and high prices of operating-period changes. In 2011, the UK Government Treasury Committee reported that (PFI is the UK name for PPP finance):

> Private Finance Initiative (PFI) funding for new infrastructure, such as schools and hospitals, does not provide taxpayers with good value for money and stricter criteria should be introduced to govern its use.

PPP is easier to use on simple infrastructure, such as linear transport systems, where a single payment can collect tolls direct from individual users, and major changes in future use are unlikely. It is much harder to get PPP right in complex structures involving changing user needs and behaviour, such as the schools and hospitals mentioned in the quote. If you are considering using this financial model, examine the report for the necessary changes.

You must deal with one key sustainability issue: given a 15–25 year operating life, how will the energy cost and global warming risks be covered, and whose liability are they? Gradual temperature rise will add new risks – higher energy costs, more air-conditioning use in buildings, buckling rails, melting roads, more ecosystem die off in watercourses, and higher processing temperatures – and you should consider and deal with these risks in PPP contracts. Because of the large uncertainly (which will generate high risk prices), it is better not to place all this risk on the financier, but to share it, to incentivise resilience in design, and to adopt a joint 'adaptive management' approach between the financier and the user. Where you can meet these deficiencies, and private project finance is the best funding source for your project, PPP can be a good model for sustainable project delivery.

These forms of modern procurement have much potential to help you deliver sustainable infrastructure. Many engineers working in 'alliancing' teams now say that their cultures are more innovative and multidisciplinary than those of their 'home' companies, and combining the depth of knowledge of the assets of the client's staff with the wider experience of the supply chain staff also contributes to this. You can use target-cost contracts and programmes to incentivise 'doing more for less'. This is a fundamentally more sustainable service business model (see Section 2.3) for the supply chain, and using gain/pain shares with targets can commercially incentivise meeting other sustainability objectives too.

So, improved, effective practices are available, tried out, and being continuously developed. Use them wherever possible. They will work best when introduced as part of an overall procurement strategy.

6.5. Integrate it all within a procurement strategy – for development

How does the engineering project meet the clearly defined needs of all project proponents and the end users?

Ineffective infrastructure delivery directly limits and delays tackling principle A2 – Socio-economic sustainability – 'development'. The World Bank found that the competitiveness of its regional strategy for Africa is: 'impeded by poor public investment choices, weak budget management, and corrupt or lethargic procurement practices'. And it identified, as one of the three main risks facing Africa in realising its full potential for sustained growth and poverty reduction, 'inadequate resources available to implement the strategy' (World Bank, 2011).

The challenge is not just poor outcomes from individual projects, but the ability to invest all the money available. The same report quoted earlier World Bank work by V. Foster in 2008, who examined the infrastructure in 24 countries, which together account for 85%

of GDP, population and infrastructure aid flows in sub-Saharan Africa. It was found that, among other things, 'countries typically only manage to spend about two thirds of the budget allocated to investment in infrastructure' (ICE, 2010a). Many urban areas in developing countries have too few, too scattered and too undertrained people to keep up with infrastructure investment demand; rapid improvements are critical, and programme management can help. The example given in Box 6.4 comes from Kansas City, Missouri, in the USA, but the lessons could be applied widely.

Box 6.4 Kansas City's CIMO greatly improves project delivery performance

In the 2000s, Kansas City, Missouri, USA, was suffering from many typical problems. The city strategy was to revitalise the city centre, through public investment, triggering private money too. However, each separate city department had its own project management and contract systems and people, operating inefficiently, and they also fought with each other about 'permitting'. As a result, the city had a large backlog of undelivered projects, and a reputation of being a poor client, resulting in higher contractor prices and fewer bids. To achieve credibility with politicians, the public, contractors and the supply chain, the city urgently needed to make improvements in its project costs, time, quality and on-costs. The solution was to set up a single Capital Improvement Management Office (CIMO), to manage all projects within a programme (Kansas City, 2006).

CIMO was set up as an integrated team of 140 people, 115 of whom were seconded from four city departments. The rest were key staff in leadership and project management positions, from outside or consultants. They worked with the in-house staff to catalyse the use of new methods and to instil a new culture, and to 'train themselves out of a job'. They set up programme management processes, improved and re-engineered city procurement and project management, and transformed the culture into a collaborative one.

This team managed and delivered a 3-year capital improvement programme, initially of $210 million worth of projects. The number of projects under management has risen to 340, worth $1.2 billion, as other departments have added theirs in since the success of the approach became evident.

CIMO results 2 years in, reported in 2006

- Halved project delivery time, to about 1.5 years
- cut project management costs by at least 40% – down from about 18% of total project value to around 10%
- reduced the time from bid to award by 30%, and the time taken to pay contractors by more than 50%, garnering cost savings and other 'preferred client' benefits for the city
- reduced project backlog, expanded project capacity and delivered more than $210 million of public infrastructure.

(Source: CIMO and personal communication)

The culture change CIMO made internally is illustrated by one of their Project Manager's comments:

The most important improvement to project delivery that I have experienced as part of CIMO is the project team approach. Before, it was me against the world – against the consultant, against other City departments, against other City agencies. Now I have a team of people helping and working on my project before I even ask. ... We are all focused on the same goal, working together to determine what needs to be done to complete projects. Now, there are accountability measures in place to ensure we are being effective.

(Kansas City, 2005)

The CIMO programme management model was, in some ways, less ambitious than the UK example given in Section 6.4, as CIMO retained more conventional contracts with designers and contractors. But its success lies in the same key ideas – combining internal and external people into a single 'virtual' self-accountable team, and application of the programme to more projects and over a longer timescale, to allow investment in innovation, improvement and learning. The programme greatly improved project delivery for the city, but also included local socio-economic benefits; it built on the expertise of the city's infrastructure staff, and the capability of contracting companies located in Kansas City, so widening the sectors that benefited from the investment in public infrastructure and redevelopment. It shows the adoption of an integrated procurement strategy, and is an example of how you can put into practice principle O4.4 – Integrate working roles and disciplines.

6.5.1 Use procurement strategy and practice to reduce corruption in construction

Corruption is a sustainability issue, because it so greatly affects the development of infrastructure. Some see the construction sector as the most corrupt sector in the world (Transparency International, 2010), and in 2005 the American Society of Civil Engineers (ASCE) stated:

The annual expenditure worldwide on construction was estimated in 2004 at $3.4 (USD) trillion. Transparency International estimates that 10% of that sum is lost through bribery, fraud and corruption, resulting in about $390 (USD) billion diverted annually from projects that provide water, pollution control, electricity, roads, housing, and other basic human needs, Fighting corruption can mitigate against poverty, disease and famine and can assist in building a fair and civil society.

Corruption has many causes, some of which are nothing to do with procurement methods, but one reaction to prevent it has been to avoid any need for (possibly suspect) judgement in evaluating tenders: chop investment into small pieces; use lump-sum contracts with anyone allowed to bid; and buy only on a single number – lowest cost. However, this notionally less suspect tendering process can produce the unrealistically low prices that can actually cause many of the corrupt practices and poor outcomes. These include collaboration to fix prices between tenderers, poor quality materials and people, bribery of inspectors to turn a blind eye on quality, and artificially induced changes and claims (Global Infrastructure Anti-Corruption Centre, 2008; Transparency International, n.d.; U4 Anti-Corruption Resource Centre, n.d.; United Nations Global Impact, n.d.).

To counter this simplistic thinking, one outcome of the CIMO programme was rewarding. The greatly improved procurement management lowered bid costs, increased genuine competition and made the city a favoured client for more competent and creditable contractors; and this itself reduced the likelihood of corruption. So, dysfunctional procurement management – poor clients and procurement processes – can itself increase corruption. In this light, the modern procurement innovations discussed above favour and help to develop competent contractors, and to remove some temptations for corrupt practice. They use transparent costing, reward added value and honest collaborative behaviour, and give productive ways of making more profit.

So you can use such approaches to contribute to reducing corruption in construction. Some may have concerns that the earlier, more complex, team tendering and evaluation processes needed may be more open to corrupt manipulation. South Africa's experience so far has been that this can be avoided by setting out very clear criteria, and making the evaluation fully transparent.

More widely, you need to adopt good anti-corruption practice in procurement. Good advice and case studies are available from the UK Anti-Corruption Forum (n.d., 2011). Transparency in procurement and contracts, allowing public scrutiny, is a powerful preventive mechanism. The international Construction Sector Transparency Initiative (CoST) aims to improve transparency in the delivery of construction projects (see Construction Sector Transparency Initiative, n.d.).

6.5.2 Use an integrated procurement strategy, with considered objectives and appropriate forms of contract

The eThekwini Water Mains Replacement programme (see Box 6.2) enabled its innovative socio-economic benefits by using partnered target cost contracts, and programme management. The details are shown in Box 6.5. This required an integrated procurement strategy to make both aspects work.

Box 6.5 Partnering and programme management in the eThekwini water mains replacement programme

- Four designer/main contractor teams worked in parallel, each within one area. A single programme manager compared performance across the four areas, and shared the lessons learnt from each.
- With this competitive pressure, the client could award successive 'packages' to the same teams in each area (selected in 'framework agreements'). The longer timeframes allowed collaborative relationships to grow, and allowed successive packages to build in lessons learned in the previous one.
- Target cost contracts (NEC3) incentivised all parties to collaborate to save money.
- During mains replacement, each contractor team also had to take responsibility for running that part of the water network. Crucially, this aligned the team's interests with those of the client (rather than exploiting interface clashes to claim delays).

(Source: Ron Watermeyer, 2010. Reproduced courtesy of Ron Watermeyer)

Table 6.1 Changes needed in procurement (Jowitt, 2010)

From	To
Master–servant relationship of adversity	Collaboration between two experts
Fragmentation of design and construction	Integration of design and construction
Allowing risks to take their course, or extreme and inappropriate risk avoidance, or risk transfer	Active, collaborative risk management and mitigation
Meetings focused on the past – what has been done, who is responsible, claims. etc.	Meetings focused on 'How can we finish project within time and available budget?'
Develop project in response to a stakeholder wish list	Deliver the optimal project within the available budget
'Pay as you go' delivery culture	Discipline of continuous budget control
Constructability and cost model determined by design team and cost consultant *only*	Constructability and cost model developed taking into account the contractor's insights
Short-term 'hit-and-run' relationships focused on one-sided gain	Long-term relationships focused on maximising efficiency and shared value
Procurement strategy focused on the selection of the form of contract	Selected packaging, contracting, pricing and targeting strategy and procurement procedure aligned with project objectives
Project management focused on contract administration	Decisions converge on the achievement of the client's objectives
Training is in classrooms unconnected with work experience	Capability building is integrated within infrastructure delivery

At the end of the deliberations at the Second Middle East and Africa Convention (ICE, 2010a), ICE issued an agreed statement on procurement and project delivery, recommending the changes listed in Table 6.1. Among other things, the statement said:

> Performance in the delivery of infrastructure needs to improve. ... This necessitates the following culture changes. The contract terms – roles, performance measures, risk allocations, pricing, incentive 'carrots', legal 'sticks' – must be aligned with project objectives, to drive effective behaviour. The form of procurement and contract matter. The NEC3 family of contracts integrates risk and project management processes, and provides a wide range of contracting strategies including priced based, cost reimbursable, target cost and management contract. This family of contracts is accordingly well placed to support the required culture change and broader project objectives.

So, through an effective procurement strategy, you can put into practice principle O4.4 – Integrate working roles and disciplines – to improve delivery effectiveness. By adopting collaborative, longer term relationships with your supply chain you can drive sustainable sourcing (see Section 9.5), and serve principles O1.1 – Set targets and measure against environmental limits – and O2.1 – Set targets and measure for socio-economic goals – for greater sustainability. You can incentivise your teams of people, to work together, with opportunities to put into practice both individual principles: I.1 – Learn new skills – and I.2 – Challenge orthodoxy and encourage change – and actively seek innovation. Use such approaches in all your projects.

Procurement: summary

Make your supply chain choices within an integrated procurement strategy, using considered objectives and appropriate forms of contract.

Within this structure you can provide extra socio-economic benefits – training, jobs and income, wealth – to the locality, region or country; use longer programmes to allow local supply chains to respond; and reduce corruption in construction.

Use modern collaborative working relationships, enabled by the right forms of incentivised contract, to deliver more ambitious sustainability objectives. This can include a 'service' business model, and you can add sustainability objectives into contract cost targets.

REFERENCES

Ainger C (2010) UK Water Sector Development 1989–2010 – lessons from experience; presentation to Indian Ministry of Urban Development at British Water. London, 23 November 2010.

ASCE (2009) Policy Statement 510 – Combating Corruption. Available at: http://www.asce.org/Public-Policies-and-Priorities/Public-Policy-Statements/Policy-Statement-510 – Combating-Corruption/(accessed 9 September 2013).

Construction Sector Transparency Initiative (n.d.) Available at: http://www.constructiontransparency.org (accessed 3 August 2013).

Egan J (1998) *Rethinking Construction*. The Report of the UK Construction Task Force. Department of Trade and Industry. Available at: http://www.constructingexcellence.org.uk/pdf/rethinking%20construction/rethinking_construction_report.pdf (accessed 30 July 2013).

Environment Agency (2011) Broadland Carbon Reduction Case Study. EA Case Study IEM01/2011/001. Available at: http://www.environment-agency.gov.uk/static/documents/Business/BFAP_Carbon_Reduction_Case_Study.pdf (accessed 3 August 2013).

Global Infrastructure Anti-Corruption Centre (2008) *Examples of Corruption in Infrastructure*. Available at: http://www.giaccentre.org (accessed 3 August 2013).

ICE (Institution of Civil Engineers) (2010a) *Accelerating Infrastructure Delivery – Improving the Quality of Life*. Second Middle East and Africa Convention, Cape Town.

ICE (2010b) An engineer's toolkit. Available at: http://www.ice.org.uk/patoolkit (accessed 3 August 2013).

ICE (2012) 2012 Learning the Legacy. Available at: http://www.ice.org.uk/topics/ Learning-Legacy (accessed 30 July 2013).

Jowitt P (2010) Accelerating infrastructure delivery – improving the quality of life. Statement issued by the ICE President arising from the deliberations at the Middle East and Africa Convention, Cape Town.

Kansas City (2005) Kansas City's Innovative Approach to Tackling Capital Improvements Backlog. CIMO, City of Kansas City, MO.

Kansas City (2006) Capital Improvement Management Office. Available at: http://www. kcmo.org/CKCMO/Depts/CapitalProjectsDepartment/CityImprovementsManagement Office/index.htm (accessed 3 August 2013).

KwaZulu-Natal Department of Transport (n.d.) *Zibambele. An initiative of the KwaZulu-Natal Department of Transport.* http://www.kzntransport.gov.za/programmes/ zibambele/index.htm (accessed 3 August 2013).

MOST Clearing House Best Practices (n.d.) *Soweto: Mobilising the Community.* Available at: http://www.unesco.org/most/africa16.htm (accessed 3 August 2013).

NEC (2009) Procurement and Contract Strategies. Thomas Telford Ltd. Available at: http://www.neccontract.com/documents/Procure_press.pdf (accessed 9 September 2013).

Sakal MW (2005) Project alliancing: a relational contracting mechanism for dynamic projects. *Lean Construction Journal* **2(1)**: 67–79.

Schefer C and Barker C (2010) A 20 year perspective on municipal service delivery models in the UK water sector – lessons for South Africa. MWH presentation to CESA Conference and AGM.

Sydney Water (n.d.) Northside Storage Tunnel. Available at: http://www.sydneywater. com.au/SW/water-the-environment/how-we-manage-sydney-s-water/wastewater- network/northside-storage-tunnel/index.htm (accessed 3 August 2013).

Transparency International (n.d.) Available at: http://www.transparency.org (accessed 3 August 2013).

Transparency International (2010) U4 Expert Answer. Available at: http://www. transparency.org/files/content/corruptionqas/254_Approaches_to_adressing_corruption_ in_construction.pdf (accessed 3 August 2013).

Treasury Committee (2012) *Private Finance Initiative.* Treasury – 17th report UK Government. Available at: http://www.publications.parliament.uk/pa/cm201012/cmselect/ cmtreasy/1146/114602.htm (accessed 3 August 2013).

U4 Anti-Corruption Resource Centre (n.d.) Available at: www.u4.no (accessed 3 August 2013).

UK Anti-Corruption Forum (n.d.) Available at: http://www.anticorruptionforum.org.uk (accessed 3 August 2013).

UK Anti-Corruption Forum (2011) *Guidance on the Bribery Act 2010 for the Infrastructure Sector,* abridged version. Available at: http://www.britishexpertise.org/bx/upload/ Events/ACF_Abridged.pdf (accessed 3 August 2013).

United Nations Global Impact (n.d.) Available at: http://www.unglobalcompact.org (accessed 3 August 2013).

Vaskimo J (2012) Jose Reyes and Martin Barnes deliver keynote presentations at the 26th IPMA World Congress. *PM World Journal.* Available at: http://pmworldjournal. net/jose-reyes-and-martin-barnes-deliver-keynote-presentations-at-the-26th-ipma-world- congress (accessed 30 July 2013).

Watermeyer R (1997) Job creation in public sector engineering and construction: why, what & how. Commonwealth Engineers Council 50th Anniversary Conference, Fourways.

Watermeyer R (2011) Standardising construction procurement systems. *The Structural Engineer* **89(20)**: 4–8.

World Bank (2011) *Africa's Future and the World Bank's Support to It.* Available at: http://siteresources.worldbank.org/INTAFRICA/Resources/AFR_Regional_Strategy_ 3–2–11.pdf (accessed 3 August 2013).

Sustainable Infrastructure: Principles into Practice
ISBN 978-0-7277-5754-8

ICE Publishing: All rights reserved
http://dx.doi.org/10.1680/sipp.57548.141

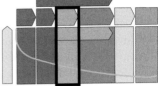

Chapter 7
Outline design

7.1.   What this stage involves

> By the time the [outline] design for most human artifacts is completed ... 80–90%
> of their life-cycle economic and ecological costs have already been made
> inevitable.
>
> (Hawken, Lovins and Lovins, 1999)

Outline design, also called a 'feasibility study' or 'optioneering', is the second key part of
the project development stage. The owner's capital investment department takes the
project scope performance brief – which defines the problem – and chooses for it a
preferred solution. They may well involve some early-appointed supply chain members
in this process, and, in best practice, other stakeholders too (see Chapters 5 and 6). If
you are a supply chain employee, this may be the first formal opportunity you have to
influence the infrastructure sustainability.

At this stage, the asset owner is much more in control of their own choices. However,
it may still be necessary to gain critical external approvals such as planning permission
and social and environmental consents and permits, or these may come later. The
sustainability opportunities, however, go far beyond just meeting any compliance
requirements. Many of the sustainability principles discussed in Part I need to be
applied, but now in a more detailed way. So, bring targets and choice criteria from
A1 – Environmental sustainability, within limits – and A2 – Socio-economic sustainability – into the development of your outline design. Similarly, apply the insights and
approaches within A3 – Intergenerational stewardship – and A4 – Complex systems –
in how you approach the outline design and make choices.

The extent of outline design varies from one organisation and sector to another,
but essentially it will include decisions on the main *components needed* to achieve
the required performance, where they will be *located*, how they *fit together* and
what they will *cost*. For example, for a public transport performance scope it might
involve

- deciding to use light rail rather than rail, road bus, guided bus or trams
- confirming the outline location, land take, the scale of the track, termini and
  stops, and their traffic capacities and transit times
- defining other component performance criteria

- attaching constraints governing the geographical, social and environmental interfaces of the project
- finalising the agreed cost budgets.

It will not go into detail about the materials or design of each component.

Within this outline design output you will have broadly decided on the *technology* to be used to supply the infrastructure service. This will broadly fix the project's *impact* on the environmental limits, at a global, regional and local scale (see Section 2.3). So when making choices during the outline design phase, it is critical to consider sustainable options and to apply sustainability targets.

7.2. Sustainability in outline design – and what can engineers do?

What mechanisms are used to encourage creativity and innovation?

Outline design is a structured decision-making process. (The process described here is also used in other delivery stages. It may be used upstream in a scoping or a pre-feasibility study; and it will certainly be used several times, for different components, in formal or more informal and intuitive ways, in detailed design. We describe the process in detail at this stage because it is essential here.) It typically starts by brainstorming a range of possible solution options (here the term 'option' is used to mean any possible solution that is considered, and the term 'solution' is used to mean the chosen preferred option). It then evaluates the feasibility of these options to solve the problem with respect to how well they meet the performance required.

Next, the options are compared on some basis of value, this being typically done by using a value management process to look at performance versus cost. ('Value management is a structured, facilitated, process in which decision-makers, stakeholders, technical specialists and others work collaboratively to bring about value-based outcomes in systems, processes, products and services' (Barton, 2000). Value management techniques are widely used in feasibility studies (Institute of Value Management, n.d.).) One option is chosen as the best solution, and a capital, operating and whole-life-cost budget are drawn up. The solution will then be authorised by a management decision 'gateway', authorising release of the solution for detailed design and construction. This sequence of activities is shown in Figure 7.1.

In earlier stages, the strategic objective(s) (1 in Figure 7.1) have been translated into a project scope, defining size and performance (2). The process now is for you to construct and fill in an *evaluation matrix* by the following.

- Brainstorming possible *options* (and often sub-options) that may satisfy the scope (3). Typically this will involve a range of components interacting with each other. These form one side of the evaluation matrix.
- Setting out the range of *choice criteria* (4) – cost, various aspects of performance – against which you will compare the options, along the other side of the matrix.

Figure 7.1 The outline design decision-making process

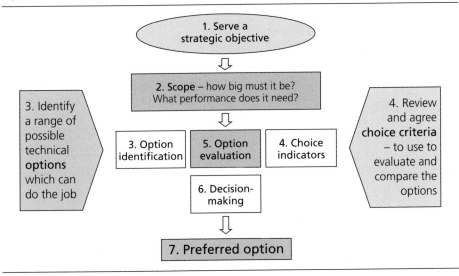

- Using a range of methods by which you evaluate or score each option (5), against each critieron, in each box of the matrix. These should vary with the type of criteria.

The scores in this evaluation should be as objective as possible. The information that you obtain can then be used in the separate decision-making stage (6), to arrive at a solution (7). The decision-making itself is subjective, because it depends on the teams' decision on how to apply the criteria, and any trade-offs between them, which brings in the values of the organisation and its stakeholders. There are many approaches available. Each has good and bad aspects, and pitfalls; some are better than others at accommodating sustainability objectives. An example of a simple decision matrix is shown in Figure 7.2.

A focus on sustainability principles O1.1 – Set targets and measure against environmental limits and O2.1 – Structure business and projects sustainably, will have added more environmental and socio-economic objectives (better ecosystem protection, less land take, lower CO_2e emissions, more community socio-economic benefits) to the project scope, to be dealt with in the outline design, in addition to the standard pressures for effective user service and cost reduction. These just widen the definition of the project 'value' that you seek to optimise. In the past, all these factors have been assumed to add extra cost, thus challenging user affordability, which requires more service for less money. You need to harness these pressures to drive creativity for more innovation in your outline design – find a cleverer solution.

7.2.1   What can engineers do?

Most infrastructure organisations are likely to use some form of this decision-making process at the outline design stage, so you are unlikely to have to introduce the idea

Figure 7.2 Decision matrix for a sewage treatment plant for a small UK community. (Source: Fenner *et al.* (2008))

Matrix has options 1–6 as columns, and criteria 1–16 as rows

| 1 – Best |
| 2 – 2nd best |
| 3 – Average |
| 4 – Bad |
| 5 – 2nd worst |
| 6 – Worst |

Uses 6 degree ranking scoring system, added with equal weight for all criteria for decision-making

		Treatment and transfer		Treat locally (levels of treatment)			
		Option 1 Septic tank and reed bed	Option 2 Septic tank and SAF	Option 3 MBR	Option 4 Septic tank, SAF and sand filter	Option 5 Septic tank and SAF	Option 6 Facultative lagoon
Environmental indicators	1 Cultural heritage and archaeology	1	1	1	1	1	1
	2 Biodiversity on species and habitats – priority with action plans	1	1	1	1	1	1
	3 River water quality	1	1	1	4	5	6
	4 Area of brownfield land used for construction	6	6	6	6	6	6
	5 Minimise greenhouse gases emissions – transport	2	3	4	6	5	1
	6 Minimise greenhouse gases emissions – electricity	1	3	6	5	4	2
	7 Chemical usage	1	4	6	3	4	1
	8 Types and volume of construction materials used	3	5	6	4	2	1
	9 Visual impact	3	2	5	5	1	3
Social	10 Reversibility – decommissioning	1	3	5	6	2	2
	11 Nuisance events – odour and noise	1	4	3	4	4	2
	12 Accidents during the construction process	5	5	2	3	3	2
	13 Planning policy context	1	1	1	1	1	1
Econ	14 Use of local workforce	1	3	6	3	3	1
	15 Origin of construction materials	1	2	6	2	2	2
	16 Type/volume of construction material to landfill after construction	6	2	2	2	2	2
	Total score	35	44	59	56	46	34
	Sustainability ranking	1	3	6	5	4	1

for the first time (but, if no such process is being used, do so). The established method-ologies in common use can accommodate this sustainability focus, but you must expect to have to push the innovation boundaries in all aspects of the process.

So your opportunities to put sustainability principles into practice need innovative action at each part of the outline design process, as listed here with cross-reference to the operational principles set out in Chapter 2.

- Work in a multi-disciplinary team – to ensure that all interfaces are considered (O.4.4 – Integrate working roles and disciplines).
- If sustainability is not reflected in the performance and scope, review it and widen the range of options considered to include real sustainable alternatives (O1.2 – Structure business and projects sustainably; O4.3 – Consider integrated needs).
- Ensure that novelty, uncertainty and lack of direct control do not eliminate promising more sustainable options during evaluation (O4.2 – Deal with uncertainty).
- Include sustainability-related choice criteria, and take care in their evaluation (O1.1 – Set targets and measure against environmental limits; O2.1 – Set targets and measure for socio-economic goals).
- Use decision-making processes that reflect sustainability principles (O3.2 – Consider all life-cycle stages; O2.2 – Respect people and human rights).

The first of these was raised in Section 6.4; we discuss the last four below. This could be the most critical stage for you to act on innovation for sustainability, because it may be the first time you can influence the project, and it is the stage at which your opportunities become most 'closed down'. Very early, right at the start of this stage, you may still be able to rethink some of the things decided upon, or missed, in upstream strategy and scoping (see Chapters 3 and 4). By the end of the stage, to achieve the management function of the project, the project focus will have been narrowed and firmed up in order to make the detailed design efficient, as it must be. So you should take every opportunity to act in this stage.

7.3.   Review the scope, and widen the range of options

Have an extended range of options been examined?

Your sustainability opportunities here depend on how performance is defined in the project scope, and what sustainability objectives have already been included. If sustainability has been effectively addressed in project scoping (see Chapter 4), then no sustainability review may be needed, and you can go straight on to brainstorm options. If sustainability has not yet been addressed effectively, this is your last chance to introduce the missing elements, but this must be done right at the start of the stage.

This may sometimes be difficult. Under pressure from time and 'engineering on-cost', many clients want to reduce time spent on outline design, and will aim to consider fewer options. So they will standardise solutions for typical types of problem into a simple choice matrix. This could be between, say, three options, depending on three main criteria. This efficiency is desirable, particularly for multiple smaller scale asset renewal or capital maintenance projects. But preparing such a choice matrix itself requires carrying out an outline design decision-making process on the typical project type, so you can still apply the scope-review and options-widening ideas in that process. This is a brief window, but is a key opportunity.

7.3.1   Review the scope

Most outline design processes are kicked off by assembling the team together with the client to review the brief and to set out a management plan. The time to ask about scope is at this first meeting, as part of interpreting and understanding the brief. The questions you should consider raising, to open up more sustainable solutions, reflect the principles in Chapter 2 and the project scoping practice in Chapter 4.

- What additional sustainability objectives can be added, including socio-economic ones (see Section 4.3)? And, should they be absolute targets, or criteria for ranking alternatives (see Sections 2.3 and 2.4)?
- Can the need that the project is to meet be defined as a *service*, and not a *product* (see Section 2.3)?
- Do the project boundaries include the in-use phase (see Section 2.5)?
- What are the real service performance requirements, on a whole-system basis? Test any 'standard' assumptions. For instance, the RMI (2003) quotes a Factor 10 example (Factor 10 refers to the target of reducing infrastructure impacts by 90% from the business-as-usual case), in which the assumed very demanding computer and cooling energy needs for computer data centres were reduced by 66–89% by first redesigning the way in which the client's servers worked (for another data centre example see Eubank *et al.* (2003)).
- Is the scope defined in terms of a required performance, not as an assumed solution (see Section 4.6)? If not, discuss and agree what the owner 'really, really, wants' in terms of *performance*.

The possible arguments that you may use to raise these issues are many. Taking an 'opportunity' perspective, you might quote public sustainability commitments by the client, or the sector, which have not yet been translated into project targets. You may find it powerful to quote examples of similar projects done by the client or in the sector that include more sustainable goals or outcomes; in particular, use examples of projects done from any organisation that is seen as a major competitor or comparator. Alternatively, taking a 'risk' perspective, you may quote existing or possible new regulatory or planning demands, or a failure to create adaptability and resilience in the light of climate change, or a huge potential rise in energy prices. The right mix will depend on the project, the client and the context; and the arguments must be real, relevant and made specific to the project, so they cannot be dismissed as generalised 'pie in the sky', or 'good ideas but not relevant here'.

Do not accept at this stage that any redefining of scope for more sustainability must be excluded on the grounds of an assumption that it will cost more and be unaffordable.

Collect and quote examples showing that this is not true (as in Box 11.7). You can also deflect that argument by saying: 'That is exactly what the outline design process will quickly show; if that is true, then we can drop it.' Your aim is to keep the options open.

7.3.2   Widen the range of options

Are adverse impacts only accepted reluctantly?

Most outline design sequences will use some form of workshop, early on, bringing your project team and key stakeholders together. It will typically include using creative thinking techniques to brainstorm a first long list of possible options. Sometimes just setting a new hard sustainability target (such as the 'halve embodied CO_2e' target in Box 11.7) is itself enough for you to get teams to be much more creative in coming up with innovative options. But, in many cases, the teams may find it hard to think laterally enough about possible solutions, so you will need to require a review of their first list. The benchmark sustainable option (BSO) methodology (UKWIR, 2004) does this partly by testing your initial options to see how far they fit with a list of key sustainability principles – even if these are not within the brief.

A range of generically more sustainable options form the six levels of the 'hierarchy of sustainability' (see Section 2.3); you can start by comparing where your brainstormed options lie against these. Many items on your first list may come in at the lower levels – 5 and 6 in Figure 2.4. If you can, start by eliminating level 6, in which environmental impacts are considered inevitable, and outside the boundary of the project. Make level 5, in which any environmental impacts must be compensated for, the minimum acceptable. An example was given in Box 4.3, and another is given in Box 7.1. Just the *requirement* to include, and cost, environmental mitigation at this early stage helps change mindsets and generate more imaginative options.

Box 7.1 'No net environmental impact' – rules for SUDS in the UK

Sustainable urban drainage systems (SUDS) are increasingly used in many countries to control the rate of stormwater runoff from urban developments to streams and rivers, and improve its quality.

In the UK, this uptake has been driven by a simple rule, applied to developers by the Environment Agency as a permit condition: *the net flow rate in the stream or river receiving the rainfall outflow must not be greater than it would be in the 'green field' condition, before development.*

Taking this further, you can suggest that the team comes up with options up to the top of the hierarchy – 'multiple small projects', 'do nothing' (reuse existing structures, perhaps?), or 'reduce or manage demand'. These may seem completely inappropriate to the brief, but asking the team to consider them and prove this assumption, can be very creative. It is a way of indirectly reviewing the brief, and applying 'whole systems engineering' (see RMI, n.d.) using principle O4.1 – Open up the problem space (see Section 2.6). You should only accept a long list of options, to complete this part of the process, when this review has been exhausted.

7.4. Ensure that promising options are not eliminated during early evaluation

Can you guarantee it will work?

Typical barrier to innovation, quoted in Ainger (2012)

After brainstorming a possibly long list of options, outline design will typically then

- identify and agree choice criteria for selecting the preferred option
- carry out an initial first filter evaluation to reduce the long list to a manageable shortlist of practical options
- have the team develop these options further and undertake a detailed evaluation
- use this analysis to select the preferred outline design option.

Hopefully, the long list will now include some more radically sustainable options. These may be unfamiliar to most workshop members, and outside their technical expertise, and contain uncertainties in data and performance. They may also be seen as outside the teams' control, and as having a number of perceived barriers to acceptance. So they are often likely to be eliminated in the initial crude first filter evaluation, one of the criteria of which is 'practicality'.

Even if some get through this filter, they may then be compared poorly with conventional options, because teams make overly conservative assumptions about performance or data, due to uncertainty. For instance, in the water sector, reduction of water use through metering might be assumed to be a lower end figure of 5%, whereas several studies show that a 10–15% reduction may be equally likely.

Such risk-averse concerns are understandable, but their effect is unacceptable for innovation: they eliminate most new options from further consideration, while everyone can say 'well we tried'. To avoid this situation you can ensure that at least one of the most sustainable options survives the first filter and is assessed fairly during development and final evaluation and comparison, by using the 'benchmark sustainable option' (BSO) approach (UKWIR, 2004). This requires that you

- identify at least one (possibly more, if there are varieties of more sustainable option types) BSO from the long list – representing the likely most sustainable solution against the choice criteria
- include this (these) in the shortlist, regardless of any 'practicality' concerns or uncertainties there may be about it – even if someone claims that there are insuperable barriers to its adoption

■ evaluate the BSO through to final decision-making, together with the other shortlisted options, as well as you can, and see how it compares. Deal with uncertainties by specific investigations, questions and answers, to provide an audit trail, and if there are doubts about performance, include the range of uncertainty, not just the worst case.

This gives you three possible outcomes.

(i) The BSO is both possible and the preferred option – the barriers are not real, some possible myths are dispelled – so it can be adopted immediately as the solution.
(ii) The BSO is attractive and competitive, but requires some key change – perhaps the reduction of some performance uncertainty, or the removal of some specific regulatory barrier – before it can be adopted, next time around.
(iii) The BSO is just wishful thinking for this application, with some real drawbacks that make it uncompetitive; so it can be discarded as an option (for this application) next time.

If the outcome is not (i) but (ii), you have still challenged, and removed for next time, some assumptions that acted as barriers to innovation. Even if the outcome is (iii), you have eliminated one idea from next time's long list. And, if challenged about accepting a less sustainable option in this case, you have an argument explaining why. Box 7.2 shows a success in getting past such 'not practical' assumptions in solving small community water supply in Scotland – allowing the use of a level 4, rather than level 5, solution in the sustainability hierarchy (see Figure 2.4).

Box 7.2 Community water supply – more local investigation justified a more sustainable solution

---

Initial regional hydrology and treatment process assumptions suggested that simple village-scale works would not work. But local knowledge suggested that more water was actually available; and local gauging of flows showed available low flows were 70% higher than assumed. In addition, pilot trials showed that activated carbon was an effective treatment, avoiding the need for membranes. So local works were chosen, with much lower CO_2e emissions, cost and environmental impact than a new centralised treatment.

(Source: adapted from MWH)

Adding a BSO option is a minor addition to a well-established methodology. It is an example of creating 'experimental space', to get past risk-averse mindsets (see Section 11.4). It is better tactics to say to your team 'Let's just test "what-if" we tried the BSO, even if we don't expect it to succeed', rather than just insisting that a more sustainable solution must be found. By investigating the BSO together, you and the team are learning, challenging the 'business-as-usual', and developing a more sustainable mindset for next time. Use the process every time if you can, so that on each project you can make some sustainability progress.

7.5.   Include sustainability-related choice criteria; take care in evaluation

How do the choice criteria needed to evaluate decisions reflect sustainability issues?

Delivering sustainable outline designs requires that you include choice criteria that follow through (see Section 3.6) on the set sustainability business objectives. There are very many sustainability choice criteria now coming into use. A summary of sustainability indicators and themes that are being developed is given in Table 7.1. It shows which are used within various project rating tools (see Section 13.5), in some academic texts (MacAskill, 2011) and in CIRIA's guidelines (Berry and McCarthy, 2011).

The indicators vary greatly. Some measure project outputs/impacts that have different significance (scale × duration); some are direct and some are indirect effects. Some measure inputs and some relate to how the project is carried out, not what it achieves. Indicators are also measured in a range of ways, from objective scientific numbers, to opinions of relative ranking.

These large differences mean that you need to take great care when using them to evaluate each option. You should apply some of them as absolute targets, and some as criteria for ranking alternatives (see Section 3.3). We discuss these issues below.

7.5.1   Distinguish between different types of indictator, for evaluation

The criteria can be categorised by the different ways in which they can be quantified to evaluate the options; they vary from objective numbers to subjective opinions. This partly depends on the amount of data available at this stage in the project.

- **Type 1**: Factual objective assessments, of output or input, expressed in numbers (e.g. amount of service or product provided, cost, CO_2e emissions, land area taken). These express the *absolute* score of the option, independent of other options. You can use these as absolute targets, or to rank and compare options.
- **Type 2**: Partly 'objective' assessments, independent of other options but expressed in words by experienced people (e.g. regarding the technical functionality of a process: 'familiar to operating staff'). Such statements can be turned into a 'score', by a disciplined process of allocating a number (say 1 to 5) to a series of descriptive statements about the topic, such as: from 'never used before' (score 1), through 'some operators are familiar with it' (score 3), to 'everyone is familiar with it' (score 5).
- **Type 3**: Less objective ranking opinions, such as 'high', 'medium', 'low'.
- **Type 4**: Simple 'yes'/'no' answers to questions about following *processes* (i.e. not about output) (i.e. 'Can it meet planning requirements?').

These four types are identifiable in Table 7.1.

Table 7.1 Typical sustainability indicators and their use in rating tools (© MacAskill 2011)

T = Rating tool R = Reporting A = Academic research I = Industry guidance ▢ = theme included ▣ = theme implied	AGIC T	CEEQUAL T	BREEAM T	LEEDS (US) T	GRI CRESS R	Fern and Rod A	Sahley et al. A	Koo et al. A	Ugwu and Haupt A	Willetts et al. A	CIRIA I
Social											
Culture/communities	▢		▢		▢				▢		
Accessibility	▢										
Health and safety and/or security	▢		▢						▢		
Participation/inclusiveness	▢										
Integration/skills/equity/acceptability	▢										
Heritage	▢		▢								
Stakeholder satisfaction/positive legacy	▢								▢		
Landscape/visual impact	▢										
Human rights						▣					
Environmental											
Soil	▢	▢	▢	▢	▢	▢	▢		▢	▢	
Water	▢	▢	▢	▢	▢	▢	▢	▢	▢	▢	
Atmosphere	▢	▢	▢	▢	▢	▢	▢		▢	▢	
Biodiversity/ecology	▢	▢	▢	▢	▢	▢	▢	▢	▢	▢	
Resource efficiency/materials use	▢	▢	▢	▢	▢	▢	▢	▢	▢	▢	
Minimise residuals/waste	▢	▢	▢	▢	▢	▢	▢	▢	▢	▢	
Energy	▢	▢	▢	▢	▢	▢	▢	▢	▢	▢	
Noise/dust	▢	▢	▢			▢	▢		▢	▢	
Climate change	▢	▢	▢	▢	▢	▢	▢	▢	▢	▢	
Land management	▢	▢	▢	▢	▢	▢	▢	▢	▢	▢	
GHG management	▢	▢	▢	▢	▢	▢	▢	▢	▢	▢	
Economic											
Costs/cost-effectiveness	▢				▢		▢				▣
Indirect cost/impact							▢				▣
Bureaucracy	▢										▣
Innovation investment											▣
Growth/competitiveness							▢				▣
Climate change	▢										
Technical/project execution											
Technical functionality/requirements											
Performance of operation (reliability/resilience/vulnerability)											
Contract and/or procurement method	▢						▢				▣
Supply chain	▢										
Reporting and responsibilities/knowledge sharing	▢										
Making decisions/responsible practice	▢										▣
Transport (materials/construction)	▢				▢		▢				
Context/political											
Regional priority/general interest				▢		▢					▣

7.5.2   Use customer-based measurement units

You need to choose carefully the *units* chosen for measuring output or impact criteria, as they affect the comparison of options. Consider the example of CO_2e emissions from the water sector.

- '$kg\ CO_2e/Ml$ of water provided' measures 'efficiency per unit of product'. It favours conventional technical-efficiency-based approaches, because it does not register the direct CO_2e emissions reduction that a more sustainable demand-reduction option, reducing water use per customer, would give. (Actually, pumping and treating less water with existing assets may slightly raise the kg CO_2e/Ml.)
- '$kg\ CO_2e/customer$' is an 'efficiency per customer' measure. It will include any direct reduction in CO_2e achieved by a water-demand-reduction option, as well as registering fairly the reductions achieved by technology efficiency options.

The same effect will apply in transport, buildings or other utilities. For all kinds of project outputs or impacts, you can use 'efficiency per unit of product' units when comparing technology choices at the detail design stage, or performance improvements during operation. But, for outline design, you must use 'efficiency per customer' units, to give a level playing field for evaluating non-assets-based solutions as well as assets-based ones. Check the units conventionally used for such a criterion – Will it measure the provision of less product?' – because many of them unconsciously assume only technology-based solutions.

7.5.3   Apply consistent sustainability indicators: take account of their significance

What distinction is made between actions that lead to large, irreversible and uncertain impacts (e.g. climate change) and actions that lead to smaller reversible impacts?

To embed sustainability in your preferred project solution, you must apply choice criteria that reflect the sustainability targets included in the project scope (see Section 4.3). If these were not already acknowledged, hopefully you managed to include some during the scope review (see Section 7.3).

All projects will have their own set of appropriate local or project-specific criteria, but to be consistent with sustainability principles O1.1 – Set targets and measure against environmental limits, and O1.2 – Structure business and projects sustainably, you should expect to include the generic criteria suggested in Table 7.2. As well as varying in type (column 3), they also vary greatly in significance (e.g. scale × duration, in the case of environmental criteria), and this determines whether you should apply them as specific *targets* to be met by all options, or as comparative ranking criteria within a commitment to minimise or maximise them (column 4).

Table 7.2 Generic sustainability indicators for sustainable option choices

Choice criteria or indicator	Comments on significance	Type – units and evaluation	Use in decision-making – target or comparator
Infrastructure service – the reason for and cost-effectiveness of the project			
Socio-economic or environmental need being met	Primary target – the main driver of the project	Direct output; units vary with sector – type 1 number	All options to meet target; exceedance can be extra ranking comparator
Whole-life cost (capital and operating costs)	Key criteria for cost-effectiveness	Direct input: £ SUM as net present value (NPV) – type 1 number	Key ranking comparator – may also have absolute affordability target
Resilience	Key secondary target – how the whole system responds	Direct outcome; measure by objective statement – type 2 number – or H/M/L – type 3 number*	All options meet target, as secondary target; and use as ranking comparator
Other environmental indicators (see Section 2.2 and Tables 2.2 and 2.3)			
Global warming control – minimise carbon emissions	Global/irreversible or VLT impact – set absolute target	Direct output: GHG emissions, in tonnes of CO_2e, from construction and operation – type 1 number	All options meet target; and use as ranking comparator
Land take and biodiversity	Regional (maybe global)/irreversible or VLT impact – set absolute target	Direct input: land take in hectares – type 1 number. Biodiversity value will vary: consider also ecological services value,† or eco-footprint (hectares)	All options meet 'no net loss' target; use any direct impact on local biodiversity sites as a ranking comparator
Water use	Regional/operational life impact – set target, if clear regional limits	Direct input: megalitres per year – type 1 number, or type 2 or 3 if few data	Set 'minimise' as a clear goal; use as a ranking comparator. Set as target if regionally justified
POPs pollution discharge: persistent organic pollutants	Regional/irreversible or VLT impact – comply with standards	Direct output: kilograms per year, separately – type 1 number, or type 2 or 3 if few data	Set 'minimise' as a clear goal; use as a ranking comparator. Set as target if standards or limits have been set

Table 7.2 Continued

Choice criteria or indicator	Comments on significance	Type – units and evaluation	Use in decision-making – target or comparator
Nitrogen and phosphorus load use/discharge	Regional/ operational life impact – set target, if clear regional limits	Direct input or output, tonnes kilograms of nitrogen (or phosphorus) per year – type 1 number, or type 2 or 3 if few data	Set 'minimise' as a clear goal; use as a ranking comparator. Set as target if regionally justified
Ocean acidification; atmospheric aerosol load; stratospheric ozone	Global/irreversible or VLT impact. Comply with any country limits	Direct output – type 1 number if enough data, or type 2 or 3 if few data	Set 'minimise' as a clear goal; use as a ranking comparator. Set as target if country standards or reduction limits have been set
Other socio-economic indicators (see Section 2.4 and Table 2.4)			
Social equity, gender equality	Key secondary target – ensure fair access for all	Direct outcome; measure by objective statement – type 2, or H/M/L – type 3	All options meet, as secondary target; and use as ranking comparator
Create jobs and income Create wealth, by ownership	Regional/local secondary target – maximise use of the opportunity	Indirect outcome; measure by objective statement – type 2 – or H/M/L – type 3	Set 'maximise' as a clear goal; use as an extra ranking comparator
'Voice'	How local stakeholders are heard – helps acceptability	Direct input: measure by objective statement – type 2 – or H/M/L – type 3	Set 'maximise' as a clear goal; use as a project risk ranking comparator

GHG, greenhouse gas; H/M/L, high/medium/low; VLT, very long term
* This may be difficult to measure in number terms, but it should be possible to establish relative risk differences in outline solution options. For work on this see Croope (2010)
† The 'value' of ecosystems is complex and very specific to a locality. The exact location of a new feature, such as a wetland or forestry strip, relative to existing features (perhaps to connect up a wildlife corridor) may transform its ecological contribution. See the Ecosystem Valuation website. Unless detailed 'conservation' projects are involved, it may be possible to use only an overall average, or an equivalent land area in hectares; both may be misleading. Using an eco-footprint in hectares may be simpler

We considered suggesting a standard 'traffic-light' ranking for each criterion, according to its relative criticality to sustainability, but criticality is very specific to the type and context of an individual project. Clearly, the scale of these impacts will vary greatly from sector to sector and from project to project. The approach we suggested to choosing objectives in Section 3.3 is equally relevant to choosing criteria here. Take a common-sense approach, accordingly, as to how much effort to put into assessing each of the criteria. At outline design stage, few data may be available, and type 2 or 3 evaluations may be necessary. The key point is that you should consider all the criteria, not ignore any just because they are unfamiliar or difficult to measure.

7.5.4 Use 'whole-life' impacts in option evaluation

It is an almost standard approach that in outline design you assess costs on a whole-life basis (see Section 13.7). This includes the initial capital cost, the operating and maintenance cost over the design life, the intermittent refurbishment cost and the final decommissioning cost. It combines these into a calculated, discounted, net present value (NPV).

In the same way, evaluating any other impacts and benefits against choice criteria such as CO_2e emissions or pollution discharges, should be done on a whole-life basis, covering the same stages of project life. Such methods are typically referred to as forms of 'life-cycle assessment' (see Section 13.2), even if the full methodology is not followed. This reflects, too, the 'extend the boundaries' ideas discussed in Section 3.3, and 'work across the interfaces' ideas in Section 4.4. Box 7.3 shows one way of applying a whole-life cost approach to highway design (through the form of the contract specification). Calculating whole life impacts for non-cost items does not involve discounting (but see Section 13.6.4).

Box 7.3 How whole-life approaches to highway design can influence solution choice

A UK highways project was being privately financed with a 25-year contract life. The contract would be paid for through annual instalments to cover the whole-life cost. The standard 'public finance' road base design life was 20 years. It might have been thought that a private finance mindset would take a shorter term design view, but the outline design team actually chose a 40-year design life, at a higher initial capital cost.

This was because a 20-year life road structure would have to be replaced within the 25 years of the contract, and the contract specification included a high future 'lane rental' cost, representing the lost transport capacity value that would occur while the owner closed lanes to replace the road. This quantified high future cost, even when discounted, meant that the 40-year design life gave the lowest whole-life cost.

(Source: authors' personal experience)

7.6. Use decision-making processes that reflect sustainability principles

What mechanisms are used to encourage creativity and innovation?

You will now have an evaluated matrix of options and choice criteria. As you start decision-making, first check that the ideas mentioned above have been reflected and evaluated, so that

- the key service-need outcome that any option must meet to be considered as 'preferred', reflects any wider scope definition that you have established (see Section 7.3)
- representatives of most sustainable options, such as a BSO, are still being considered (see Sections 7.3 and 7.4)
- the choice criteria reflect sustainability issues; their differences, such as measurement type and significance (scale × duration), are recognised and applied appropriately; and the units used relate to service, not just to the product (see Section 7.5.2).

Then you can compare the criteria scores of the options and identify the preferred solution. There are many different decision-making approaches, and none is necessarily better than others in enabling more sustainable solutions. Your best approach will be to use the system that your organisation is familiar with. However, within that system keep in mind some key points that will give the best chance for innovation for sustainability. (If the familiar structure proves to prohibit this, then you might suggest adding a more sustainability-friendly approach, possibly just as a 'what-if' in parallel with the established method, to see what difference it makes.) Some approaches within 'multi-criteria decision-making' (see Section 13.10) may be less useful in outline design, because the methods can be both conceptually difficult and very demanding of data. Life-cycle assessment (see Section 13.2), and particularly a range of building and project rating systems (see Section 13.5) and cost–benefit analysis (see Section 13.6) are more popular, but are still somewhat 'black box', which means they can hide detail and assumptions, and be contentious. The ideas included in a paper by Jowitt *et al.* (2012) on 'protocols' for carbon accounting are more widely applicable in evaluation and decision-making.

You should use a transparent decision-making process, involving a series of comparison steps that are carried out in the right order.

1 **Meet targets**: compare evaluations against any targets, and eliminate options that fail to meet them (but also consider choices just either side of targets – see below). Do this first for the key project-driving service needs, and then for the other criteria that have been set as targets, such as CO_2e.
2 **Do a comparative ranking**: rank the surviving options, using the comparative ranking criteria and separate groupings, in order, for:
   (*a*) **Cost** – to see which solution is most 'cost-effective' in meeting the target (use whole-life costings).
   (*b*) **Other project output/input criteria** – recognising different significance and measurement types in any weighting or combinations.
   (*c*) **'How the project is done' criteria** – separately from inputs/outputs.
3 **Review comparative patterns and relationships, and decide**: this may suggest better refined or combined options, so iterate and recompare if necessary.

As you work through these steps, consider and apply lessons learned from experience, as described in the following sections.

7.6.1 Consider choices just either side of targets – for the 'best bang for the buck'

How is it recognised that best value is not always lowest cost?

All schemes are driven by some performance target, defining the key social or environmental need; and normally only outline design options that fully meet that need will be considered. Although this will always be the first approach, sometimes it may be worth considering preferring solutions that either do not quite meet the target, or just exceed it, to get the 'best bang for the buck'.

Figure 7.3 shows an example of the first, from the water sector. A new 'consent' (defining the quality of sewage effluent allowed to be discharged) is being applied, which is more stringent than the existing one (x axis). However, the 'breakpoint' shows that the final part of the reduction in consent to the target requires a disproportionate increase in CO_2e emissions (y axis), perhaps because an extra energy-intensive treatment process is needed. In this case, the 'better bang for the buck' – measured as 'maximum progress towards the consent threshold at minimum carbon emissions increase' – might involve negotiating an agreement for a reduction in new consent to below the breakpoint, with a much lower carbon impact (Barker, 2011). This might also involve agreeing a compensating tighter consent at another sewage discharge.

A preferred solution might also be justified that delivers a result better than the target. This applies the same kind of logic as before, but in reverse. Figure 7.4 is adapted from Brunei Shell, in the oil and gas sector, where the 'as low as reasonably practical' (ALARP) principle is often applied (Shamhary and Mustapha, 2009). Here the performance goal is to reach a CO_2e reduction target (grey dashed line, and left-hand axis). The chart shows how four options measure up to this. Option 1 fails to reach it, but 2 does; so 2 would be the lowest cost (black line and right-hand axis) acceptable solution. The

Figure 7.3 Applying a carbon breakpoint analysis. (Source: © MWH)

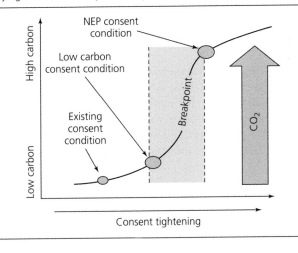

Figure 7.4 Applying the ALARP principle (adapted from oil industry sources)

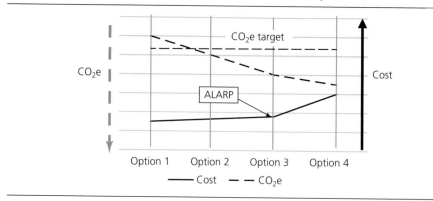

chart also shows why applying the ALARP decision-making principle would lead to a different option being chosen. Option 3 delivers further reductions in CO_2e, for a modest increase in cost; whereas option 4 does even better, but with a large cost increase. The principle defines that option 3 is ALARP, and is preferred because it is where there is maximum extra risk reduction for minimum extra cost, i.e. the 'best bang for the buck'.

Application of this idea is worth considering once the outline design evaluation process has shown the pattern of the performance of the best solutions in relation to the key targets.

7.6.2 Be careful in weighting and adding evaluation criteria

Like cost–benefit analysis, the practice of weighting and adding evaluated impacts turns them all into a single number that apparently tells you the answer. This can appear to be an attractive outcome, and is in common use. One critique of many of the available assessment tools (see Section 13.5) is that many of them do just this, and thus hide the important details and weightings contained within them.

There is nothing wrong with weighting, combining and aggregating information, but do it carefully, with these rules.

- Only weight, combine and aggregate information within a 'comparative ranking', after eliminating any options that do not meet the targets. (However, criteria that set targets can still also be used for comparative ranking, for all options that meet or beat the target.)
- Give extra weight for criteria having a greater significance (see Table 2.3) – do not give a global/irreversible impact the same weight as a local/construction stage impact.
- Never weight and add fundamentally different types of criteria – project results (the 'what') with any aspect of 'how' it is being delivered – or combine an output result with a 'yes'/'no' answer to 'Did you follow the process?'. Both are nonsensical in logic terms – and represent a 'category error'.

- To combine environmental impact criteria, find a summary measure based on science – for instance by turning them all into an eco-footprint – rather than by using an arbitrary weighting.
- If combining two weighted group totals, just adding them can allow a very poor score from one to make up for a very good one on the other. If you want to recognise the importance of both, consider multiplying them instead (see Box 8.1).
- Do not spend a lot of time arguing about the 'correct' weightings; it is much better to use one of the presentational multi-criteria software tools (e.g. Criterium Decision Plus 3 (Haerer, 2000)) that allow your team, possibly together with stakeholders, to try a range of weightings. A good test of a robust preferred option is that it remains the best across a range of weightings.
- Remember that the relative weighting factors that you decide on for any two criteria are actually defining your trade-off value judgements between them. Two weighting factors might be determining, for instance, that 'We are prepared to spend an extra £1 million to save another 1000 tonnes of CO_2e emissions per year'. It is important to extract these explicit trade-off values, and examine them, before agreeing weightings. Most presentational software tools can do this.

7.6.3 Avoid optimism bias, so as not to discredit more sustainable solutions

Have plans and proposals been prepared that reflect the true position, not an idealised one?

Optimism bias reflects a tendency to overestimate future project benefits and to underestimate future risks and costs:

'Optimism bias' is the term used to describe the demonstrated, systematic tendency for project appraisers to be overly optimistic about project costs, duration and benefits.

(DCLG, 2007)

Your evaluation of more innovative and sustainable options may involve a relatively high level of uncertainty, and there may be more potential for such bias. This perception may be reinforced by others' knowledge that you are an advocate of such solutions. It is equally likely, actually, that conventional solutions will be subject to the bias. Examples include a tendency to underestimate both the risks of future rise in energy costs, and the scale of the combined impacts of long-term climate change on infrastructure.

Guidance on correcting for optimism bias is given in government publications. One UK report suggests raising pre-feasibility (or 'scoping' stage) costs by as much as 60%, and even increasing it at the detailed design stage by 30%, to counteract this bias (Defra, 2003). These are extreme examples, perhaps reflecting poor experience of cost estimating. Another approach is to avoid bias by recognising the uncertainties, and carrying out the sensitivity analysis needed to reflect it. In either way, it is important that you avoid optimism bias, and such accusations in decision-making, by being even handed in assessing both conventional and innovative solutions.

7.6.4 Engage with the ideas and opinions of stakeholders in their local context

As part of community stakeholder engagement, it is useful to give local stakeholders, including infrastructure users and operators, a 'voice' (see Table 2.4) in the outline design process and in contributing to decision-making. Local stakeholders can provide more local knowledge than the project team has, and may come up with additional options, or different opinions about choice criteria, because of their understanding of the local context (see the note about different rural attitudes to smell in UKWIR (2004)). In addition, the involvement of these stakeholders will increase their understanding and appreciation of the issues and constraints, and improve community buy-in to the chosen solution.

> The implication is clear: the preferred options are not determined by somehow capturing stakeholder values and opinions *a priori* and then using these to calculate the most preferred alternative, but instead by keeping the points of agreement and disagreement very much in the open as each option is assessed. It might seem a more political and less objective process, but it is much more likely to lead to acceptance.
>
> (UKWIR, 2004: Vol. I)

While continuous engagement is ideal, it can be a complex process, partly because of the sheer amount of detail that may be involved (Box 7.4). In practice, use the following practical steps, which have emerged from experience (extending those in Section 5.4), to involve the community at the right times in the process.

Box 7.4 Water treatment study: managing complex stakeholder engagement

---

How and when do you consult stakeholders on complex projects with many component combinations? This project had

- water demand reduction: 5 options
- treatment technologies: 6 options
- potential development areas: 17 options
- possible storage locations: 19 options.

At an early stage in the evaluation, these multiply to give 9690 possible combinations – far too many to 'consult' in detail on.

Solution

Engage early to obtain local knowledge on components; work up combinations into a shortlist of alternative solutions; then engage on the final evaluation and preferred options.

The project solution gained planning permission at the second attempt.

(Source: personal experience)

- Get local knowledge and ideas early (as in Box 7.4), by discussing local needs, constraints and knowledge, and include them in the brainstormed range of options.
- Include all these options in the 'long list'.
- Engage early with stakeholders on the choice criteria – an interesting example is the engagement with Maori environmental philosophy done in the Hastings, New Zealand, sewage treatment plant (Fraser and Bradley, 2007).
- Then consciously 'take on' these concerns of community stakeholders, through the detailed stages, keeping in touch with key community representatives.
- Include key stakeholder representatives – to contribute opinions, not (probably) as decision-makers – in the decision-making process to select the final solution.

Finally, making good, acceptable, choices needs good communication, as well as good information and processes. Taking time and care to present complex information in an understandable way is worth the effort. For instance, methods such as the radar plot shown in Figure 7.5 can show complex patterns in an understandable way. They can be used to combine information on separate multiple criteria, using a traffic-light scale, to give an overall comparison of performance.

The outline design stage gives most engineers their first, best chance to frame more sustainable projects. By taking the opportunities to ask better questions at the right time, you can serve principles O1.1 – Set targets and measure against environmental limits – and O2.1 – Set targets and measure for socio-economic goals – for greater sustainability. This stage also provides good opportunities to put into practice both

Figure 7.5 A radar plot presentation of multiple project performance criteria. (Source: Author's own presentation, Christchurch, New Zealand, 2008)

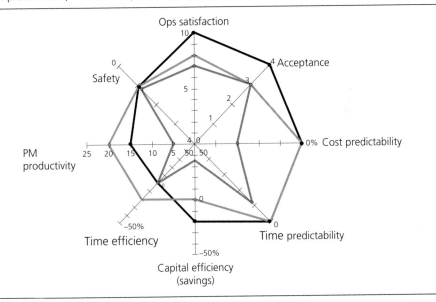

individual principle I.1 – Learn new skills – competences for sustainable infrastructure – and I.2 – Challenge orthodoxy and encourage change.

Outline design: summary

This may be your first chance to influence a project – take it.

Work in a multidisciplinary team – to ensure that all interfaces are considered.

Review the project scope to ensure that sustainability is reflected in it; and widen the range of options considered, to include real sustainable alternatives.

Identify a benchmarked sustainable option (BSO), and take it through to the final shortlist – so that more sustainable options are not eliminated during evaluation.

Include sustainability choice indicators; distinguish between different types and their significance; use customer-based measurement units and evaluate whole-life impacts.

Use decision-making processes that reflect sustainability principles; consider choices just either side of targets – for the 'best bang for the buck'; use 'weighting' with care.

Avoid optimism bias so as not to discredit more sustainable solutions.

Engage with the ideas and opinions of stakeholders in their local context.

Suggested Further Reading is listed after References.

REFERENCES

Ainger C (2012) Briefing: Speeding up innovation by better 'first use' reporting. *Proceedings of the ICE – Engineering Sustainability* **166(1)**: 8–10.

Barker C (2011) *Sustainable Environmental Regulation*. MWH, Leeds.

Barton R (2000) From Value Analysis to Soft Value Management – A Learning Journey. Quotation is included in: http://www.slidefinder.net/b/barton/barton/684788 (accessed 10 September 2013).

Berry C and McCarthy S (2011) Guide to sustainable procurement in construction. CIRIA, London. Available at: http://www.ciria.org/service/Web_Site/AM/ContentManagerNet/ContentDisplay.aspx?Section = Web_Site&ContentID = 19278 (accessed 5 August 2013).

Croope SV (2010) Managing critical civil infrastructure systems: improving resilience to disasters. PhD Thesis, University of Delaware, Newark, DE. Available at: http://www.ce.udel.edu/UTC/Presentation2010/Silvana_Croope_Final-Disseration_100504.pdf (accessed 5 August 2013).

DCLG (Department for Communities and Local Government) (2007) *Adjusting for Optimism Bias in Decent Homes Standard Investment Programmes*. Guidance Note for Housing Revenue Account Private Finance Initiative, Product Code 06HC04379. DCLG, London.

Defra (2003) *FCDPAG3 Economic Appraisal, Supplementary Note to Operating Authorities,*

March 2003 – Revisions To Economic Appraisal Procedures Arising From The New HM Treasury 'Green Book', Annex 2. Available at: https://www.gov.uk/government/uploads/system/uploads/attachment_data/file/181443/fcd3update0303.pdf (accessed 3 August 2013).

Eubank H, Aebischer B, Lewis M *et al.* (2003) High Performance Data Centers. Presented at Building Performance Congress. Available at: http://www.cepe.ethz.ch/publications/Aebischer_EUBANK_RMI.pdf (accessed 3 August 2013).

Fenner RA, Ainger CM, Cruickshank HJ and Guthrie PM (2008) Widening horizons for civil engineers. Presented at the Institution of Civil Engineers, London.

Fraser D and Bradley J (2007) Cultural dreams become a technical reality with innovative wastewater treatment. Ingenium Conference 2007, Invercargill, New Zealand. wat50507319. Available at: http://www.gdc.govt.nz/assets/Wastewater-Plant/Library/117192-David-Fraser.pdf (accessed 3 August 2013).

Haerer W (2000) Criterium Decision Plus 3.0. Versatile multi-criteria tool excels in its ability to support decision-making. *ORMS Today*, February. Available at: http://www.orms-today.org/orms-2-00/swr.html (accessed 3 August 2013).

Hawken P, Lovins A and Lovins H (1999) *Natural Capitalism – The Next Industrial Revolution*. Earthscan, London.

Institute of Value Management (n.d.) Available at: http://www.ivm.org.uk/whatisivm.php (accessed 3 August 2013).

Jowitt P, Johnson A, Moir S and Grenfell R (2012) A protocol for carbon emissions accounting in infrastructure decisions. *Civil Engineering* **165(CE2)**: 89–95.

MacAskill K (2011) Risk management as a framework for applying sustainability concepts on infrastructure projects. MPhil Dissertation, University of Cambridge, Cambridge.

RMI (Rocky Mountain Institute) (n.d.) Available at: http://www.rmi.org (accessed 5 August 2013).

RMI (2003) *Design Recommendations for High-Performance Data Centers*. Report of the Integrated Design Charrette, Rocky Mountain Institute, Snowmass, CO. Available at: http://www.rmi.org (accessed 3 August 2013).

Shamhary PG and Mustapha PDP (2009) *Sustainable Development in Brunei Shell Petroleum – Specific Project and Operation Related Tools*. Environmental Affairs, Brunei Shell Petroleum Co., Brunei.

UKWIR (UK Water Industry Research) (2004) *Sustainable WWTW for Small Communities*, 2 vols. Vol. I: *Sustainability and the Water Industry*. UKWIR, London.

FURTHER READING

Ecosystem Valuation (n.d.) Available at: http://www.ecosystemvaluation.org (accessed 5 August 2013).

Sustainable Infrastructure: Principles into Practice
ISBN 978-0-7277-5754-8

ICE Publishing: All rights reserved
http://dx.doi.org/10.1680/sipp.57548.165

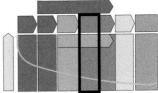

Chapter 8
Detailed design

8.1. What this stage involves

> If you always do what you always did, you'll always get what you always got.
>
> (Mark Twain)

As the quotation suggests, nothing changes unless you change what you actually do. So achieving more sustainable infrastructure will likely require some change in the detailed design process. In this process, the owner's in-house or contracted-out multi-disciplinary design team starts with the management-authorised preferred outline design solution (which describes the main components needed to achieve the required performance, where these are located, how they fit together, and what they will cost). The team takes each component, and develops its detailed design, using whatever mixture of disciplines is appropriate – civil, architecture, structural, geotechnical, mechanical, electrical, electronic and chemical/process. A core team will need to watch over the whole system, to ensure that all the intentions developed in and carried through the previous stages are applied consistently to the project detail, and that the components interact effectively to produce the required sustainable service outputs.

Some specialist design work, such as steel structural frames, cladding, foundations or MEICA (mechanical, electrical, instrumentation, controls, automation) equipment, will not be done by the team, but will be specified by them for detailed design and customised manufacture or off-the-shelf supply by suppliers. For each component the detailed choices of design and materials will likely use 'mini' versions of the decision process described in Section 7.2. The overall output will consist of drawings, specifications, quantities, costs and sustainability impact estimates. When using a design-and-build process, the main contractors and subcontractors already procured for the construction will interact with and advise the design teams, in order to optimise the design–construction interface. Some detailed design may run parallel with construction.

Many infrastructure engineers are engaged in detailed design. While much of your effort is in delivering the sustainability outputs already decided on, and managing risk, there remain new sustainability opportunities at this implementation stage. It is at this design stage that you finally decide on the detailed **technology** to be used to supply the infrastructure service – and so finally fix the project's **impact** – that is, the environmental limits at the global, regional and local scale (see Section 2.3). So it is critical to maintain sustainability objectives as key criteria when making design choices.

8.2. Sustainability in detailed design – and what can engineers do?

What clear responsibilities to the client and to society and the environment have been identified?

As you move through the successive stages of project delivery, the choices about technical aspects, and their associated sustainability opportunities and impacts, become ever more sector specific. In detailed design, these choices are hugely varied, and we leave it to the volumes on individual sectors to cover this in detail.

However, there are strong common themes that 'enable' design for sustainability that you can apply.

■ Regularly engage with constructors, suppliers and the views of users, to ensure that the technical interest of design does not obscure the end objective – a sustainably buildable, operable and decommissionable asset (see Chapters 9 and 10). This can be facilitated by having the design team's performance watched and assessed by the client's operations group and capital investment group (see Box 8.1 for an example). You should be aware of Chapter 10 – In-use to end of life – in order to pick up early on the issues that will be critical to success at those stages. This applies operational principle O3.2 – Consider all life-cycle stages.

> Box 8.1 Driving the design team to consider operating-stage convenience and costs
>
> In the UK, Thames Water's Aspect investment programme utilised a key performance indicator (KPI) system to score the delivery performance of designers and constructors with regard to how well they satisfied both capital investment and operations staff. These two client departments can operate within their own silos, blaming the other if something is not working right; and it is hard for the supply chain, usually contracted to the capital investment group, to serve two masters.
>
> In this case, 'joined-up thinking' was used to score and add up the design team's KPIs as usual for delivery on capital expenditure, but the operators/maintainers also scored the team too. The latter score was converted to a mark between 0 and 1, and was used to *multiply* (not add to) the capital expenditure score. So, to score well overall, the team had to satisfy *both* the capital investment group and the operations group.
>
> (Source: Author's direct experience)

■ The design and specification of some components will be developed within a single discipline. You can however, require that the multidisciplinary group together frequently reviews the whole design and its interactions, to ensure that the system as a whole is optimised and that interface opportunities (see Section 8.6) can be taken. This applies operational principle O4.4 – Integrate working roles and disciplines.

■ Finally, it goes without saying that applying an approach of 'safety in use, through design' is an essential basis for achieving a socially sustainable project (for examples see Queensland Government (2012) and Design Best Practice (n.d.)). This applies operational principle O2.2 – Respect people and human rights.

8.2.1 What can engineers do?

Beyond these basic expectations of good detailed design practice for sustainability, there are more opportunities for you to intervene for greater sustainability. They are, with cross-reference to the principles set out in Chapter 2.

■ During design keep sustainability metrics and climate change in mind (O1.1 – Set targets and measure against environmental limits).

■ Challenge traditional approaches and design standards (I.2 – Challenge orthodoxy and encourage change).

■ Explore design-life and reuse options, and 'off-site' implications (O3.1 – Plan long term).

■ Combine the functions of components to achieve more sustainable systems (O4.3 – Consider integrated needs).

■ Choose more sustainable materials (O1.1 – Set targets and measure against environmental limits).

■ Keep in mind biodiversity and wildlife support when making detailed decisions about siting and landscaping (O1.1 – Set targets and measure against environmental limits).

We describe these in detail in the following sections.

8.3. Keep sustainability metrics and climate change in mind

How has the engineering process shown respect for people and the environment?

Two key sustainability issues during detailed design are: (i) to keep checking design options and choices for their impacts against the sustainability objectives; and (ii) to build in provision for realistic long-term adaptation to climate change, as an external design driver. The latter is very important; because things change relatively slowly, it is easy to underestimate the scale of the *cumulative* impacts of climate-change adaptation (see Section 6.4.3). The most extreme requirement with regard to this issue is to check that the location and site of your infrastructure project, within its longest likely 'use' life, will not be lost to future long-term sea-level rise. (Formal government predictions may be too optimistic to be taken seriously, partly because they rely on IPCC reports, which tend to become quickly outdated. It is also difficult for governments politically to admit to the real scale of the cumulative future changes. For instance, credible work summarising experts' opinions and published data in 2010 (Dyer, 2010) would suggest that the eventual loss of the whole Greenland ice sheet is probably already unavoidable, and will trigger by about a 7 m rise in sea level over some hundreds of years. This means that central low-lying areas of most coastal cities in the world, and all their assets, may need to be abandoned within this period.) An example of a project choice partly dictated by this is given in Box 8.3.

8.3.1   Use sustainability metrics

It is normal to keep asking the question 'What will it cost?' as a key check on the implications of the accumulating detailed design choices. In order to pay equal attention to sustainability impacts, you must also measure 'sustainability' regularly. The tools described in Section 13.5, such as BREEAM or LEED for buildings, and CEEQUAL for other infrastructure, are becoming increasingly popular for doing this.

Figure 8.1 shows how BREEAM can be used to weight and combine scores against the range of sustainability targets for a building. Such tools also have the advantage of now being widely recognised 'brands' that are able to demonstrate the achievement of pre-established levels of sustainability performance, and can be used to benchmark your project against others. They also help in setting a predetermined target for reaching an ambitious level, which can then drive innovation during detailed design. For these reasons, if specific separate sustainability goals have not already been set for your project, suggesting using one of these tools as a target is a credible way of adding them in.

However, the tools should be used with caution, having carefully considered the concerns about combining different types and scales of impact, and of weighting (see Section 7.5). Overall, make sure that the tools you use include, exclude and emphasise correctly all the right impacts that you have been trying to apply systematically through each successive project stage (see Section 3.6). Getting this right may require some customisation, and

Figure 8.1 Sustainability metrics – BREEAM weighting for a building. (Source: MWH)

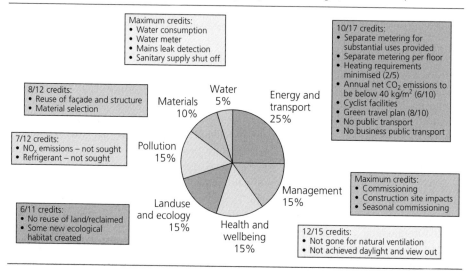

some iterative reviews, which will involve learning by all members of the combined design team (see Section 13.5).

8.3.2 Measure embodied CO₂e emissions as a parallel separate target

Assuming that the outline design stage (see Chapter 7) has settled the *operational* energy use and CO_2e emissions, independently of the detailed design choices, then detailed design can focus on minimising the CO_2e emissions embodied in the construction. There is great merit in measuring project embodied CO_2e emissions as a separate key sustainability goal in design, because

■ climate change is of a larger scale and irreversibility than are other impacts (see Table 2.3)
■ emissions from all CO_2e sources in the project can be added together to represent the whole project, based on scientific fact, without any weighting based on a value judgement
■ because the embodied CO_2e emissions originate in resource extraction, manufacturing, energy use and transport, minimising them provides a good proxy for minimising many other environmental impacts associated with these activities
■ setting ambitious targets for reducing embodied CO_2e emissions has been found to spur creative design and construction innovations that cost less too (see Box 11.7).

This proxy measure is also useful in comparing materials choices for sustainability (see Section 8.7).

8.4. Challenge traditional approaches and design standards

How is careful and informed material selection ensured and overspecification avoided?

Detailed design involves a series of decision-making routines, similar to but narrower and probably more informal than those described in Section 7.2 for outline design. Driving more sustainable detailed design also requires challenging the traditional range of options at this stage too. You should consider new detailed approaches (see Section 8.5), role combinations (see Section 8.6) and materials (see Section 8.7), but also you can challenge the conservatism of design standards and codes, and use approaches that minimise the amount of materials or energy used, to meet the design goal.

Two examples of this are taken from concrete design.

■ For beam or slab design, use post-tensioning or pre-stressing, rather than simple reinforcement, to reduce the use of in-concrete material required to span the design distance under the design load. For example, post-tensioned slabs will carry the same load as a conventional reinforced concrete slab, but use about a third of the steel and reduce the depth of the concrete by about 50 mm (for flat slabs) (Institution of Structural Engineers, 2012).

■ For the design of a water-retaining concrete structure, the concrete wall thickness and the density of reinforcing steel – and so the amount of materials used – depends more on designing to limit crack width to maintain impermeability than on the structural loadings. Typical practice is to design for a 0.2 mm crack width (in BS 8007: 1987), but, as concrete is a 'self-healing' material, where cracks fill up with cementitious material, you could instead design for a 0.3 mm or 0.35 mm crack width (particularly for water-retaining structures that are not full all the time). This can save 30–50% of reinforcing steel and concrete (Ainger, 2005).

In road construction, a good example is not to excavate cuttings to shallow slope angles as dictated by worst-case assumptions about soil strength and geotechnical codes, or pre-assuming that retaining walls are needed. Rather, excavate first to the steepest likely angle that might stand up, and then design on-site to 'treat as found', and stabilise the slope using a range of soil-engineering techniques, applicable to the detail of the varying soil conditions found along a cutting.

Another example is the specification and design of pumping equipment packages, such as are used extensively in buildings and process plants. These often use small-diameter pipe-work, and include many tight bends, with the assumption (often not actually specified) that saving space is the overriding requirement. But the friction energy of pumping increases as approximately the inverse fifth power of the pipe diameter, so small increases in pipe size (say from 25 mm to 40 mm internal diameter) and the use of less acute bends could reduce the pumping energy, and thus CO_2e emissions and operating costs, by about 50%, with little real increase in the space used (the same also applies to larger pumping stations, but with a less proportional impact).

Many such opportunities exist in the detailed design of infrastructure components. Applying them requires you to specify explicitly the key sustainability performance that you require (e.g. a water-retaining concrete tank in which self-healing of cracks is acceptable, or a pumping package for which lowest possible energy use is more important than the space used) rather that just refer to standard sector design codes.

Challenging hallowed design codes is no longer just an occasional need during design. Organisations will do well to review *all* standard design codes and manuals, as called for in Box 8.2. Such a review can be driven by the need for cost savings, but it can be equally driven by a demand for a reduction in embodied CO_2e, which often can be a stronger driver for reducing both CO_2e and cost, than cost alone (see the examples in Boxes 8.4 and 8.6).

Organisations can use CO_2e reduction to challenge design codes, by first writing a new, quite independent 'low carbon design manual', and requiring all designers to use this. When the low carbon manual clashes with existing standards, set the rule that the low-carbon approach always takes precedence, until it comes up against a demonstrably absolute design safety limit.

Box 8.2 Make a 'bonfire' of the standards

There is a genuine problem – specifiers are habitually over-prescriptive, and this is because they do not differentiate between what really needs to be mandatory to achieve their stated performance aims, and what is advisory or even simply suggested ...

First, specifiers should explicitly consider the optimum level of prescription for their project – (less for highways, more for nuclear).

Second [all] should be encouraged to challenge specifications and standards in a constructive manner, We found several stimulating examples of standards being amended to result in outstanding savings and performance improvements.

(Source: Terry Hill, Chairman of Infrastructure UK's Industry Standard Group, in *New Civil Engineer*, 19 July 2012)

8.5. Explore design life and reuse options, and 'off-site' implications

What flexible and adaptable designs have been developed to allow for extended useful life?

Often, decisions about the design life of major infrastructure are defined at the project scoping stage. Sustainability raises our consciousness of increased future risk and uncertainty, and the need for adaptation to climate change, and makes 'adaptive management' an attractive proposition (see Section 4.5). So we are forced to think more carefully about the appropriate design life of infrastructure.

The issue arises again in the detailed design stage, as you chose materials. Box 8.3 shows a project that consciously chose refurbishment, rather than new build, with a shorter design life, to match the expected life of the other parts of the plant on the site. After this time a new build for the whole plant is likely to be required. The refurbishment solution has correspondingly lower cost and embodied CO_2e emissions too. Had a new filter plant with a 40-year life been built, the possible loss of 25 years (60%) of its design life could have made it a bad financial investment, as well as a less sustainable option.

Box 7.3 showed another example of a conscious choice of a different design life for road pavements – in this case doubling it to 40 years. The point is not that longer or shorter design lives are better, but that you should make a conscious 'most sustainable' choice to fit the context. That said, it seems likely that the sustainable response will tend to one of the two extremes.

- Applying a shorter design life, using reusable or recyclable parts or units, and more use of steel, plastic, precast concrete and timber.

Box 8.3 Match the design life to suit external factors and constraints

Essex & Suffolk Water, UK, chose to refurbish a water-filter plant for 15 years more design life rather than build it anew.

(Source: Essex & Suffolk Water)

The structural team determined the residual structural strength, and worked with the client supply chain and industry concrete specialists to develop innovative and successful concrete repair techniques.

The refurbishment solution cost about £7 million less than a new plant, with less embodied CO_2e. This solution provided a further 15 years of life for the existing structure, which corresponds to the expected life of other assets on the site. After this time a new build is likely to be required due to deteriorating raw water quality.

(Source: MWH, UK)

■ Applying a long design life, but building in flexibility in use, allowing for whole-system re-specification, additions, or reuse for other purposes.

By the detailed design stage, the major questions of overall design life may have been settled, but if they have not, re-examine them now, before the opportunity is lost.

8.5.1 Design for reuse of components – and off-site construction
Once the design life has been decided on, there is scope to actively design for *easier reuse*. The preferred 'decommissioning hierarchy' at the end of life of the infrastructure (see

Section 10.7) emphasises that dismantling and reusing as many as possible of the individual components has a much lower processing energy and CO_2e emissions impact than demolishing and recycling the materials.

So the aim is to design infrastructure that can be dismantled easily, and the components rearranged and reused. Sometimes this may involve providing two components to do a job that was done by a single component in the standard 'business-as-usual' approach. For instance, one traditional way to provide watertight water-storage tanks has been to specify concrete designed to a water-retaining code (see Section 8.4). This requires in situ concrete, which can only be demolished, not dismantled. If the water-tightness is provided separately by a lining membrane, the structure can be composed of precast concrete units, with tensioned hoops if needed, and these units can be dismantled and reused. An example of gaining first-use savings in cost and embodied CO_2e emissions in this way is given in Box 8.4 (Riley, 2011).

For tanks, this type of design can also employ reusable timber components (e.g. Timber-tank, n.d.) with separate watertight membrane linings, as used for New York's famous roof-level water tanks. Such tanks can be up to 6000 m^3 in size. An example is shown in Figure 8.2.

You can assist the future reuse of components by providing clarity on design drawings about how a structure is to be put together and how it can be dismantled. It can be made a requirement that components are marked up to allow them to be identified.

Once you look to employ more reusable components, this drives the construction towards more prefabrication or 'off-site construction'. For example, 'modern methods of construction' (MMC) (Building Research Establishment, 2009) is a general term applied to innovative construction techniques that shift a great deal of on-site work to an off-site factory – a more tightly controlled environment. This can reduce waste and materials use, and may allow a reduction in design safety factors (which allow for less control over in situ construction) and overdesign. Issues such as material selection and waste handling, material transportation, and embodied energy or CO_2e of different designs are likely to become important to the structural designer.

These are ways to help use less material and reduce waste on site (see Section 9.6), and may reduce overall sustainability impacts:

> As we continue to drive down lifetime energy consumption, the embodied energy starts to become more significant. Offsite can really score here, with reduced waste, fewer transport miles and material substitution to reduce reliance on high-energy materials, all very feasible. Interestingly, this flies in the face of the general perception that somehow offsite creates lots of CO_2 emissions. Modern factory manufacture is the de-facto lowest cost and energy balance for producing mainstream products in all other sectors, so why not construction?
>
> Mtech Consult, n.d.

Box 8.4 Alternatives to in situ reinforced concrete for watertight storage tanks – saves cost and CO_2e

Conventional reinforced concrete techniques are labour, materials and carbon intensive. Alternatives include:

- Precast concrete units with a separate liner (in the right ground conditions): saved cost (28%) and embodied CO_2e (19%); 50% with cement replacement) and is reusable.

- Structured plastic pipes: saved cost (34%) and embodied CO_2e (39%).

(Source: © Anglian Water, UK)

Figure 8.2 Timber water tanks with reusable timber components. (Figure reproduced courtesy of Timbertank Enterprises Ltd)

8.6. Combine components' functions, for more sustainable systems

While the concrete tank example given above required *separating* previously combined functions, some emerging more sustainable systems designs involve *combining* previously separate functions. The many opportunities vary in their applicability between sectors. They all require that design teams take a multidisciplinary, whole-system approach right from the start of design in order to identify the opportunities. One good example of this is the Rocky Mountain Institute's (RMI) Factor 10 approach, the final principle of which is: 'capture multiple benefits from single expenditures' (see Section 4.4). We summarise some opportunities below, under three headings.

8.6.1 Integrate structure, energy and environment

In the search for lower energy consumption, and the desire to use decentralised, site-based renewable energy sources, opportunities are emerging to integrate site structures and energy. For instance

- appropriate selection of building orientation in order to be able to exploit 'solar gain' for heating
- building thermal mass and insulation into structural components and cladding

- using chimney-type designs, from traditional Middle Eastern practice, to induce natural ventilation
- integrating solar photovoltaic generation into building cladding, covers and roofs
- exploiting underground structures, such as piles, retaining walls and tunnels, to act as sources for ground source or geothermal heat pumps (Box 8.5) (for an example in commercial buildings see Smith (n.d.) and for an example in a residential building see Wood (n.d.)).

Box 8.5 Piled foundations also serve as ground source heat pumps

Images reproduced courtesy of Skanska, UK.

A similar trend exists for gaining environmental benefits from structures, and this is referred to as 'green infrastructure' (see Section 13.6).

- Seeding vertical walls with green 'vertical gardens' (Fedele, 2013), for better air quality and working conditions, and even for useable crops.
- Building 'green roofs' of a variety of kinds, for ecosystem provision, wildlife support and/or reduction of rainwater runoff.

New applications of these kinds of ideas are emerging, and you should search for them early on in the detailed design stage.

8.6.2   Recover energy or resources from infrastructure 'throughput'

A key sustainability principle is to use cyclical rather than linear systems, so that useable resources or energy are recovered from what we thought of as 'waste', and even from what we had not categorised as a possible resource. For instance

- recovering water for air-conditioning/cooling in buildings in water-short areas by treating grey water or sewage in the basement
- extracting energy, phosphate, organic soil conditioner and, perhaps, 'biochar' and algae growth media from sewage sludge and municipal waste, using various processes, including anaerobic digestion and pyrolysis
- recovering large quantities of heat from sewage flows, using water source heat pumps, these usually running on the effluent (the Helsinki district heating system described in Box 4.2 is a large-scale example)
- in buildings, using the body warmth of occupiers and heat generated by equipment such as computers to contribute to heating (as in the Passivhaus (n.d.) design)
- in urban areas, generating useable energy from the force of walkers' footfall on pavements or stairs.

These technologies are at very different stages of development, from about 100 years of experience (anaerobic digestion) to just being developed (footfall energy), but these and other new ideas will be increasingly available to infrastructure designers. You should search for them early on in the detailed design stage.

Interestingly, one example of the most complete application of these ideas of zero waste and recycling systems is not in an urban area but in the latest cruise ships. The largest of these may constitute a population of up to 4000 people, enclosed in the equivalent of an urban system, with a profligate lifestyle in terms of quantities of energy, resources and waste. However, these ships cannot discharge any waste at all in the most pristine environments that they visit (Wein, 2011). This makes them quite a test bed for applying technology to solve these problems, and are worth exploring for ideas.

8.6.3 Integrate temporary and permanent works

The final area to consider – integration – is not new, but has been enabled by the increasing use of combined 'design and construct' teams, and has been incentivised by target-cost contracts. The idea is to minimise the use of materials and energy – and hence minimise cost – by not building separate 'temporary works' just for construction purposes.

- Building basements and many other forms of underground construction where the piled walls or linings that keep the excavation open for construction are also part of the permanent structure.
- Matching the structural strength of a pipeline as it passes through a range of soil types to allow maximum refilling using the as-excavated material as the compacted 'surround', to provide the as-laid pipeline's overall structural strength (which is a function of the interaction of the pipe and its 'surround'). This ends the traditional practice of disposing of all excavated material as 'unsuitable', and having to import special surround material.
- The above is a special example of a broader trend, where you design so as to allow all excavated material on a site to be reused on site, aiming at zero exports of soil off site.

■ Going even further in optimising the pipeline example, some are now using special trench excavation buckets to match the exact size and curve of the pipe to be laid, to absolutely minimise the need to excavate and replace soil at all. The CO_2e and cost savings made using this approach are illustrated by the example given in Box 8.6.

Box 8.6 Curved-bottom trench excavation to match pipe design – reduces cost and CO_2e emissions

---

This extends the given in Box 8.4 of using structured plastic pipes for water storage.

A special bucket is used to excavate a curved-bottom trench, saving on imported pipe bedding and surround materials. This makes further cost (4%) and CO_2e (11%) savings, over the plastic pipe alternative in Box 8.4.

(Source: Anglian Water, UK)

Many of these more sustainable ideas are being generated in areas of standard and regular design, in which one might not have expected there to be much improvement possible. Within these ideas the designed components are relatively low-tech, but the creative innovation is often possible by using new, high-tech analysis and design tools. This was the case in the underground construction, slopes engineering and pipeline examples given above. Furthermore, creative innovation in such areas is often driven by setting strongly disciplined and difficult to deliver CO_2e reduction targets, rather than by just the traditional search for cost savings. You can do the same.

8.7.   Choose more sustainable materials

Choosing the design life, designing for reuse and integrating roles may already have dictated many of your materials choices, but this remains another area in which you can take sustainability into account. The example given in Box 1.4 of comparing Roman and Peruvian bridges demonstrated that you cannot entirely specify 'sustainability' directly. The most sustainable material for a project depends on the project context, and local availability and familiarity. So consider using local supply chains for construction, and choosing materials that are locally sourced.

8.7.1   Measuring embodied CO_2e in practice

As stated above, the embodied CO_2e emissions of a material *are* a good proxy for sustainability impacts, but you need to take care in estimating them. Some sources (e.g. Green Spec, n.d.) tend to assume that *embodied energy* is a good proxy for *embodied CO_2e*, but these two are only actually proportional if the electrical energy sources involved in producing the materials you are comparing have the same average CO_2e emissions factor (EF). This is often not the case. The average EF of a country's power generation, or the EF of the actual power source used (e.g. aluminium smelters favour direct use of low CO_2 hydropower), vary widely with the power generation mix involved. So new steel manufactured in, say, Norway, using almost 100% low-carbon hydropower at about 0.1 t CO_2/MW h (2010 figures: generated 377 TW h, CO_2 emissions 39.2 Mt (International Energy Agency, 2011)), may have lower CO_2e emissions than recycled steel manufactured in the UK. Although the embodied energy of recycled steel may be only a third of that of new steel (BMRA, n.d.), in 2011 the EF of the UK's average mains electricity was about 0.5 t CO_2/MW h, i.e. five times that of Norway. There is an increasing number of embodied CO_2e EF sources for construction materials. A well-used one is the University of Bath database (Hammond and Jones, 2011). The ICE *CESMM3 Carbon & Price Book 2010* (ICE, 2010) now quotes EFs for materials as well as costs, and a new guidance document for carbon accounting for the UK water sector (UKWIR, 2012) is compiling a 'meta-database' of EF sources. An increasing number of tools is available for CO_2e measurement (see Section 13.3).

So, when comparing materials, use embodied CO_2e, not just embodied energy, and also consider the overall installation, not just the main material. For instance, for insulation, it may be attractive to use natural material such as sheep's wool (EF practically zero) rather than a manufactured rigid foam sheet of polyurethane insulation (EF 3.48 kg CO_2/kg (Green Spec, n.d.), but the foam sheet is also self-supporting. If the application is cavity filling between structural walls and cladding, the sheep's wool has no extra penalty. But if it is to line the underside of a roof, the extra impact of an extra supporting or containing structure to hold up the sheep's wool may cancel out its apparent advantage. Again, the context determines the overall most sustainable solution.

It is for this reason that we do not attempt to suggest here a table or hierarchy of more or less sustainable materials. Checking the CO_2e/kg of the material is an important start, but gives only part of the information for making the best choice; you have to work out the weights of material needed in the actual application. (What is needed, ideally, is a measure of intrinsic sustainability per unit of performance – perhaps for structures, a measure of 'CO_2e per tonne of loading per metre of span' for different structural materials.)

8.7.2   Minimise embodied CO_2e in materials in other ways

A good overview of materials and their sustainability, including new alternatives, is given in a recent report by the Institution of Structural Engineers (2010). A forward-looking review of the sustainability current and potential materials is given by Allwood and Cullen (2011).

Apart from choosing a 'better' material, another way to minimise CO_2e emissions is to reduce the content of the high-CO_2e ingredient. For instance, this can be achieved by reducing the cement content in concrete. A typical $1:1.5:3$ mix concrete for an in situ floor slab would have an EF of 0.159 kg CO_2/kg, whereas an alternative with 50% ground granulated blast furnace slag to RC40 would reduce this by a third, to 0.101 kg CO_2/kg. Other emerging cement alternatives are to use a magnesium-based cement rather than a lime- (calcium) based one, and to make cement using captured CO_2 from power-plant emissions (Bielo, 2008). You can also specify the use of recycled aggregates, or other recycled materials.

Finally, while your overall aim will be at least to minimise the overall CO_2e, you might also deliberately use materials that actually sequester CO_2 – that is, that are 'carbon positive'. Timber is the obvious one of these, as it sequesters the CO_2 absorbed during its growth (although it is not often given credit for this in EF tables: many say 'excludes sequestration', which is unfair). An emerging possibility is the design of concrete to be CO_2 absorbing, which might be given credit for an annual rate of CO_2 sequestration throughout its design life (SCIN, n.d.). It is clear that such new opportunities and materials will continue to emerge, and all designers should be looking out for them.

8.8.   Keep biodiversity and wildlife in mind in detailed design landscaping

As development reduces the size of individual wildlife habitats in many places, there is evidence that providing 'wildlife corridors' connecting habitats can support their biodiversity and the resilience of their wildlife and plant populations:

> We show that corridors are not simply an intuitive conservation paradigm. They are a practical tool for preserving biodiversity.

> (Roach, 2006)

As long infrastructure projects (roads, railways, pipelines) cut across many environmental habitats, their detailed site and landscape design, plus careful provision and siting of crossing points, can actively damage or support biodiversity and wildlife. So, when making the detailed design, you have a responsibility to minimise damage and prevent the cut-off of habitat connections; but you also have a large opportunity to enhance them, by choosing the right detailed landscaping layouts, sizes and planting. Some tools are becoming available for doing this (Beier *et al.*, n.d.).

All these ideas can significantly reduce the environmental impact set by the design of a project. You can optimise them through close designer–constructor–supplier collaboration, and this forethought – on how the project will be constructed – can also minimise the impact of the construction stage itself.

Detailed design: summary

Ensure that all the sustainability intentions carried through the previous stages are consistently applied now to the project detail, and seek more opportunities.

Use sustainability metrics, and keep climate change in mind. Measure embodied CO_2e emissions as well as costs.

Challenge traditional approaches and design standards.

Explore design-life and reuse options, and off-site manufacture; combine the functions of components, for more sustainable systems; design for reuse; integrate structure, energy and environment; recover energy or resources from infrastructure 'throughput'; integrate temporary and permanent works.

Choose more sustainable materials, by measuring embodied CO_2e, and minimise embodied CO_2e in materials in other ways.

Keep biodiversity and wildlife support in mind when detailing the design of landscaping.

REFERENCES

Ainger C (2005) Climate change and 'UK Water', presentation to Water UK Innovation Conference, May 2005.

Allwood JM and Cullen JC (2011) *Sustainable Materials – With Both Eyes Open.* UIT, Cambridge.

Beier P, Majka D and Jenness J (n.d.) Conceptual steps for designing wildlife corridors. Northern Arizona University. Available at: http://corridordesign.org/dl/docs/ConceptualStepsForDesigningCorridors.pdf (accessed 3 August 2013).

Bielo D (2008) Cement from CO_2: a concrete cure for global warming? *Scientific American*, 7 August. Available at: http://www.scientificamerican.com/article.cfm?id = cement-from-carbon-dioxide (accessed 3 August 2013).

BMRA (British Metals Recycling Association) (n.d.) *About Metal Recycling*. Available at: http://www.recyclemetals.org/about_metal_recycling (accessed 3 August 2013).

BSI (1987) BS 8007: 1987 Code of practice for design of concrete structures for retaining aqueous liquids. BSI, London.

Building Research Establishment (2009) *Modern Methods of Construction (MMC)*. BeAware Supply Chain Resource Efficiency Sector Report. Available at: http://www.bre.co.uk/filelibrary/pdf/rpts/BeAware_MMC_Sector_Report_02Mar09.pdf (accessed 3 August 2013).

Design Best Practice (n.d.) *Sustainability*. Available at: http://www.dbp.org.uk/sustainability.htm (accessed 3 August 2013).

Dyer G (2010) *Climate Wars*. One World Publications, London.

Fedele A (2013) *Singapore's Sky High Vertical Garden Opens to the Public*. DesignBuild. Available at: http://designbuildsource.com.au/singapores-sky-high-vertical-garden-opens-to-the-public (accessed 3 August 2013).

Green Spec (n.d.) *Embodied Energy*. Available at: http://www.greenspec.co.uk/embodied-energy.php (accessed 3 August 2013).

Hammond G and Jones C (2011) *Inventory of Carbon & Energy, ICE V2.0*. Sustainable Energy Research Team (SERT), Department of Mechanical Engineering, University of Bath. Available at: http://www.siegelstrain.com/site/pdf/ICE-v2.0-summary-tables.pdf (accessed 3 August 2013).

ICE (Institution of Civil Engineers) (2010) *CESMM3 Carbon & Price Book*. ICE, London.

Institution of Structural Engineers (2010) Sustainability: responsible sourcing. *The Structural Engineer* **88(6)**.

Institution of Structural Engineers (2012) *The Value of Structural Engineering to Sustainable Construction*. Available at: http://www.istructe.org/webtest/files/ff/ff7707de-d68f-47f2-b002-b472d90861d8.pdf (accessed 3 August 2013).

International Energy Agency (2011) *Key World Energy Statistics 2011*. International Energy Agency/OECD, Paris. Available at: http://www.iea.org/publications/freepublications/publication/key_world_energy_stats.pdf (accessed 3 August 2013).

Mtech Consult (n.d.) Originated at: http://mtech-consult.com/editorial/editorial/more-offsite-less-energy/, but domain has terminated. For a 2004 presentation, see: http://www.cibse.org/pdfs/1%20Martin%20Goss%20-%204a.pdf

Passivhaus (n.d.) Available at: http://www.passivhaus.org.uk (accessed 3 August 2013).

Queensland Government (2012) *Making Public Spaces Safer*. Department of Local Government, Community Recovery and Resilience. Available at: http://test.dsdip.qld.gov.au/indigenous-councils/making-public-spaces-safer-through-design.html (accessed 3 August 2013).

Riley T (2011) Embodied carbon – the journey through measurement and reduction. Presented at Anglian Water's 'Carbon' Event.

Roach J (2006) First evidence that wildlife corridors boost biodiversity, study says. *National Geographic News*, 1 September. Available at http://news.nationalgeographic.co.uk/news/2006/09/060901-plant-corridors.html (accessed 3 August 2013).

SCIN (n.d.) *Carbon Dioxide Absorbing Concrete*. Available at: http://www.scin.co.uk/material.php?id = 92&PHPSESSID = 7be6a34def45a77975044b13a1d1df28 (accessed 3 August 2013).

Smith P (n.d.) *Energy Piles® – Renewable Energy from Foundations*. Available at: http://www.ice.org.uk/getattachment/33f6e360-0301-45f7-b187-f4ba4e4368e1/Energy-Piles – renewable-Energy-from-Foundations.aspx(accessed 3 August 2013).

Timbertank (n.d.) Available at: http://www.timbertanks.co.nz (accessed 3 August 2013).

UKWIR (UK Water Industry Research) (2012) *A Framework for Accounting for Embodied Carbon in Water Industry Assets*. 12/CL/01/15. UKWIR, London.

Wein A (2011) Innovative water technologies on a cruise ship – the ultimate smart water community. In *Innovations for Smart Water Communities Workshop* (Wein A and Scanship AS (eds)), NTNU, Trondheim.

Wood C (n.d.) Energy Piles for Residential Installations and Other Low Rise Buildings. Roger Bullivant/University of Nottingham. Available at: http://www.gshp.org.uk/GroundSourceLive2011/ChrisWood_Piles_gsl.pdf (accessed 3 August 2013).

Sustainable Infrastructure: Principles into Practice
ISBN 978-0-7277-5754-8

ICE Publishing: All rights reserved
http://dx.doi.org/10.1680/sipp.57548.185

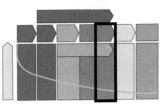

Chapter 9
Construction

9.1. What this stage involves

> The greatest problem with communication is the illusion that it has been accomplished.
>
> (George Bernard Shaw)

Many problems in the construction stage occur because of poor communications – hence the above quotation. 'Construction' builds the project to the detailed design, using a set of (mostly) contracted out supply chain companies working together. It involves managing a complex mixture of people, equipment, suppliers, materials and logistics, with a combined focus on quality, cost, time, safety and environment, and not allowing trade-offs between these. The core project skills of managing risk, change and information must be enabled by excellent and timely communication.

Construction involves interaction between the owner client and all members of the supply chain – designers, key suppliers, programme and project managers, and constructors. It includes sourcing subcontractors, materials and equipment, and effective handover to operations and users. Continued practical interaction with the local community (Chapter 5) is also important. Typically, the modern 'design and construction' team may well include detailed design, with a range of contracted companies acting as a 'virtual team', with some design and supplier procurement continuing in parallel with construction, working to a partnering ethos, with joint risk management. Modern construction management guidance will refer to these, and to sustainability (Harris *et al.*, 2013).

Many infrastructure engineers are engaged in construction. Most of your effort is in delivering the designs already decided upon (hopefully, with some input from you), and managing risk, time and quality. However, you still have further sustainability opportunities with regard to how you construct the project. So it is important to understand the sustainability objectives and to serve them during construction.

9.2. Sustainability in construction – and what can engineers do?

> No construction project is risk free. Risk can be managed, minimised, shared, transferred, or accepted. It cannot be ignored.
>
> Sir Michael Latham, 1994 (quoted in Godfrey, 1996)

Successful construction of more sustainable projects requires that you ensure that the full planning and design intent, including all the sustainability initiatives discussed in previous stages, are delivered on site. Dealing with risks is a key part of this. Including sustainability objectives adds pressure on good project and construction management, because sustainability adds extra issues and concerns. These include adding the risks of environmental impact and social/community relations into standard risk management. Also, in addition to addressing the usual targets of cost, time and quality, sustainability concerns add a strong focus on managing reductions in energy use and waste production and disposal. It can also require using unfamiliar materials or new methods (e.g. sealing buildings to airtight standards), which challenge the attitudes and skills of construction staff and require additional training.

Construction management traditionally focuses mainly, quite naturally, on *getting it built*. That process ends with the handover of the project to the 'in-use' stage; and this requires the final consistent transfer of intent, from one stage to the next (see Section 3.6), in the project delivery process. So, as construction completion approaches, you also need to look forward, to handle all the different issues of successful commissioning into operation and maintenance. This adds a separate *handover management* function, to coordinate and manage all the things that the construction team and equipment providers/installers must pass on to the operators, so that they can operate and maintain the assets effectively and sustainably. To match the sustainability objectives, handover must include training in any new technologies, post-commissioning testing for lowest possible energy use in operation (see Section 10.3), and providing information to facilitate future dismantling for reuse in the operating and maintenance manuals.

So constructing sustainable projects requires using the best available modern project, risk, health and safety, construction, environmental and handover management practice. However, paying attention to risk management, less energy use and waste, good training, a better handover and good community relations is also likely to save money, so you should not assume that construction costs are higher for a more sustainable project.

9.2.1   What can engineers do?

Beyond these basics, there are clear specific opportunities to improve sustainability during the construction stage. Four key actions are listed below, with cross-reference to the principles set out in Chapter 2.

- Manage risks to time, quality, sustainability and cost, as a single team (O4.4 – Integrate working roles and disciplines).
- Purchase all materials and operating equipment on a whole-life-cost basis (O3.2 – Consider all life-cycle stages).
- Source materials and inputs sustainably (O2.2 – Respect people and human rights).
- Proactively reduce construction energy, carbon dioxide equivalent (CO_2e) and waste (O1.1 – Set targets and measure against environmental limits).

We describe these in detail in the following sections.

9.3. Manage risks to time, quality, sustainability and costs as a single team

What steps are actively taken to minimise pollution and negative visual impact?

Many of the traditional risks to successful project completion are to do with issues concerning internal 'interfaces'. These include failures in communications or procedures, such as the design approvals process and programme integration (approvals, design and construction), clashes of interest (such as client scope changes and their consequences), and problems in designer–constructor–client roles and procedures. This has led to the development of comprehensive risk management processes (Godfrey, 1996). Another effective response has been the development of modern partnering (see Section 6.4), this being a 'single team' approach that contributes to handling these issues by setting up a joint and collaborative working culture with joint incentives.

It goes without saying that you will set high expectations and standards, and use excellent methods, for health and safety. (See, for example, the UK Construction Industry Training Board's website and manuals on construction site safety (CITB, n.d.). Equally, as standard practice you should set up and run a project environmental management system, and work to ISO 14001, Environmental management (see Section 13.4), and proactively manage local community relationships through a framework such as the UK's Considerate Constructors Scheme (n.d.).

9.3.1 Manage external social and environmental risks jointly

Sustainability brings a stronger focus on managing and mitigating more external risks, which can have a huge delaying impact on projects, with consequential costs. These might include

- achieving timely planning permission (see Section 5.4)
- gaining local community collaboration on heritage and environmental concerns (see the example in Section 5.3.2), and maintaining good local community relations
- gaining stormwater and pollution discharge consents
- managing interference with, or from, the infrastructure owner's operations (see the example of the eThekwini water mains replacement in Box 6.5)
- complying with wildlife-protecting seasonal permits.

In the past, most of the above were handled by the client or by separate specialist companies. The initial reaction of a typical design-and-construct team to being required to include and manage these risks too was to say 'Those risks are outside our control, and are not acceptable'. However, their actual experience of taking these risks into their 'single team' management has usually been to reduce, not increase, project risk. This is because that inclusion allows proactive management of their previously separate work and timing within the project programme (in the same way as for all 'internal' risks), and provides the incentive to collaborate to achieve that programme.

9.3.2   Integrate the management of critical environmental permitting

Complying with environmental permits to protect wildlife can present large time and cost risks on projects. These risks can be largest on projects comprising multiple or extended sites that may cross many habitats (e.g. many distributed projects, or pipelines or transport routes). Box 9.1 illustrates some of the issues arising on a multi-project utility programme, and demonstrates the value of integrating environmental risk management with all others.

So you can better deal with the external sustainability issues and risks as well, by adopting this single team, inclusive approach, and integrating the risks into overall project programme and risk management.

9.3.3   But take calculated risks, for innovation

Much technical innovation for greater sustainability and lower costs in processes or equipment for infrastructure is first done by small start-up companies. In Schumpeter's 'creative destruction' theory of innovation (Schumpeter, 1942), these would be successful in ousting the established, less sustainable suppliers. The biggest practical threat to this is crossing the gap from a successful research funded proof-of-concept pilot plant (even if at a realistic large scale), to taking part in competitive procurement processes for the first time. In practice, conservative risk clauses in purchase contracts for equipment can shut out small new players, who are too small to take on the transferred financial risks.

Box 9.2 tells a story of the difficulty that an alternative new technology had in breaking into the market. The actual risk of failure was small, and the cost of replacing the equipment if it did fail was not a large part of the project cost. So you can adopt a 'risk guarantee' approach, by getting an outside innovation-supporting party to provide a guarantee in case of failure. Alternatively, have the risk self-insured by the infrastructure client or, if the contract is under a large partnered, target-cost-incentivised programme, have the programme team accept the risk within their overall programme target.

This is a good example of how overly risk-averse contract standards (as with design standards – see Section 8.4) can unintentionally inhibit the innovation needed for more sustainable solutions. You need to look out for these and, with care, challenge them.

9.4.   Purchase all materials and operating equipment on a whole-life-cost basis

Are costs minimised only where all costs over whole life are included?

It is often still common practice for operating equipment and plant, particularly smaller items, to be purchased on the basis of lowest capital ('first') cost, like other materials and components of construction. This habit comes out of a purchasing tradition the only interest of which is lowest first cost. This is in spite of the fact that the choice of the preferred solution is (nearly) always made on the basis of whole-life cost (see Section 7.5), and that the use of energy and other consumables in the operating stage will cost

Box 9.1 Environmental permitting – lessons learnt on major utility programmes

Managing environmental issues on large utility programmes, composed of hundreds of projects, has to deal with a fast-moving, time-constrained and ever changing programme. One example is given below.

Key environmental permitting risks include:
- Protected species – great crested newts, natterjack toads, trees, birds, badgers.
- Invasive species – giant hogweed, Japanese knotweed.
- Consents for river crossings, planning, discharges.
- Third parties – residents, regulators.

Typical time and season constraints and costs:
- Avoid badgers between December and June.
- Avoid bird breeding season, February–August.
- Toads – must survey between February and April.
- Consents to discharge – may take 4 months.
- Planning consents may require work before starting on site.
- Residents may object to plan applications.
- Great crested newts – potential £500 000 delay costs.
- Invasive species regulations – potential £250 000 costs.

Manage all this by:
- Identifying issues early enough to avoid impact.
- Good early consultation about concerns with third parties.
- Manage environmental permit risks within programmes.

Proactive management has reduced risks – one year, e.g.
- Numbers of projects and their risk status being managed each month (red = high, yellow = medium, green = low).

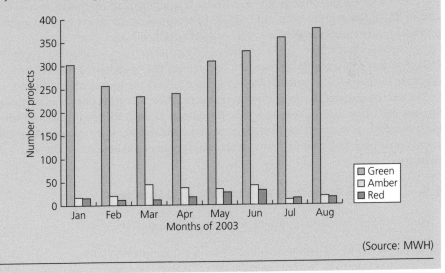

(Source: MWH)

> **Box 9.2** Supply contract risk terms prevent technical innovation for sustainability
>
> A new electrolysis-based nitrate removal treatment process for drinking water was developed, pilot tested (with water company support) and proven by a small start-up company. It was cost competitive, and was included on tender lists.
>
> However, the small company could not accept the water company's contract conditions – to be liable for the full replacement cost if the process did not work. So it licensed its technology to an established larger supplier, who could afford the contract risk; but their interest was diluted, being spread over a wide range of older technologies, some of them threatened by the new process.
>
> Five years later, no full-size plants had been built, and the start-up went bankrupt; but the IP was bought up by a privately owned larger group, and the inventor joined them too. This new combination is now beginning to succeed in the US market.
>
> (Source: Waite, 2012)

the operator much more than those 'embodied' in the first purchase of the equipment (see Section 2.5). So, to be consistent with the preferred solution (and even where you have already specified an energy performance standard), it makes sense to base your equipment purchase choice on offers that have a *projected lowest whole-life cost*, including operating and repair costs, reflecting the working life of the equipment. This will require you to carry out operating efficiency tests as part of commissioning (see Section 10.3).

There may be a practical difficulty of trust when you first tell your suppliers that you are now changing to buy on a lowest whole-life-cost basis, as they may not believe that you will follow through on that. Suppliers will suspect that some competitors will not believe you either, and will still offer their lowest first cost equipment, so that when your purchasers see those lower prices they will fall for the temptation and revert back to your old practice. In the authors' experience, this is a well-justified fear.

One good way to overcome this issue, as you change your purchasing practice, is to ask suppliers for two offers – one of their equipment having the lowest whole-life cost, and one of their equipment having the lowest first cost. Then, all competing suppliers can put up their best offers against both. It is to be hoped that you will stick to their lowest whole-life cost intent, as this matches your real cost and sustainability interests. But, even if you are tempted back to lowest first cost, at least all suppliers will have a fair chance on that basis too.

9.5. Source materials and inputs sustainably

How is the exploitation of distant resources and people minimised?

'Sourcing' refers to the procurement of all services, equipment and materials for design and construction. The particular issue of proactively serving *extra* socio-economic sustainability objectives through procurement has already been emphasised (see Section 6.3) and you can optimise this potential by considering it in advance, throughout construction, and at all scales of purchasing.

Beyond this, as a standard 'corporate responsibility' expectation, infrastructure clients and stakeholders increasingly demand that the ideas of 'do no harm' and 'encourage good practice' are embedded throughout your supply chain, and are applied to all purchases. It includes issues such as the treatment of workers, and materials impacts and source traceability: 'If I buy those cheaper pumps from X, what worker conditions am I tacitly supporting?' To address this, you can follow published advice: a useful summary of what is required and how you can do it is available from the Institution of Structural Engineers (2010), and a comprehensive guide is provided by CIRIA, UK (Berry and McCarthy, 2011):

> A growing world population, depletion of resources, greater interest by stakeholders of how the products they consume are created, and the increasing importance of making wise financial investments in projects, all point to the importance of sustainable procurement.

In addition to *voluntarily* exercising 'buyer power', some country strategies, such as the UK's *A Strategy for Sustainable Construction* (DETR, 2000) (which has established a target that, by 2012, 25% of products used in construction projects should be sourced from schemes recognised for responsible sourcing), and international standards such the ISO 21930: 2007 (ISO, 2007) (which provides the principles and requirements for type III environmental declarations of building products) are reinforcing the drivers. A number of sustainable scoring systems, such as BREEAM and the UK's 'Code for Sustainable Homes', now award additional points for using such sourcing practice. The UK Building Research Establishment's responsible sourcing standard (BES 6001), published in 2008 (BRE, 2008), makes this link, and this is reinforced by the subsequent British Standard BS 8902: 2009 (BSI, 2009).

So you can serve sustainability goals beyond your specific project objectives, and (as an extra argument for business leaders) increase your project sustainability scores, by requiring your supply chain to follow such codes as standard good practice.

9.6.   Proactively reduce construction energy, CO_2e and waste

> Every little helps
>
> (Tesco)

Your other main opportunities to make a difference lie in reducing – by actively 'managing down' – all your construction *energy use* and *waste*. The key drivers for doing this are that you are thereby actively reducing the CO_2e emissions embodied in your construction, and reducing the cost of energy and wasted materials too. A UK-wide study in 2008 found that the construction stage influences the total CO_2 footprint of a project in the following proportions (BIS, 2010)

- 15% in the purchase of manufactured materials and equipment
- 1% through 'distribution' – transport to and from site
- 1% through on-site energy use in construction

the other components were 82.6% in operation and maintenance, and 0.4% in demolition.

Construction energy environmental impacts are beginning to be assessed as an issue (Sharrard *et al.*, 2007), and construction waste may be around a third of national waste totals. (According to Defra (2006), the UK generated 335 Mt of waste, of which 32% (107 Mt) was from the construction and demolition industry. Construction and demolition contributed 11% of the UK's GDP, so our sector's waste is disproportionately large, compared to its value to the economy.) Furthermore, waste may cost 0.3–0.5% of a project's overall value (and this does not take into account the large cost of wasting purchased materials that are thrown away). On both energy and waste, construction attitudes have started to be transformed in recent years, and there are increasing examples of both new regulation and good practice, partly because of the realisation of the cost savings that can be achieved. There is no single 'magic bullet', but a range of small actions all add up to less sustainability impact and, probably, lower cost – 'every little helps'.

9.6.1 Minimise the CO_2e emissions of all purchased materials and equipment

To minimise that 15% of the CO_2 footprint in the materials and equipment, your design process will have already aimed at low CO_2e emissions (see Section 8.7), and CO_2e is also a useful proxy for other environmental impacts. Now you must carry this through into purchasing. The best way to do this is to require, as a condition of tendering, that all suppliers tell you the CO_2e footprint – both embodied and operational – of their materials, products and services, as well as their cost. In this way, you can get a whole-life carbon measure (see Section 13.7) for purchases, which parallels the whole-life-cost approach (see Section 9.4 above). You can then tell suppliers that you will take this comparative measure into account, as well as cost, in your purchasing choices.

Within the general business world there are several good examples of driving down carbon footprints in the supply chain, from companies such as Walmart (Wong, 2012), and advice from the World Business Council for Sustainable Development (WBCSD, n.d.) and Greenhouse Gas Protocol (n.d.). Infrastructure still has much potential for better practice. One pioneer, particularly for CO_2e, is Anglian Water in the UK (Anglian Water, 2012) (see Box 11.7). Such new client demands require suppliers to generate new environmental information about their businesses and to share this with competitors, and so it can be uncomfortable for supply chains to respond to this request. Clients need to drive this, and you can do it most effectively within longer term, collaborative, supply chain relationships (see Chapter 6).

9.6.2 Manage down construction carbon – through energy efficiency

There remains the 2% of CO_2e emissions from transport to and from site, and on-site energy use. This requires proactive energy management. Methods to minimise this energy use include

- optimising construction plant use and movement (particularly for large-scale excavation and filling), driver training in efficient plant operation, and use of only highly efficient plant
- optimising *transport* efficiency for materials delivery, waste disposal and construction workers' travel – by minimising all the quantities needing transport and their distances travelled, and insisting on weight- and fuel-efficient vehicles – 'An efficient concrete mixer truck is capable of transporting nearly a cubic metre of additional concrete in one load' (MAN Transportefficiency, n.d.)
- minimising energy used in heating temporary site buildings – 'Temporary cabins are identified as one of the top sources of carbon emissions and wasted energy on construction sites' (Buckler, 2010).

On the best-practice sites, this approach has the potential to reduce energy use and CO_2e emissions considerably, as illustrated by the example in Box 9.3 (Ko, 2010).

Box 9.3 Reducing construction energy use

The UK's Strategic Forum for Construction developed an action plan on carbon reduction (Ko, 2010):

15% reduction in carbon emissions from construction processes and associated transport compared to 2008 levels

The 2008 baseline assessment identified the largest sources of carbon emissions. This Action Plan focuses on the top four:

- On-site construction (plant and equipment) and site accommodation
- Transport associated with the delivery of materials and removal of waste
- Business travel
- Corporate offices etc.?

They also have plans for reducing waste and water use (Strategic Forum for Construction, n.d.).

9.6.3   Manage down all construction waste

Over 90% of non-energy minerals extracted in the UK go into construction. Every year some 70 million tonnes of construction and demolition materials and soil end up as waste, and around 13 million tonnes of this is discarded each year without even being used (DETR, 2000). In addition, the issue is made more complex when further creators of waste are added during the subcontract and construction phases; the wide range of possible waste causes is shown in Figure 9.1 (Keys *et al.*, 2000).

The minimisation or elimination of construction waste requires proactive management. This includes

- all demolition waste, and components and materials from it – by minimising disposal off-site, maximising reuse or recycling (see Section 10.7)

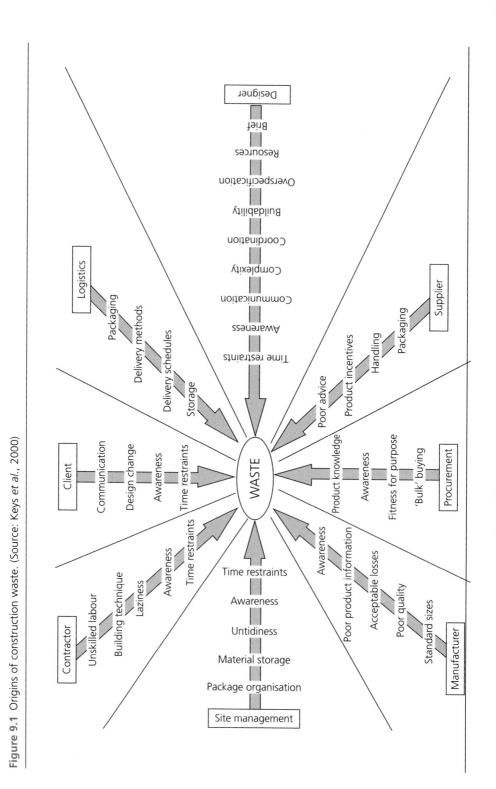

Figure 9.1 Origins of construction waste. (Source: Keys *et al.*, 2000)

- all waste of bought-in materials – by not over-ordering, and taking great care to avoid damage in storage and overspecification
- all waste excavated material, and the need for imported filling material – by optimising the balance of both, and maximising reuse on site.

Clients can allow more time for waste recovery within project timetables, and can allow innovation in the novel reuse of materials, and contractors can ensure efficient ordering processes and implement waste-segregation policies and waste auditing as a key part of their project management.

The additional effect of minimising waste is, of course, to reduce energy use as well, because it directly reduces the quantities you purchase, remove and transport. For bought-in materials such as concrete, reducing waste also depends on what the supplier does with over-ordered returned material. One new technical solution for waste concrete may be to treat it in-lorry, to extract the aggregate for reuse (*New Civil Engineer*, 2013). You can use the lists of waste causes in Figure 9.1 as a framework for joint waste audits.

Best practice, and in some cases regulations, requires you to make and execute a site waste management plan. The Site Waste Management Plan Regulations 2008 came into force in England and Wales on 6 April 2008, and seek to reduce the quantity of construction waste that is sent to landfill. Good guidance on how to go about this is given by the UK's Waste and Resources Action Programme (WRAP, n.d.). As a stretch target, and to get the site design and construction team in the right frame of mind, you may want to set a target of 'zero waste'. This may be more practically defined as 'zero net waste' (i.e. waste-neutral construction projects) (DTI, n.d.), where the value of materials reused or recycled in a construction project at least equals the value of materials delivered to site that are wasted:

Waste neutrality therefore depends on reducing waste, segregating material for re-use and recycling, and using more recovered material.

In selecting new materials you should make maximum use of renewable resources such as wood from properly accredited sources, and greater emphasis should be placed on locally available materials. You should minimise the use of those materials with poor environmental performance, such as metal derivatives or synthetic compounds. In general, materials should be used in such a way that they can be beneficial for many generations; for example, a structure built with design life of 500 years should be designed to allow for adaptation within the shell to suit a variety of potential future uses.

Another new tool in the UK is the National Industrial Symbiosis Programme (NISP), which has been operating since 2005. This the only national programme of its kind. It facilitates mutually profitable links (synergies) between a network of 12 500 companies, so that previously unused or discarded resources such as energy, water and materials, from one organisation are recovered, reprocessed and reused by others elsewhere in the network. The state of development of markets, systems and technologies for reusing and recycling materials varies greatly between countries, and is rapidly changing.

The materials that are used most in any civil engineering works are aggregates, and the UK is fortunate to be at the leading edge of aggregate recycling in Europe (WRAP, 2005). Specifying that a percentage of any project should comprise using recycled materials can provide the incentive and flexibility for contractors to look for effective economic solutions. Government organisations can take a lead here, through their procurement strategies, with, for example, the Environment Agency requiring 60% recycled materials in its capital programmes for flood defence and other projects. The use of pulverised fuel ash and ground granulated blast furnace slag as cement replacements was pioneered in the 1980s, and these recycled materials have been included in British Standards for over 20 years. Not too long ago, recycled aggregates were excluded from some standards and specifications, but now aggregates are covered by joint British and European Standards, which mean that they specifically allow the use of recycled materials (e.g. EN 933-11: 2009 Classification test for the constituents of coarse recycled aggregate, and EN 1744-6: 2006 Determination of the influence of recycled aggregate extract on the initial setting time of cement). Recycled materials can perform as well as primary materials, and in some cases they enhance performance.

In new road construction, the engineering elements that can contain recycled materials include asphalt in the road, safety barriers, top soil, drainage channels, filter materials and geotextiles. In the UK, the asphalt base of the A21 refurbishment was constructed using a recycling process. This resulted in a direct saving of £525 000 through the use of recycled compared to primary materials. Additional indirect savings in avoided transportation and extra work were also considerable (approximately £480 000). Similarly, the construction of a large sewage works in the north-east of England used 270 000 t of locally sourced coarse pulverised fuel ash to raise the level of the sewage works. A direct saving of £108 000 was achieved by using this recycled material in preference to a primary material. Other examples include the laying of water distribution pipelines, which can contain recycled materials in the asphalt road materials, fill material, water pipes and pipe bedding. Further good ideas that you can adopt come from the 'lean construction' movement (Constructing Excellence, 2004); these are best applied in a joint 'design and construct' team. The UK Environment Agency's Burrowbridge river engineering project (Box 9.4) is an example of using many of these ideas.

Box 9.4 Reducing waste and CO_2 emissions in the £270 000 Burrowbridge bank project, UK

The project was to repair a failing river bank, part of the River Parrett tidal reach. The bank had eroded, lost many of its timber piles and been affected by landslides.

The solution was to trim the riverside slope to a more suitable profile, while maintaining crest width by filling at the rear. The toe of the embankment was protected by new timber piles and stone, while a soft engineering solution was used for the bank slope.

The initial design predicted a total CO_2 footprint of about 140 t, mostly from import of clay, stone and timber to site, and export of trimmed material off-site. By eliminating the stone access track material, reusing some bank 'cut' as fill (by challenging the original specification), and using reclaimed timber piles from a nearby demolition, the project reduced its total predicted footprint by over 40%, and avoided more than 100 lorry movements.

(Source: Environment Agency, 2011)

Proactively managing down your construction energy and waste – and thus your CO_2e – will serve your sustainability goals, and also save you money. You can use your project environmental management system as one monitoring and improvement target-setting tool for this. By adopting the practices suggested, you can deliver your project to fulfil the sustainability objectives that have been carried down through all the delivery stages. This sets up the best possible potential for a successfully sustainable in-use stage.

Construction: summary

Focus on delivering the sustainability features of the designs already decided on, but also do the extra things you can to minimise impacts, and meet sustainability targets.

Add environmental impact risk and permitting, and social/community relations into standard risk management, but take calculated risks, for innovation.

Purchase all equipment on a whole-life-cost basis.

Source all materials and inputs sustainably.

Proactively manage down construction energy, CO_2e and waste in all purchased materials and equipment, and through on-site energy efficiency.

REFERENCES

Anglian Water (2012) *Innovating Collaborating Transforming.* Love Every Drop 2012. Available at: http://www.anglianwater.co.uk/_assets/media/Love_Every_Drop_2012.pdf (accessed 3 August 2013).

Berry C and McCarthy S (2011) *CIRIA Guide to Sustainable Procurement.* CIRIA C695. CIRIA, London.

BIS (Department for Business, Innovation and Skills) (2010) *Estimating the Amount of CO₂ Emissions that the Construction Industry can Influence*. Supporting Material for the Low Carbon Construction IGT Report. Available at: https://www.gov.uk/government/uploads/system/uploads/attachment_data/file/31737/10-1316-estimating-co2-emissions-supporting-low-carbon-igt-report.pdf (accessed 3 August 2013).

BRE (Building Research Establishment) (2008) BES 6001 Framework Standard for the Responsible Sourcing of Construction Products. BSI, London.

BS EN ISO 14001:2004 (2004) Environmental management systems. British Standards Institution.

BS EN 933-1:2012 (2012) Tests for geometrical properties of aggregates. Determination of particle size distribution. Sieving method. British Standards Institution.

BS EN 1744-8:2012 (2012) Tests for chemical properties of aggregates. Sorting test to determine metal content of Municipal Incinerator Bottom Ash (MIBA) Aggregates. British Standards Institution.

BSI (2009) BS 8902:2009 Responsible sourcing sector certification schemes for construction products. Specification. BSI, London.

Buckler S (2010) *Green My Cabin: Site Offices Key to cutting carbon emissions in Construction*. Housing and Planning, Local Government. Available at: http://www.govtoday.co.uk/local-government-news/21-housing-planning/4491-green-my-cabin-site-offices-key-to-cutting-carbon-emissions-in-construction (accessed 3 August 2013).

CITB (Construction Industry Training Board) (n.d.) Construction site safety. Available at: http://www.ge700.co.uk (accessed 3 August 2013).

Considerate Constructors Scheme (n.d.) Available at: http://www.ccscheme.org.uk (accessed 3 August 2013).

Constructing Excellence (2004) *Lean Construction*. Available at http://www.constructingexcellence.org.uk/pdf/fact_sheet/lean.pdf (accessed 3 August 2013).

Defra (Department for the Environment, Food and Rural Affairs) (2006) *The Environment in Your Pocket*, 10th edn. PB. Available at: http://www.lbp.org.uk/downloads/Publications/CommunitiesLAs/eiyp2006.pdf (accessed 3 August 2013).

DETR (Department for the Environment, Transport and the Regions) (2000) *Building a Better Quality of Life – A Strategy for Sustainable Construction*. Available at: http://www.berr.gov.uk/files/file13547.pdf (accessed 3 August 2013).

DTI (Department for Trade and Industry (n.d.) Waste. In *Strategy for Sustainable Construction – Consultation Events*. Available at: http://www.bis.gov.uk/files/file37173.pdf (accessed 3 August 2013).

Environment Agency (2011) *Burrowbridge Carbon Reduction Case Study*. Available at: http://www.environment-agency.gov.uk/static/documents/Business/Burrowbridge_Carbon_Reduction_Case_Study.pdf (accessed 3 August 2013).

Godfrey PS (1996) *Control of Risk – A Guide to the Systematic Management of Risk from Construction*. Special Publication 125. CIRIA, London. Available at: http://www.ciria.org/service/free_publications/AM/ContentManagerNet/ContentDisplay.aspx?Section = free_publications&ContentID = 19459 (accessed 3 August 2013).

Greenhouse Gas Protocol (n.d.) Available at: http://www.ghgprotocol.org (accessed 3 August 2013).

Harris F, McCaffer R and Edum-Fotwe F (2013) *Modern Construction Management*, 7th edn. Wiley-Blackwell, Chichester.

Institution of Structural Engineers (2010) Sustainability: responsible sourcing. *The Structural Engineer* **88(6)**.

ISO (International Organization for Standardization) (2007) ISO 21930: 2007 Sustainability in building construction – Environmental declaration of building products. ISO, Geneva.

Keys A, Baldwin AN and Austin SA (2000) Designing to encourage waste minimisation in the construction industry. *Proceedings of CIBSE National Conference, CIBSE2000, Dublin*.

Ko J (2010) *Carbon: Reducing the Footprint of the Construction Process. An Action Plan to Reduce Carbon Emissions.* Strategic Forum for Construction/Carbon Trust. Available at: http://www.strategicforum.org.uk/pdf/06CarbonReducingFootprint.pdf (accessed 3 August 2013).

MAN Transportefficiency (n.d.) *More Payload Due to Lightweight Construction and High-strength Materials.* Available at: http://www.transportefficiency.com/en/technology/lightweight_construction/Lightweight_construction.html (accessed 3 August 2013).

NISP (National Industrial Symbiosis Programme) (n.d.) Available at: www.nisp.org.uk (accessed 3 August 2013).

New Civil Engineer (2013) Spotlight concrete: clean solution. *New Civil Engineer*, January, p. 40.

Schumpeter JA (1994) [1942] *Capitalism, Socialism and Democracy*. Routledge, London.

Sharrard A, Matthews H and Roth M (2007) Environmental implications of construction site energy use and electricity generation. *Journal of Construction Engineering Management* **133(11)**: 846–854.

Strategic Forum for Construction (n.d.) Sustainability. Available at: http://www.strategicforum.org.uk/Sustain.shtml (accessed 3 August 2013).

Waite M (2012) Innovation and water: an entrepreneur's adventure. Presented at CMS (Communications and Management for Sustainability) Conference, London. Available at: http://www.coastms.co.uk/conferences/466 (accessed 3 August 2013).

WBCSD (World Business Council for Sustainable Development) (n.d.) *GHG Management.* Available at: http://www.wbcsd.org/work-program/capacity-building/ghg-protocol.aspx (accessed 3 August 2013).

Wong KA (2012) Walmart commits to scale sustainability of global supply chain. Available at: http://www.greenbiz.com/news/2012/10/25/walmart-commits-scale-sustainability-global-supply-chain (accessed 3 August 2013).

WRAP (Waste and Resources Action Programme) (n.d.) *The Site Waste Management Plan Template.* Available at: http://www.wrap.org.uk/sites/files/wrap/WRAP%20SWMP%20Template%20v2.0%20Information%20Sheet.pdf (accessed 3 August 2013). *Site Waste Management Plans.* Available at: http://www.wrap.org.uk/content/site-waste-management-plans-1 (accessed 3 August 2013).

WRAP (2005) *Opportunities to Use Recycled Materials in Preliminary Building Works and Civil Engineering: Quick Wins Guide.* WRAP, Banbury.

Sustainable Infrastructure: Principles into Practice
ISBN 978-0-7277-5754-8

ICE Publishing: All rights reserved
http://dx.doi.org/10.1680/sipp.57548.201

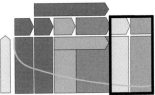

Chapter 10
In-use to end of life

10.1. What this stage involves

The challenge, then, for civil engineering, is to be concerned [with] how such technology sits within the public realm and how it will affect the communities it touches.

(HRH The Prince of Wales, 2012)

The in-use stage is the longest stage. In it, infrastructure operation makes use of the project output, which are the built assets, to provide the intended service to users. It also includes maintenance, and periodically replacing worn-out or inefficient 'capital' assets to keep them in effective use for the design life of the structure. A fundamental test of sustainability success is how well the infrastructure serves its communities, as the above quote identifies. However, all assets eventually come to their end of life (see Section 10.7), either by reaching the end of their original design life, or by having it cut short or being in need of adaptation due to changes in circumstances, technology or need; sometimes the life of an asset may be extended beyond the original intent.

Infrastructure may last in-use for 40–100 years or more. The in-use stage involves three separate but interacting sets of actions: actions by operators, who run and maintain the fixed infrastructure assets; actions by mobile vehicles (in some sectors), which are operated to use them; and actions by end-users themselves, whose behaviour interacts with the other two. These interactions determine how well you deliver the socio-economic service, while also minimising the environmental impacts. The combinations vary from sector to sector.

- In buildings, the users *occupy* the assets, and their behaviour *directly* influences how the assets can be operated.
- In transport, the users ride in mobile vehicles (car, lorry, bus, train, ship, plane) that have their own extra impacts, and use the fixed assets (road, rail, port, airport) only *indirectly*. The overall sector impact depends more on the way in which the vehicles are designed and driven, and how effectively their passenger/goods capacity is utilised, than on how the fixed assets are operated. In most cases, the specification and design of the vehicles is outside the infrastructure engineers' control, but it can be influenced through sector relationships between infrastructure owners, vehicle/service providers and government policy-makers.

■ In utilities (e.g. gas, electricity, water) the user uses only the *product* of *the assets*, entirely *separately from* their operation. The operation of the assets, and the user's behaviour with the product, have entirely separate impacts, but the utility service provider can still influence the product use in various ways. So the interactions between assets, operators and users are critical to sustainable operation.

The decisions made in all the previous project delivery stages (through detailed design and construction) will have set out your scope for sustainable operation and then decommissioning. However, optimising these stages still requires applying the key sustainability objectives and principles discussed in Part I. By putting these principles into practice, operators, users and decommissioners can deliver the sustainable service outcomes that were intended.

10.2. Sustainability in operation and use – and what can engineers do?

Does the engineering product provide value and satisfaction to meet the needs of end users and the general community?

Operation and maintenance engineers have always been driven by their *operational* responsibility for cost-efficiency and reliability, in keeping the asset working. A strong sustainability focus needs you to also pay attention to optimising its use by all customers, which determines how well it actually delivers its socio-economic service – that is, the *user outcome*. It also adds minimising, and absorbing, global warming impacts into your efficiency and reliability objectives.

The UK high-level study first referred to in Section 9.6 (BIS, 2010) estimated that the carbon dioxide equivalent (CO_2e) emissions from the in-use stage are at least five times those embodied in the project assets (in-use and end-of-life stages comprised 84% of the whole-life CO_2e emissions, compared with 16% embodied CO_2e in the construction stage). However, this depends heavily on the sector and the power generation assumptions used. The authors' project experience suggests that a ratio of 10 times or more is frequently found. So, this in-use stage has by far the largest global warming impact and, within the constraints of the assets, you have many opportunities to help control it. At the same time, the threats that it brings (see Chapter 1) require you to build asset *resilience* to accommodate them.

Dealing with these threats depends on the *efficiency* of each of the three factors described above – asset operation, maintenance and replacement, vehicles use, and user behaviour – and also on optimising their interactions. Table 10.1 summarises the factors that control the in-use socio-economic service, and environmental impacts, and the ways in which these can be minimised, in each sector.

The table demonstrates some common features across sectors, in spite of the large differences between them. Similar influences include government sector policies, and vehicle or appliance specifications. Part of your role is to influence these over time, to

Table 10.1 Factors that control in-use sustainability impacts and their minimisation

Sector	Factors that control the sustainability impacts of the in-use stage – for effective operation (i.e. minimising all types of impact)
Buildings	■ Building *design*, including heating and ventilation, can enable efficient operation, user behaviour and service; it is affected by government *planning and regulations* ■ Individual user's *behaviour* is also critical to efficient operation; it can be influenced by employer's *incentives/rules, culture and leadership*
Transport	■ Road/track and port/airport and user access *design* can enable efficient operation, affected by government *planning and transport regulations* ■ Vehicle/train and ship/plane *specification* and *design* can contribute, affected by government *vehicles (all types) regulations* ■ Individual user's behaviour is critical to efficient occupation of available capacity and quality of service; it is influenced by *tolls, tariffs, business model, schedules, convenience, education*
Utilities	■ Facility, treatment and network *design* directly enables efficient operation, influenced by government gas/electricity/energy/water/telecoms *regulations* ■ *Appliance* specification, design and business model can contribute; it is affected by government *energy, water and industrial policy*, which can indirectly or directly reduce gas/electricity/water impacts (including energy and water efficiency) ■ Individual user's behaviour is critical for efficient use of gas/electricity/water and quality of service; it can be influenced by *tariffs, business model, legislation, direct assistance with appliances, insulation, public education*
All	■ Government *energy and fuel policy*, particularly on decarbonisation, fundamentally dictates, and can reduce CO_2e impact of infrastructure energy use ■ Operators must operate and maintain assets optimally; this can be incentivised by tenant/facilities management/operator/franchise *contracts* ■ Clever *management* can help match user demand with available infrastructure, industrial or environmental *capacity*

increase sustainability (see Section 3.4). Also, the behaviour of the *users*, rather than of the asset *operators*, can dominate impacts and affect access and service. You need to play close attention to 'enabling' this behaviour. Lastly, the terms of operating contracts, and the tariffs, business model and incentives used are critical to maximising user service, with minimum impacts. Your 'operation' includes these aspects, as well as your vital role in efficient operation of the assets.

10.2.1 What can engineers do?

The main task that you have during the in-use stage is to optimise the asset operation and its use, in order to deliver (at least) the sustainability performance it was designed for. Beyond this, you will find new opportunities for improvements, as needs change and technology develops. The four key things you can do to create and use these opportunities are summarised below, with cross-reference to the principles set out in Chapter 2.

- Hand assets over effectively from construction (O4.4 – Integrate working roles and disciplines).
- Incentivise operational and maintenance teams, and user behaviour (O1.2 – Structure business and projects sustainably; O4.3 – Consider integrated needs).
- Keep social needs in mind (O2.1 – Set targets and measure for socio-economic goals).
- Add environmental performance and resilience into effective maintenance (O1.1 – Set targets and measure against environmental limits; O3.2 – Consider all life-cycle stages).

Some of these interventions may not seem to be in the traditional job description of operations and maintenance (O&M) engineers. Using all the opportunities to maximise sustainability in-use may require you to extend those roles, and raise new questions, whenever possible. We describe these interventions in detail below.

10.3. Hand assets over effectively from construction – include efficiency trials and training

How is performance benchmarked as a precursor to seeking continual improvement?

At the end of construction, the formal handover of the asset to the operational team needs to demonstrate that it works according to its specification. This will usually include final inspections for quality, equipment testing and commissioning, and training in O&M, including providing technical manuals.

For sustainability, you should also ensure that handover includes demonstration, training and trial operation in *optimum efficiency* mode – delivering the specified benefits, while operating with the lowest possible use of electricity and other inputs and resources, and the lowest possible emissions of CO_2e and pollutants. Ideally, the specification and performance guarantees should have included a requirement for this efficiency demonstration. The demonstration must be long enough for the constructor to show that the asset is delivering its specified service benefits, within its expected impacts. It can thus test performance against an accepted whole-life-cost tender offer (see Section 9.4), and this can be incentivised as part of a target cost contract (see Section 6.4). You are thereby 'proving' the constructed solution against the criteria that were used to choose it (see Section 7.5).

There is evidence that some infrastructure assets for which sustainable credentials are claimed do not perform to the designed standard:

Design intent does not always translate into real-world performance ... GSA must take into account the way its buildings perform on the ground. Upfront investments in sustainable measures need to be matched by sustainable O&M practices.

(GSA Public Building Service, 2011)

So your other key action, for sustainable operation in line with design, is to provide appropriate training to both operators and also to users, where needed.

Through this process you complete the transfer of the original project scoping intent into the in-use stage (see Section 3.6). You will have used operations people to help choose the solution and design (see Section 7.6), and have now required the construction team and equipment providers/installers to pass on to the operators and users the knowledge they need to operate, maintain and utilise the assets effectively, to deliver the intended performance.

10.4. Incentivise O&M teams, and user behaviour, for sustainable performance

How is resource and energy efficiency optimised over the whole life of the project?

All O&M teams will have 'efficiency' as a performance target, usually defined mainly in terms of 'defined performance at minimum cost'. So you need to include the sustainability objectives alongside the other performance criteria, just as they were included at earlier stages of project delivery. It is helpful to the cause of reducing CO_2e emissions that energy costs are now an ever-increasing operational cost, so reduction of energy use is a strong 'efficiency' focus in most infrastructure operations.

For example, in existing buildings that have not recently had an energy audit, potential achievable energy savings may be of the order of 10–40% of total energy consumption. A typical split is shown in Figure 10.1, and some famous examples are given in Box 10.1. So, if you are dealing with existing infrastructure, using LEED or BREEAM type tools and doing an energy audit is a good starting point.

Because sustainability objectives may be new targets, some organisations create new roles to specifically focus on them. For instance, this can involve the appointment of a 'carbon manager' or 'sustainability manager', as well as the more usual 'energy manager' or even just 'operations manager'. Whether or not you choose to suggest a special sustainability role may depend on your organisation's culture (see Section 11.4). In some organisations, a newly named or created appointment will be needed to gain the right authority and attention (and funding); in others, the most effective approach may be to add in the new sustainability performance measures, as further criteria that you are now going to measure.

The concern raised in Section 7.5.2 about measuring sustainability and cost performance to include user response also applies here. You will still use impact *per unit of product* to measure operational efficiency, but you must also measure customer-based units – that is *impact per user* – to drive optimum customer use of the assets. A near empty train will have much the same CO_2e emissions per kilometre travelled as a full one. Measuring *emissions per customer per kilometre* will put pressure on the operator to fill the train up.

Figure 10.1 Typical potential energy savings in commercial buildings – split between sources. (Source: MWH)

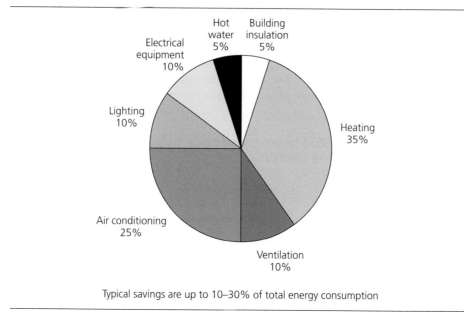

Typical savings are up to 10–30% of total energy consumption

10.4.1 Incentivise O&M operations and facilities management teams/ contracts

To further complete the transfer of the original project scoping intent into the in-use stage, and the training of O&M staff (see Section 10.3), you can ensure that the terms and targets driving O&M performance also reflect the target and choice criteria figures that were used in the outline design to choose this solution (see Section 7.5). The meeting of these targets may be ensured by offering performance/reward terms for an asset owner's in-house team, or through an out-sourced contract, and should incentivise the operator to operate the asset as sustainably as possible. For example, if the outline design choice for a building, with a set level of occupier service, was for a whole-life solution with a higher construction cost (due to building in passive solar gain, thermal mass and much insulation), then you can use the expected resulting low running cost and low CO_2e operational emissions to set tight performance targets for facilities management. This training is likely to need updating at intervals during the design life; and it needs funding as part of O&M.

Whatever were the key sustainability criteria and targets set in scoping and design, you can use all of them as performance measures for rewarding operational success, with cost impacts attached to each. This may perhaps be aspects of the asset performance, as measured by BREEAM, or LEED or CEQUAL, and such tools usually provide for actual in-use testing of results (see Section 13.5). In this way, the O&M contract terms become *service*, rather than *product* based, with a built-in drive for more sustainable, lower resource use, lower impact operation (see Section 2.3). Even better, you can

Box 10.1 LEED for existing buildings certification drives retrofit energy use reduction

This new (2011) data mark the first time that LEED-certified existing buildings have surpassed LEED-certified new construction cumulatively.

Projects worldwide are proving that green building does not have to mean building new. By undertaking a large renovation, the recently LEED-certified Empire State Building has predicted it will slash energy consumption by more than 38 percent, saving $4.4 million in annual energy costs, and recouping the costs of implementation in three years.

Empire State building   Transamerica Pyramid

San Francisco's Transamerica Pyramid also earned LEED Platinum as an existing building, 39 years after it was originally built. The landmark's onsite co-generation plant saves an average of $700 000 annually in energy costs.

(Source: Planet Profit Report, 2011)

provide an incentive to beat the targets, with incentivising gain/pain share arrangements (see Section 6.4).

10.4.2 Incentivise user behaviour too

In all the sector examples given in Table 10.1, the behaviour of the *user*, interacting with the operation of the assets, also greatly affects the sustainability of the operation. Once the infrastructure asset has been built with a certain capacity, the sustainability objective can be thought of as this 'service' model, so the goal is to:

serve the maximum number of users/customers, providing just the level of service they really require (but no more), while using the smallest possible amount of energy, water and other resources, and emitting as little as possible CO_2e and other pollutants.

So, for example:

- **For a building**: user education prevents opening of the windows when the heating is on; internal temperature is set as low, and lighting as suitable as is tolerable; all systems are operated with smart meters and controls, and the maximum number of occupiers is spread evenly over as many hours of use as possible. All of this is incentivised through the terms of the tenancy, and you could include rental and service charge target prices, with gain/pain share arrangements for beating them, to encourage user–service provider collaboration to achieve this (see Section 3.5).
- **For transport**: use automatic signals, and speed controls for minimum fuel use and maximum vehicle throughput, with penalties, based on traffic prediction and modelling; and then time-varied ticket prices and toll structures can encourage the maximum number of users to occupy the available vehicle capacity evenly over as many hours of use as possible.
- **For a utility**: smart meters and structured tariffs can incentivise the least use really needed, and help smooth out time of demand within production capacity. Direct customer assistance with appliances and insulation, etc. can also help minimise use. Again, you could include target product use prices, with gain/pain share arrangements for using less, to incentivise customer–service provider collaboration to achieve this (see Section 3.5).

The other way in which you can efficiently match operational supply to user demand is to use a demand modelling tool. These are particularly used for utilities, but are available for user demand in transport too (Mathew and Krishna Rao, 2005).

10.5. Keep social needs in mind, for users and operators
And not just infrastructure, but infrastructure that delivers real, pro-poor outcomes.
(Ainger and Macabuag, 2012)

In the technical and managerial challenge of effective asset operation, it can be easy to forget your ultimate purpose – providing all users with their fair service outcome. For socio-economic sustainability (see Section 2.4) this could include, for example, the following.

- **Care with pricing/tariff structures**, to assist take up by the poor, as in Southern Water's metering scheme in the UK (Groundwork South, n.d.).
- **Easy public transport accessibility** and affordability, to maximise usage, as for Curatiba's buses in Brazil (Goodman *et al.*, 2006).

■ **Gender equality**, in designing urban areas for safety (ACPO Crime Prevention Initiatives, 2004; Women's Design Service, 2007; Jagori, 2010), or even just providing enough toilets so that women do not have to queue.

■ **Enabling disabled access in urban areas**, for example through a GPS-based software tool, as in the AMELIA (A Methodology for Enhancing Life by Increasing Accessibility) project (Pratt, 2010).

Your other key opportunity is to use local people or organisations to carry out O&M functions. Apart from being cost-effective, and minimising travel impacts (see Section 5.3), this can also serve a socio-economic objective (see Section 6.3). Box 10.2 summarises the poverty-reduction effects of the 60 million Rand/year (£5.3 million/year), 2500 km long, Zibambele road maintenance project around Durban, South Africa.

Box 10.2 Poverty alleviation and road maintenance – the Zibambele project, Durban, South Africa

---

The challenge in this project was to maintain steep rural roads that were strongly affected by regular rainfall. The solution met these technical needs, and at the same time provided a local poverty-reduction programme.

Brushing, digging mud and cutting vegetation to maintain the roads is the most technically effective way to clear these roads and drains between rainfalls. The poorest 'head of household' local women were selected and registered as hand labour maintenance contractors. They were given the tools and training to maintain a 0.5–1.0 km stretch of road close to where they live, and were paid for 2 days work each week. The project employs 30 000 women, who are managed, at low overhead, using 'high-tech' mobile phones, GPS and computer systems.

(Source: Tom Wilcock)

Their small extra wages allow the women to send their children to school. They were helped to register as voters, open bank accounts, form savings clubs and apply for loans to start their own businesses. The women provide a road maintenance service that could not be delivered in any other way, and their families' lives are improved.

(Source: ICE, n.d.-b)

209

Most infrastructure maintenance involves a mixture of low- to medium-tech work on landscaping, civil structures and simple machinery, and some high-tech expert work on MEICA, automation systems or specialist structures. Even though the second category may need centralised, travelling experts, there is much to be said for the first category of work to be done by local, perhaps part-time, people and companies, to help maintain local skills and employment. This can apply just as much in 'developed' as in 'underdeveloped' countries.

Your opportunities and concerns will differ on each project, but you need to pay attention to them when measuring, and improving, socio-economic sustainability performance, throughout the in-use stage.

10.6. Add environmental performance and resilience into effective maintenance

Maintenance is perhaps the single most important element of government's stewardship obligation.

(South African National Council on Public Works Improvement, 1988)

The least sustainable form of maintenance is to do nothing but an occasional greasing or cleaning until failure occurs, and then to replace the failed assets, accepting the cost and loss of service involved. This is 'reactive' maintenance; it may be inadequate to maintain 'serviceability', and can lead to serious deficiencies ('30% of sub-Saharan Africa infrastructure assets are in need of rehabilitation' (ICE, 2010)), and not just in developing countries (in the USA, 45% of roads are in less than good condition and 12% of bridges are structurally deficient (Council of State Governments, 2010)). 'Sustainable maintenance' involves you in much more than this. This is likely to involve

- regular or continuous condition monitoring of both fixed and moving ones, allowing preventive, not just reactive, maintenance
- equipment replacement for energy efficiency, not just failure
- and, to deal with the challenges of global warming, to adapt capital replacement projects to 're-scope' existing assets, to reduce energy use, cost and CO_2e emissions, and increase system resilience.

All of these require proper funding for O&M.

10.6.1 Ensure funding for O&M

No O&M can be done effectively if insufficient money is allocated to carry it out. This is a basic requirement for asset sustainability (ICE, n.d.-a: card 83). In developing countries (and anywhere where infrastructure funding is in short supply) there can be a tendency to divert funds intended for O&M to other purposes.

There are several ways in which you can help make sufficient funds available (ICE, n.d.-a: card 64). For example, when the outline design is chosen (see Chapter 7), the calculation of the whole-life cost includes an estimate of the future O&M and capital replacement costs. As the project is given formal authorisation to proceed, ensure that these future O&M costs are earmarked within future operational and capital budgets. O&M duties can be outsourced to a contractor, on 3- to 5-year re-awardable contracts, and this ring-fences the money as a contractual obligation, and allows you to set and incentivise improvement goals (see Section 10.4) and fixes the amount needed, leaving the risk with the contractor. One advantage of a public–private partnership privately financed contract is that it includes doing the long-term O&M, and funding it, as part of the contractual obligation and package.

All the comments about effective procurement made in Chapter 6 apply to O&M contracts too (ICE, n.d.-a: card 63).

10.6.2 Do preventive maintenance, based on monitoring

In recent years (and this has always been true for infrastructure where failure would cause loss of life), the reactive-only approach has been largely superseded (when it can be afforded) by preventive maintenance (see Reliability Web (n.d.) for a summary that is non-equipment specific). This is:

> care and servicing ... for the purpose of maintaining equipment and facilities in satisfactory operating condition by providing for systematic inspection, detection and correction of incipient failures either before they occur or before they develop into major defects.
>
> (Wu and Zuo, 2010)

This intensifies the regular attention and servicing given, predicts likely failure before it happens, and aims to replace equipment before failure.

To plan preventive maintenance you need to monitor asset condition and operating efficiency. Very small, distributed and networked continuous monitors are being developed for both fixed underground and mechanical above-ground assets (for pumps see Yatesmeter (2001)), and these are beginning to be applied in practice. To maintain service, such sensors can monitor infrastructure performance in near real time (such as pressure and flow in water distributions systems, to identify the early on-set of leaks and bursts). Sophisticated algorithms based on signal-processing techniques can interrogate the data and provide decision support tools for operational control rooms responsible for monitoring the network (Ye and Fenner, 2013). Such techniques are intended to ensure minimal disruption to customers, as the diagnostics are smart enough to enable the network to be repaired before an interruption in service becomes necessary. Similarly, sensors can continuously monitor structural condition, to predict potential failure. Such systems are referred to as 'smart infrastructure' (Bennett et al., 2010).

However, monitoring and maintenance programme need not be high-tech. The Zibambele road maintenance project (see Box 10.2) is a good example of regular preventive maintenance. Continuous monitoring (in this case by personal inspection) between rainfalls of deposits and erosion, and clearing mud immediately, restores the road's condition and slows down the wear and tear that will eventually cause failure. Thus a low-tech approach to maintenance is sufficient to extend asset life.

10.6.3   Intervene for efficiency improvements

Rising energy prices, and a concern to reduce CO_2e emissions, introduces a further sustainability driver – to minimise energy use by maintaining machinery at close to optimum efficiency, rather than accepting a slow decline in efficiency. This will likely mean more frequent intervention than would be the case with preventative maintenance. Figure 10.2 shows such a regimen applied to pumps, but you could also apply it to any other machinery. Rather than continue preventive maintenance, with a slow decline in efficiency until incipient failure occurs (upper boundary to the lower shaded area), you should do several earlier refurbishments (likely in this case to be replacing pump impellers) to maintain efficiency above an optimum minimum (upper boundary to energy saving shaded area).

The ideal contract arrangements to drive the decision-making process for this approach would be a *service* contract, say for all services to a building (see Section 10.4). This would allow the provider to optimise the balance between equipment efficiency, design life and replacement, and the costs and impacts of energy, resources and consumables use. You could also incentivise CO_2e reduction in such a contract by including a CO_2e price multiplied by the CO_2e quantity in the definition of 'costs'.

10.6.4   Make 'capital replacements' into strategic improvements

Because different types of materials and assets have different design lives (see Section 2.5), the in-use stage always needs replacement or refurbishment of some items such as ITC control systems, rotating machinery such as pump impellers, process or heating

Figure 10.2 Efficiency-driven pump refurbishment

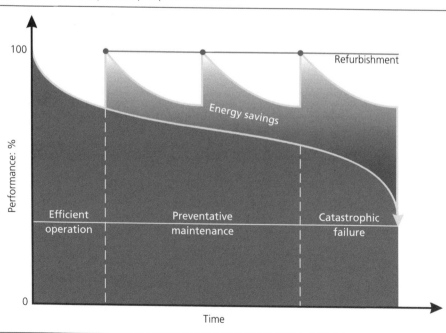

and ventilation equipment, or bridge bearings, to keep the overall asset going. This usually has been managed separately from 'new infrastructure' projects, and as a simpler process than the delivery sequence described at the start of Part II. Sometimes it has been as simple as 'read the label to see what the machinery is, and who manufactured it, and just buy another one'.

However, the rapid changes that society now faces make this simple treatment no longer sustainable. Externally, you must cope with changing demand for services, and the need for adaptation to global warming for resilience. Internally, your urgent challenges are to catch up with a backlog of over-ageing assets, and to reduce energy cost and CO_2e emissions at every opportunity. There are continuously improving technologies available – for materials, equipment, monitoring and control, and asset data and management software – which you can use.

Your first response can be to invest in equipment replacement more frequently, such as for energy efficiency, as just discussed. More widely, you can treat every capital replacement need as an opportunity to re-scope the existing assets, to meet changed demands, reduce energy use, cost and CO_2e, emissions, and increase system resilience. A diagram of this approach applied to reduce CO_2e emissions is given in Figure 10.3.

The trend is likely to be that 'new projects' and 'capital replacement' will converge, to deal with these rapid changes. You will increasingly look to meet new needs by adapting existing assets (applying level 2 of the 'sustainability hierarchy', see Figure 2.4), while capital replacements will act as a trigger for this wider re-scoping. One extra challenge of adapting existing assets is the need to build new ones around ongoing operations, while at the same time maintaining service. This adds cost and time, but infrastructure organisations are learning how to do this effectively as an extra project-management skill – one that is important for sustainable solutions. The expertise differs in each sector, and will be discussed in the sector-specific volumes.

Figure 10.3 Capital replacement as an opportunity to reduce CO_2e. (Source: Adapted from Ainger (2012))

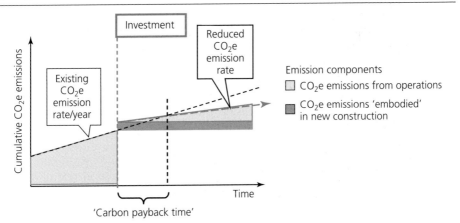

To create space for the innovation needed, you may need to go back and use the processes for scoping (see Chapter 4) and outline design (see Chapter 7), before undertaking detailed design and construction. Because adaptation projects may well be smaller and more repetitive than 'new projects', you will face understandable pressure to develop very cost- and time-effective ways of doing this, probably working on clusters of projects rather than a single project (see Section 7.3), Do not let this pressure prevent you from using the windows of opportunity that will still exist to introduce the sustainability practice ideas detailed in Part II, to deliver more sustainable outcomes.

These ideas will help you to operate and maintain your asset, to sustainably provide its socio-economic service, to minimise its environmental impact, and to seek continual improvement. Applying the ideas in practice will help your infrastructure to perform well against the more challenging sustainability targets set for it. However, finally, you will need to consider the end-of-life stage of the asset.

10.7. Decommissioning at end of life

We are not very good at taking things apart.

(Pauli, 2004)

Decommissioning and end of life is the final stage in the life of an infrastructure asset. Most infrastructure assets are fixed, and occupy a dedicated land area. In theory, all the 'constructed' parts and materials could be dismantled, demolished and taken away, and the land remediated, with holes filled in or reused as water or cliff features. In this way, the site could be returned back to something with the potential, once nature has recolonised it, to return to a 'natural' site.

For most infrastructure sites, the design life is long, and you cannot predict a preplanned dismantling process, as you might do for a car. The first choice for an end-of-use infrastructure site and its assets is usually to be changed and developed further, to serve extra or new infrastructure needs. This is a result of growth – in population to be served, and/or in the level of service expected. So, 'decommissioning' is usually actually not considered as a separate final stage, but has its characteristics put as an *input* into the 'outline design' stage (see Chapter 7) of that new upgrading or replacement project for the assets concerned.

10.7.1 Sustainability in decommisioning – and what can engineers do?

What opportunities are sought for reuse (e.g. of land, materials and building stock)?

The scope for sustainable action at this stage is large. At the very least you can see the existing assets and land as a resource and opportunity for reuse in one of various forms, rather than as a problem to be disposed of. The trend discussed in Section 10.6, to 're-scope' existing assets, can be applied even at the end of their design life. If you do this, then decommissioning is only one of the options to consider. Even when there is no prospect of the built structures being reused, the land is a precious asset that can be reused. For example, a railway line can become a walking or cycle route, a quarry, mine or industrial site might provide a wildlife haven or an urban 'green lung', and an industrial or infrastructure site has the possibility of providing remediated (brown field) land for sustainable housing or other redevelopment.

Some detailed advice is becoming available about how to assess a structure for facility decommissioning and adaptive reuse. Scadden and Weston (2001) describe this process for power plants, but their approach is applicable more widely to infrastructure containing mechanical equipment. Some facilities can have interesting potential for creative reuse, as in the well-known case of the Tate Modern art gallery in London, which makes dramatic use of the huge old turbine hall (Box 10.3).

Box 10.3 Adaptive reuse – the Tate Modern, London

The Bankside power station in London was 'adaptively reused' as the Tate Modern, a large art gallery, with the huge old turbine hall being its major attraction.

(Image © ICE Publishing)

[O]ld power plants [can be] attractive candidates for adaptive reuse ... they were constructed with large turbine-generator 'halls'. These very large open spaces represent significant opportunities for developing new functions on a grand scale.

(Scadden and Weston, 2001)

To achieve the optimum and most sustainable solution, you should apply the following *end-of life hierarchy* of approaches, which is similar to the hierarchy used for 'waste' because the end-of-life stage is, in effect, about choices made about structures and materials as 'waste'.

In order of preference:

1 **Directly re-use** as much as possible of:
    – the site (with remediation treatment if necessary)
    – whole structures
    – their main elements.

2 **Dismantle, reclaim and re-use** as many as possible of the individual *components*. This has a much lower processing energy and CO_2 emissions impact than the next level.

3 **Demolish, then reclaim and recycle** the materials, with minimum possible environmental impact (processing and transport).

4 **Demolish and dispose**, with minimum possible environmental impact (transport and 'in-tip'), and energy recovery and/or pretreatment before disposal, if these lessen overall impacts.

The extent to which you can already apply solutions at the upper end of this hierarchy is limited by past design decisions which did not take the issue into account, and the durability of the structures and their materials. As design for reuse and recycling (see Section 8.5) becomes more prevalent, your opportunities will increase. You can make demolition and disposal become very much the last resort – an example of 'bad engineering'.

In-use to end of life: summary

Deliver, and improve on, the designed sustainability performance, through measurement and follow-up action.

Hand assets over effectively from construction, including operating efficiency trials and providing training; and incentivise O&M teams, and user behaviour, for sustainable performance.

Keep social needs in mind – for user service and in O&M.

Add environmental performance and resilience into effective maintenance; ensure funding; do preventive maintenance, based on monitoring; intervene for efficiency improvements; and make 'capital replacements' into sustainability improvements.

At the end of life apply the *end-of life hierarchy* of approaches, in order of preference: directly reuse; dismantle, reclaim and reuse, demolish, reclaim and recycle; demolish and dispose.

REFERENCES

ACPO Crime Prevention Initiatives (2004) *Secured by Design Principles.* Available at: http://www.securedbydesign.com/pdfs/SBD-principles.pdf (accessed 3 August 2013).

Ainger C (2012) Speeding up innovation – for sustainability, presentation to MWH Quarterly Conference Call on Sustainability, 24th February 2012.

Ainger C and Macabuag J (2012) Engineering a better world – A toolkit for delivery. Sinclair Knight Merz, London.

Bennett PJ, Kobayashi Y, Soga K and Wright P (2010) Wireless sensor network for monitoring transport tunnels. *Proceedings of the ICE – Geotechnical Engineering* **163**: 147–156.

BIS (Department for Business, Innovation and Skills) (2010) *Estimating the Amount of CO_2 Emissions that the Construction Industry can Influence.* Supporting Material for the Low Carbon Construction IGT Report. Available at: https://www.gov.uk/government/uploads/system/uploads/attachment_data/file/31737/10-1316-estimating-co2-emissions-supporting-low-carbon-igt-report.pdf (accessed 3 August 2013).

Council of State Governments (2010) *Condition of US Roads and Bridges.* Available at: http://knowledgecenter.csg.org/kc/content/condition-us-roads-bridges (accessed 3 August 2013).

Goodman J, Laube M and Schwenk J (2006) Curitiba's bus system is model for rapid transit. *Race Poverty and Environment* **Winter 2005/2006**: 75–76. Available at: http://urbanhabitat.org/files/25.Curitiba.pdf (accessed 3 August 2013).

Groundwork South (n.d.) Southern Water Universal Metering Programme. Available at: http://www.south.groundwork.org.uk/our-projects/southern-water-ump.aspx (accessed 3 August 2013).

GSA Public Building Service (2011) *Green Building Performance – A Post-occupancy Evaluation of 22 GSA Buildings.* US General Services Administration, Washington, DC. Available at: http://www.gsa.gov/graphics/pbs/Green_Building_Performance.pdf (accessed 3 August 2013).

HRH Prince Charles (2012) Working in harmony with nature: the key to sustainability. *Civil Engineering* **165(CE3)**.

ICE (Institution of Civil Engineers) (2010) *Accelerating Infrastructure Delivery – Improving the Quality of Life.* Report of 2nd Middle East and Africa Regional Convention, Cape Town.

ICE (n.d.-a) *Toolkit for International Development. In-Use.* Available at: http://www.engineers.org.uk/taxonomy/term/16 (accessed 3 August 2013).

ICE (n.d.-b) *Maintenance for Social and Economic Gain.* Available at: http://www.ice.org.uk/topics/International-development/civil-engineers-toolkit-for-development/In-use/Procurement-and-delivery#card_65 (accessed 3 August 2013).

Jagori (2010) *Understanding Women's Safety: Towards a Gender Inclusive City. Research Findings, Delhi 2009–10.* Jagori, New Delhi. Available at: http://www.endvawnow.org/uploads/browser/files/understanding_womens_safety.pdf (accessed 3 August 2013).

Mathew TV and Krishna Rao KV (2006) Travel demand modelling. In *Introduction to Transportation Engineering, NPTEL*, Ch. 5. Available at: http://www.cdeep.iitb.ac.in/nptel/Civil%20Engineering/Transportation%20Engg%201/05-Ltexhtml/nptel_ceTEI_L05.pdf (accessed 3 August 2013).

Pauli G (2004) ZERI principles as a form of industrial ecology. In *Sustainable Development in the Chemical Industry: A Practical Approach* (Beloff B, Pojasek R *et al.* (eds)). John Wiley, Chichester.

Planet Profit Report (2011) *LEED-Certified Existing Building Surpasses New Construction. Taipei 101 And San Fran's Transamerica Pyramid Earn Platinum Status.* Available at: http://www.planetprofitreport.com/index.php/articles/leed-certified-existing-building-surpasses-new-construction (accessed 3 August 2013).

Pratt K (2010) *SUE Success Story AMELIA. Common Goal, Different Perspectives.* The ISSUES Project. Herriot Watt University/Cambridge University/EPSRC. Available at: http://www.urbansustainabilityexchange.org.uk/media/ISSUES%20Outputs/AMELIAFinal%20Report.pdf (accessed 3 August 2013).

Reliability Web (n.d.) *Preventive Maintenance.* Available at: http://www.reliabilityweb.com/art06/preventive_maintenance.htm (accessed 3 August 2013).

Scadden RA and Weston RF (2001) *Facility Decommissioning and Adaptive Reuse.* Technical Paper No. 0103. Presented at NDIA 27th Environmental Symposium and Exhibition.

South African National Council on Public Works Improvement (1988) Final Report to the President and Congress, February. Pretoria.

Women's Design Service (2007) *What to Do About Women's Safety in Parks.* Available at: http://www.wds.org.uk/www/projects_parks.htm (accessed 3 August 2013).

Wu S and Zuo MJ (2010) Linear and nonlinear preventive maintenance. *IEEE Transactions on Reliability* **59(1)**: 242–249.

Yatesmeter (2001) Available at: http://www.yatesmeter.net (accessed 3 August 2013).

Ye G and Fenner R (2013) Weighted least squares with EM Algorithm monitoring water distributions systems. *ASCE Journal of Water Resources Planning* doi:10.1061/(ASCE)WR.1943–5452.0000344.

Part III

Change

In Part I, we suggested that professional ethics now require all of us to take some responsibility for delivering more sustainable infrastructure. This needs you to play an individual active role in changing the way things are done. We suggested key individual principles to help you

- learn new skills – competences for sustainable infrastructure
- challenge orthodoxy and encourage change.

Throughout Part II, we suggested new practice to apply – new questions to ask, and ideas and solutions to try. But all engineering is done in teams, within management structures, so this requires you to persuade others to listen, and to agree to try out your ideas – to change things.

Many of us may have an inbuilt assumption that only authority – 'power over' – counts, and only the boss can tell us to change the way we do things. Experience of working in project teams, however, shows differently. Many individuals in a team can have influence – 'power to' – through technical knowledge, good timing, controlled passion and enthusiasm, careful communication and argument, and demonstrated loyalty to the project's interests. You have more power to change things than you think, if you use it carefully and at the right time.

This part of the book aims to help you do just that; to take an active role, project by project, in changing the way that infrastructure services are planned and delivered, for a more sustainable outcome. Chapters 11 and 12 work at a different level to the rest of the book; they discuss not the 'what' – the content of a change – but the 'how' – the way it happened, and how you can help.

There are two interacting aspects of this. In Chapter 11, we discuss the context – the organisational arena within which you have to act – and in Chapter 12 we discuss your individual action.

Engineers are used to taking responsibility, making up our own minds about things and acting accordingly. So influencing is mostly not 'telling people to do things' – it is helping them to develop their own innovative ideas, so that they 'own' and are keen to use the results:

Of the best leaders
When the task is accomplished
The people all remark
We have done it ourselves.

(Lao Tse)

Sustainable Infrastructure: Principles into Practice
ISBN 978-0-7277-5754-8

ICE Publishing: All rights reserved
http://dx.doi.org/10.1680/sipp.57548.221

Chapter 11
Understanding your arena for change

The best way to predict the future is to invent it.

(Alan Kay)

Alan Kay is in the IT business, and it's easy to see the reality of his statement in that sector. However, there is also a real sense in which infrastructure engineers invent the future. Every project creates assets that provide for, but also constrain, society's behaviour and choices for a long time. All projects matter.

Your ability to help invent a sustainable future depends on two interacting things: your own individual capabilities, and the characteristics of the organisation, project and sector in which you work, which comprises your 'arena', or context, for change. (Making changes for sustainable infrastructure requires innovation in everything we do. Here we treat the words 'change' and 'innovation' interchangeably, but use the latter more for the overall process.) The actions you will choose to make depend strongly on that context, which bounds an organisation's or project's capacity and opportunities for innovation – so we discuss this first, in this chapter. Understanding this will give you the best chance of applying your capabilities successfully.

There are many drivers and opportunities for creative change at all levels, and you can be a catalyst for such change – as we hope the interventions we suggested in Part II have shown – even if you are not the boss. Although we certainly do need 'an "industrial revolution" for sustainability, starting now' (John Schellnhuber, Chief Scientific Advisor to Angela Merkl, German Chancellor; quoted in *The Guardian*, 19 November 2008), we think we can achieve this through evolutionary change in our existing practices. This is not to say that it is easy; it can require a significant (and at times uncomfortable) mindset change. Box 11.1 gives a project example of this, which includes a significant change in approach. The Box describes the same project as Box 8.3, but concentrates on 'how' the change was made.

Understanding this arena requires, first, some confidence that you can act; then some thinking about the system in which change happens, the types of change and the stages it will have to go through; and, finally, how to be effective in your interventions in those stages.

11.1. Inventing the future – who, me?
Every change that has ever happened started with an individual's decision to act. Someone got up one morning, or had a 'eureka moment', decided that 'something has

Box 11.1 Changing from 'building new' to 'refurbishment' – water treatment plant

The project team changed the project output from building a new water filter, to refurbishing the existing one. This saved about £7 million and much embodied CO_2e. It provided 15 years of design life for the filter, matching the available life of the other assets on the site.

The project was 'partnered', with a joint client, designer and constructor team; the partners had established experience and trust from working with each other previously. The team suggested the refurbishment alternative right at the start of the outline design stage. The client's project leader started out being very much against refurbishment, due to an earlier bad experience of it. The technical structural leader had a passion for sustainability and experience in refurbishment; and had the trust of the project manager. Expert, detailed work was used to show evidence of the residual structural strength, and a detailed concrete specialist contractor's plan to use innovative concrete repair techniques was developed. The client accepted this option.

So, success required the right timing, enough time in the programme, technical passion, support, expertise, evidence and detailed execution plans.

(Source: MWH, personal communication)

to change', and started to take action – like the ABN – Amro man in the story in Box 11.5. But do you have the capability to do that?

Classic management writers such as John Kotter (2012) used to assume that the person starting any change had to be the CEO, or otherwise have some top-down authority – 'power over' – to require people to change. Another perspective sees an organisation as more like an organism, and naturally creative and innovative, bottom-up:

> We need to learn how to engage the creativity that exists everywhere in our organisations ... we must engage with each other, experiment to find what works for us, and support one another as the true inventors that we are.
>
> (Wheatley, 2000)

This inclusive approach feels more like the atmosphere in a good project team; the kind of creative culture that innovation for sustainability requires. In the cultures of some organisations such initiatives may be stifled by management, because they may be seen as a challenge to its authority; but they do not have to be (Box 11.2). Most organisations have a mix of both characteristics:

> Wired into the mechanism of the top-down mindset is a belief that the organisation and its people are like a machine ... (and machines are ultimately controllable and predictable).

> Buried at the root of bottom-up thinking is the belief that organisations are living systems ... change is the natural order of things ... the trick is to find ways of

---

Box 11.2 Staff as volunteers – generating innovation for sustainability

---

The company is in the infrastructure and water and environmental engineering sector. In the UK it had some staff passionately interested in sustainability, *ahead* of top management's commitment. These staff set up an informal sustainability interest group (SIG) in February 2000, initially working in their own time, 'unpaid' (i.e. after completing timesheet time). They started to develop sustainability tools and client offerings; these were shown to be useful, and in the next year a small budget was allocated. Over three years, this developed into a formal, funded, Knowledge Community – within the new 'Knowledge Management' structure that the company had just set up.

(Source: Ainger and Pickford (2007))

> unleashing people's natural ability to adapt ... The art of change management is
> to pursue both in combination.
>
> (Binney and Williams, 1997)

Change for sustainability involves both top-down and bottom-up approaches; and, of course, you need to match the right one to the right part of the process. Many of the opportunities for early action suggested in Part II involve you in bottom-up *influencing and persuading* people. This mainly uses not authority, but credibility and expertise, best described as 'power to'. The word comes from *poeir/pouvoir* (French) meaning 'to be able':

> 'Power to' is social power, experienced in relationship with others. The great lie
> in our society is that 'power over' is the route to fulfilment ... By contrast,
> 'power to' offers an attractive, abiding route out of powerlessness that everybody
> can use.
>
> Ainger (2003)

Debra Meyerson (2001) sums up the thoughts and feelings you might have when trying to act for change:

> many people who want to drive changes ... face an uncomfortable dilemma. If
> they speak out too loudly, resentment builds toward them; if they play by the
> rules and remain silent, resentment builds inside them. Is there any way, then, to
> rock the boat without falling out of it?

She confirms that there is, calling it 'quiet leadership', and goes on to say:

> I call change agents 'tempered radicals' because they work to effect significant
> change in moderate ways. In doing so, they exercise a form of leadership.

So, you do indeed have capability to act as an innovator for sustainability, particularly if you can find support (see Section 12.4). As American scientist Margaret Mead famously said:

Never doubt that a small group of thoughtful, committed citizens can change the world. Indeed, it is the only thing which ever has.

11.2. The context for innovation

Even when you have capability, the infrastructure sector is hard to change, because infrastructure assets have long lives – it takes a long time to change the *average* sustainability impact of the assets – and the sector is (properly) risk averse and in the public eye (see Section 11.3). So, you can either just say 'it's too difficult', or you can accept that, because it is difficult, you must plan innovation more carefully and strategically. Faster change needs an understanding of the systems it happens within.

11.2.1 Innovation happens within 'systems'

Just as you must scope and design the *content* of infrastructure to take account of the complex system it lies within (Principle A.4 – Complex systems, see Section 2.6), this complexity also applies to the planning and delivery *process*. The regulators, clients and supply chains involved are themselves a complex adaptive system. Changing these involves iterative learning and evolution (Clements, 2006). It is mainly about persuading people, and their reactions are hard to predict. So, in acting for change, and trying to work out what may happen as a result, you will need to 'be comfortable with uncertainty' (Wheatley, 2000) – you may make progress, but often not exactly in the way you expected.

The challenge that this makes to engineers' 'Newtonian' thinking is well expressed here in an internal report for the Sustainability Learning Network of the Cambridge Programme for Industry:

> Recent thinking around change, which is supported by research, is moving towards the idea that change cannot be managed or controlled; rather that, like in a living system, change 'emerges'. Your role [as a change agent] is therefore to *facilitate the emergence of change*. This approach to change is made even more relevant in the context of sustainable development since your destination is by definition not easy to describe or measure.

> It is worth adding that in considering the 'emergent' model of change, our group who comprise mainly engineers and scientists, struggled initially with the 'looseness' of the language and the lack of a clear action focus inherent in this approach.

> (Group Project, Leadership for Change,
> Cambridge Programme for Industry (2003))

So, the paradox is that, in the complex adaptive system of infrastructure delivery and operation, change can only 'emerge' – but, nevertheless, you can act proactively, to enable it.

11.2.2 Tensions 'pull' change, with many roles

Figure 11.1 gives a crude summary of the interplay and tensions towards sustainability that exist between *outside* and *inside* organisations in an infrastructure sector. Progress

Figure 11.1 Tensions and roles in change. (Source: Adapted from Ainger and Howards' diagrams of tensions for change in Section 2 of IEMA (2006))

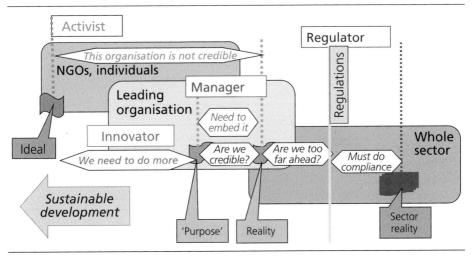

towards sustainability is to the left. Environmental activists sometimes think that change depends only on the *outsiders*: *activists and NGOs,* demanding it, and *regulators,* forcing it. But, as Figure 11.1 shows, it is the *insiders* – public or private infrastructure organisations – that actually have to change their practice in response, and they can only do so by someone *inside* inventing a new way of doing things.

External activists' demands are vital to create a 'pull' for improvement. (If you are in the supply chain, an 'activist' might be an innovative client, who demands that you improve your sustainability performance. One consultant first set up an EMS and ISO 14000 registration because it became a qualification requirement to get onto a major client's tender list (authors' experience).) But there are equally vital roles for people inside, who understand the sector and its opportunities. They include *innovators* – who first try out, and demonstrate, the new solutions – and *managers* – who then must embed these in standard practice. These key players have been labelled as 'social intrapreneurs' (for several named examples, and their stories, see SustainAbility (2008)), and have been described as:

> Someone who works inside major corporations or organizations to develop and promote practical solutions to social or environmental challenges where progress is currently stalled by market failures.

Of course, even if you work for a body that is notionally fully committed to sustainability, in an activist or regulatory role, it will still have things that need changing. Its own culture and practices can inhibit innovation, and reduce the effectiveness of the role it could be playing to encourage sustainability. So, although Figure 11.1 shows the key 'insider' roles (*innovator* and *manager*) as being within a public or private infrastructure organisation, such roles for change are needed within every 'player'

Figure 11.2 A sustainability journey for an engineering consultant

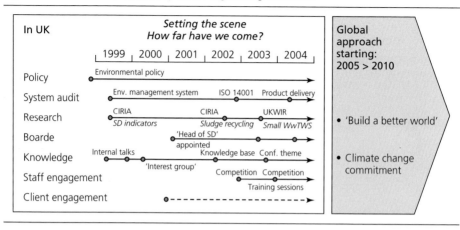

involved in a sector, and much of what we say about this context for change applies to any or all of them.

Inside most organisations there will still tend to be a gap between a declared sustainability purpose and the current reality. It is this that can create your internal business case for embedding new practices in 'good engineering'. As part of your capability for change, you can play the innovator or manager role, by creating, identifying and using the positive and negative tensions for change in your sector.

11.2.3 'It's a journey'

Changing to sustainability is often characterised as a *journey*. This is certainly true of an infrastructure organisation, and of any whole sector. The reason why it takes time, with many interacting actions, is because of the many players in the complex system, the time needed for those tensions to play their part, and the deep change of attitudes and learning involved.

A timeline of part of one international engineering consultant's such journey (not of course 'completed' yet), in the UK from 1999 to 2010 is shown in Figure 11.2. In very simplified form, each dot shows one step change. This is a point where some action with leverage produced change (see Section 11.4). The changes in each aspect of business are interconnected, with one enabling progress to another, and they combine the use of *external pressures* and *internal arguments* for change.

This is not to say that change on an individual project must be slow. Because projects are done by teams, you have already linked up most of the people to be influenced, and your single timely and influential action can create a new direction (see Box 11.1). This is an illustration of acting proactively to enable change in the whole system.

11.3. The types and 'shapes' of change

Engineers are used to accepting and delivering change, from new job administrative procedures to new clients, projects, bosses and team members. How does change

Table 11.1 Levels of change, and their implications

Level	Type of change	Who and what has to change?	Example of change
1	Within existing working practices	Front line (daily) – budgets, resources, priorities	New annual budget, job code or project start
2	Within the existing mindset and culture	Middle managers (monthly) – agendas, structures, methods, targets	New accounting system, or project management system
3	Needing a change in mindset and purpose	Top management (annually, longer?) – business purpose, strategy and language – What's our company about?	Adding 'sustainability' into standard engineering practice

towards sustainability compare with these familiar examples? Lawson and Price (2003) of McKinsey describe change as taking place at different 'levels', or depths, with very different implications (Table 11.1).

As suggested throughout this book, a goal of sustainability changes the definition of good engineering, and of 'efficiency', and thus resets the engineering ethos. So, while most of those familiar changes happen at levels 1 or 2 in Table 11.1, sustainability actually demands a level 3, deep, culture and mindset change. It requires innovation of all kinds, not just 'compliance', and probably a redefinition of purpose. The consequences of failure have personal, family and *emotional* impacts too. This is why it is hard, and takes time.

Disappointment at the rate of progress of change for sustainability in organisations can happen because the change is assumed to be only level 2. (This also can lead to too-early declarations of only superficial success.) However, if you dramatically challenge everyone to immediately embrace a deep level 3 culture change, this may well produce resistance – it can sound too daunting. You can avoid this by starting with a simpler proposal, and let the learning from it lead into a deeper change process. One example of this is given in Box 11.3.

11.3.1 Barriers to change

Because of their close interfaces with health, safety and the environment, infrastructure sectors are regulated. In such conditions, it is easy to fall back on the idea that the purpose is merely to comply with regulations. This is part of a range of barriers to innovation in infrastructure.

- Tight regulation reduces the strategic mindset, and the normal commercial incentives for competitive innovation.
- Regulators and infrastructure operators can have an adversarial, 'compliance' culture.

> **Box 11.3** Level 2 or 3 change: CIRIA and tools for sustainable development
>
> The UK CIRIA Pioneers Club project was a 2-year programme to develop and trial pilot sustainable development key performance indicators (KPIs). This was notionally a non-scary 'level 2' change; it involved leading UK construction companies that had some experience of KPIs.
>
> As the prgramme progressed, it raised deeper, wider questions, beyond just indicator tools, heading to 'level 3'. It started all the organisations involved thinking about their role and performance on SD/CR:
>
> > We recognised that by collecting the data, we had started the establishment of a unifying network or centralised sustainability management system.
>
> > Participation in the Pioneers' Club is part of our process of change, and is not seen as an isolated activity.
>
> This is a good example of starting a culture change process not by 'scaring the horses' with too deep a challenge from the start, but by letting the use of measurement tools generate its own learning about the need for deeper change.
>
> (Source: CIRIA, RP644 Project, 2001-3)

- Innovation is not a hidden internal activity; it interacts with customers, clients and stakeholders, and any failure is in public.
- Customers and the public mindset want risk averse, 'safe' solutions – which may no longer be possible, or affordable (e.g. flooding?).
- There is a mesh of professional and engineering standards and codes to work to.
- Engineers have been trained in a mindset that favours centralisation and unit efficiency, rather than system effectiveness and resilience.
- There is a large supply chain, with many players to convince.
- No one has any time; everybody is too busy with 'delivery'.

In all your actions for change, you need to identify and plan around each of these barriers.

11.3.2 Challenging the 'sustainability costs more' assumption

Most fundamentally, you often face a pernicious assumption, not explicitly expressed, about cost, namely: 'more sustainable solutions will always cost more money'. This has its origins in past environmental impact assessment practice (see Section 13.1).

Such a view can lead to the suggestion that setting sustainability goals can only be a 'values' driven, moral choice, which may be in conflict with the reality of commercial 'value':

I believe ... business is part of society ... Its social purpose is to provide products and services profitably and responsibly, ... the nature of that responsibility being

determined by society's values ... Financial failure can destroy individual
companies. Moral failure will destroy capitalism.

(Sir Geoffrey Chandler, quoted in BT, 2003)

Experience shows that, for infrastructure, a choice between 'values' and 'value' is an
increasingly false one. The examples in this book, and many others, provide evidence
that innovation can produce infrastructure services that reduce CO_2e emissions, are
more sustainable, and cost less; particularly if you address this early in the project
sequence. Part II puts much emphasis on this.

So, you can respect both aspects. *Values*, connected to purpose, often initiate innovation
for sustainability (see Section 12.1) but, to be real, this must be useful, and provide *value*
for customers and owners (both in the public or private sectors). This is the 'value-
focused' thinking we identified as one of the key sustainability competencies in
Chapter 1. You need to ask for both.

- *Customers*: How can we serve users needs, following this sustainability purpose?
- *Owners*: How can we be commercially successful (and, if privately owned, wealth
  creating) through serving these customers in this way?

To show that sustainability can be aligned with commercial value requires you to use the
sustainability challenge as a driver for step-change innovation. You can do this at two
levels within the existing commercial system. The first requires no change; the second
requires applying a broader financial approach.

1   *Directly, in separate capital and operating costs terms*: if you are innovative
    enough, it is possible, for instance, to save both money and carbon, in simple
    financial terms, as several case studies in this book demonstrate.
2   *In whole-life-cost terms*: some sustainable solutions will require higher up-front
    capital investment, with less operating cost, for a lower whole-life cost. This may
    particularly apply to reducing energy and resources use. Succeeding in this
    approach requires the *infrastructure owner* to choose investment options on the
    basis of a whole-life cost, not a narrower 'up- front cost' criterion (see Section
    2.5), and to have a budget model that allows the project to be able to benefit from
    savings in operating cost, if the owner spends extra on capital cost.

If the above points are not enough, then you need to work at sector, government or international policy level, to raise the bar. This can involve the following.

3   *In whole-life-cost terms, including sustainability 'externality' costs*: in some sectors,
    such as fossil fuel energy generation, current financial costs do not reflect real
    sustainability impacts. Here, in addition to (2) above, you need to price in carbon
    or environmental shadow pricing (see Section 2.3) to reflect this.
4   *In whole-life-cost terms, including taxes or subsidy from stronger regulation*. The
    Corporate Leaders Group on Climate Change has asked the EU and UK
    Government to regulate more strongly for climate change responses (Prince of

Wales Corporate Leaders Group, 2013). This was because commercial value and certainty for business in this area requires governments to introduce carbon taxes or levies, or other forms of quota or subsidy.

As well as the Corporate Leaders Group, business groups relevant to infrastructure who have come together to make these kinds of arguments include the World Business Council for Sustainable Development, SustainAbility and the Aldersgate Group. There is also a C40 Cities Climate Leadership Group. Web addresses for all these organisations are given in the reference list at the end of this chapter. Referring to the policies of these groups may provide you with useful business case sustainability arguments to use with infrastructure owners.

11.3.3 Processes of change – the innovation S-curve

Level 1 and 2 types of change are typically delivered top-down, through one of two 'shapes', as shown in Figure 11.3. The details have been worked out beforehand (you hope!).

- *Big bang*: 'The new project starts on Monday. Use your existing processes' (Level 1). This assumes that everyone already knows what to do.
- *Roll-out*: 'The expert team will teach you the new accounting system, one office a week, over the next three months' (Level 2). This acknowledges some learning is to be done, and allows for a little tweaking as a result of experience.

If you try to use these top-down processes for the early stages of introducing Level 3 change for sustainability, you will likely run into trouble, meeting all the barriers and 'costs more' assumptions discussed above. For this, your process needs to be different, using the S-shape innovation diffusion curve shown in Figure 11.4. You need to shepherd the new sustainability ideas out of the bottom left-hand corner, and up the curve, past an interacting 'cloud' of barriers (which are referred to by some as constituting 'socio-technical lock-in'). In this way you can help a new idea to fight its way up to

Figure 11.3 Change shape for levels 1 and 2

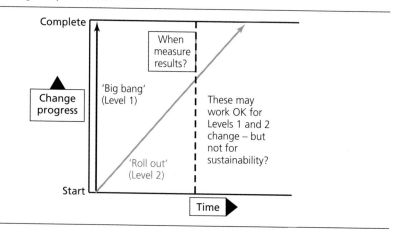

Figure 11.4 The S-curve shape of innovation (Ainger, 2012). (Source: Based on Senge *et al.* (1999) and Reason *et al.* (n.d.))

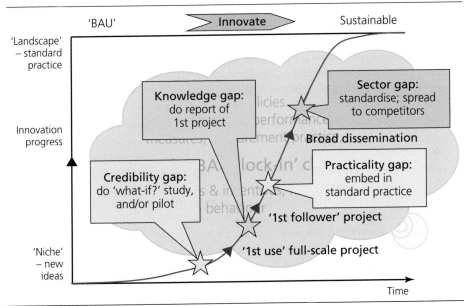

the top right-hand corner, finally becoming the new, sustainable, 'landscape' (Reason *et al.* (n.d.: 92) of standard practice. It is an influencing, (even 'marketing') process, which needs *learning*, and the development of convincing *evidence* at each stage. Once you reach the embedding stage, then a roll-out, office by office, may be appropriate.

You can think of the curve as applying to one new idea starting within one organisation, but spreading to a whole sector at its broad dissemination stage. This process, multiplied many times across all sectors, provides the journey to a new culture that embraces sustainability, as a natural part of what 'good engineering' does. Sector innovation happens like this in some form anyway, but slowly, and not nearly fast enough to be an adequate response to our urgent sustainability challenges. Innovation is drastically slowed in having to get past the four critical gaps shown in sequence in Figure 11.4.

- Credibility gap – Can you guarantee it will work?
- Knowledge gap – Who else knows about this innovation?
- Practicality gap – Is it cost-effective and proven?
- Sector gap – What? You mean we should help our competitors do it too?

You need to understand this structure, to make your change actions most effective.

11.4. Effective change actions
You can use the curve in Figure 11.4 and its stages consciously, as a road map, to help new ideas for sustainability to move up the curve faster, by using a series of facilitating, catalyst actions to get past the gaps. To plan this, consider where your innovative idea

currently is on the curve, and apply strategies for crossing each gap. Remembering the activist, innovator and manager roles in Figure 11.1, and the top-down and bottom-up routes for change (see Section 11.1), you will need to apply different combinations of these at successive stages up the curve.

Let us illustrate how you might make the journey up the curve and across the gaps by tracking a new idea – for example, measuring and pricing CO_2e emissions as a project objective, at all stages of project delivery – using the terms from these figures.

1   An *activist* or research organisation introduces the method into the sector, as a *new idea*.
2   An *innovator* in one organisation, looks *outside* their organisation, sees the new idea, and persuades a project to try it, first as a quick *'what-if' desk study*, which produces better evidence that it will work. (See the example in Box 11.4.)
3   This evidence crosses the *credibility gap* sufficiently for a brave project manager to say 'go ahead' at full scale – a *first demonstration project*. This is even though there is no formal requirement in place, i.e. it is bottom-up *in spite of* the current system. It allows the organisation to develop its own CO_2e measurement tool, at a draft level. If gaining evidence of credibility is difficult, a *pilot trial* might be a further intermediate step.

Box 11.4 Making an innovative idea more 'real', with a business case – using a 'what-if ' study

---

Heated household water runs away down the drain, and raises the temperature of the sewage. This heat can be recovered through water-source heat pump technology, and used for urban district heating systems, or other heating applications. In recent years, towns in Germany and, particularly, Helsinki in Finland have started to install such systems: Helsinki gains about 100 MW of such district heating from its central sewage treatment works effluent (see Box 4.2).

This idea was unknown in the UK. A paper on the Helsinki plant was heard of by one of the authors, who brought it to the attention of UK water utilities, but no one took it up. Later, an opportunity arose to try out the idea in a UK water utility, as the topic of a Master's degree dissertation. The author acted as a *catalyst* to bring the parties together (see Section 12.2), the utility company provided data, and the graduate student provided *time and enthusiasm* as research input (see Section 12.3).

The paper essentially asked: '*What if* we applied the Helsinki technology to four real sewage treatment plants in the UK?' It addressed the regulatory issues in the UK, estimated the heat recoverable and its possible use, discussed the technology needed, predicted the costs and benefits, and produced an outline business case (Hawley and Fenner, 2012).

With this to hand, the utility was in a much better position to decide whether to try out the technology, with much less uncertainty and risk.

(Source: Author's personal experience)

---

4   This project works, and provides harder evidence. It is written up as a case study by the *innovator*, still *in spite of* the system; bottom-up. This crosses the *knowledge gap* and the idea is taken up by a *first follower project*. This confirms that the idea has wider potential, creates an incentive to spread it, and it starts to get past a tipping point (Gladwell, 2002).

5   Then a *manager* is confident enough to decide (still with difficulty) to *embed* the new idea. It crosses the *practicality gap*, to become a new part of the organisation's standard practice. Money is found to refine the draft CO_2e measurement tool for general use; and management authority, top-down, now requires it to be applied on all projects.

6   This innovative organisation then leads a sector initiative, to cross the *sector gap*, and set this approach as a voluntary new sector standard. This can be difficult, as it requires sharing knowledge with competitors in the sector; it needs top-down authority. An example of project finance banks doing this, the adoption of the Equator Principles for social and environmental assessment in major infrastructure projects, is given in Box 11.5.

Box 11.5 Adopting the Equator Principles – sector innovation by voluntary collaboration

---

ABN AMRO and others found there was very public and very poor sustainability performance on some major projects they had helped finance. This was damaging their bank's reputation, with boycotts of their domestic banking business. This was a strong driver for action.

- The Global Head of Risk at ABN AMRO had earlier been sent down *personally* to confront a Friends of the Earth protest outside the bank's front door – and his values and judgement told him that Friends of the Earth were right. He became the 'torch carrier, committed to sustainability'. He called an old business friend at the IFC, the largest global development institution focused exclusively on the private sector in developing countries, stressing that something had to be done.

- These two persuaded ten banks and the IFC to meet in London under the Chatham House rule (which states that participants of a meeting are free to use information received, but the identity and affiliation of the participants may not be revealed). They discussed failures and successes in managing environmental and social risks in projects, and agreed that ABN AMRO, Barclays, Citibank and West LB (the 'gang of four') would take the lead in defining a set of guidelines. Instead of devising a new set of guidelines, World Bank/IFC guidelines were authoritative and available, so they adopted these, saving much argument.

- These ten large banks, enough to be significant, adopted these principles as the Equator Principles at a Washington launch on 4 June 2003. Now, more than 28 banks have adopted the principles, and this is perhaps 80% of the project finance banking market, so the initiative has passed 'critical mass'. Satisfying the Equator Principles has almost become a standard measure of an acceptable project.

- This has changed behaviour, but non-governmental organisations remain properly sceptical, and watch the process closely.

(Source: Personal communication, 2004; CPI SLN group project;
The Equator Principles Association)

Most experience shows that you will find it too difficult to go straight from (1) to (3), without the better evidence from (2) (many of you may have taken a good new sustainability idea to your manager, but have been unable to provide a guarantee that it will work, and have been turned down). Also, spreading the learning from that first demonstration project is too slow, unless you pay special attention to step (4). Step (6) makes particular demands on creating trust between competitors to discuss critical internal issues. You need to deploy the right culture and approach to support these critical steps.

11.4.1 Culture and systems favouring innovation

An instructive story of starting transformation from bottom-up driven change, and then supporting it from top-down, is Gary Hamel's account of IBM's start on the journey from a mainframe computer manufacturer to a software consulting company – requiring a huge change in purpose and culture (Hamel, 2000). The story is notable because it is one of few which acknowledge that change started from the bottom, and it provides sufficient detail to get a real idea of the subtleties of the interactions that took place. Although management did not start off with a system that encouraged such innovation, it quickly responded with just sufficient discipline to mould and support the initiative.

The 'what-if' early steps in Figure 11.4 can be cheap and quick. At project level, they are epitomised by applying the 'benchmark sustainable option' approach suggested for outline design (see Section 7.4). However, to be allowed to undertake these steps requires adopting an *experimental* mindset – an acceptance of *trying something out*, without knowing the result. Pilot-scale work requires this attitude too, and is effective for trying out new relationships or behaviour, not just for new technology.

However, this not a common feature in efficiency-driven infrastructure organisations, outside the research and development department (if there is one). It requires time, a relatively small amount of funding and, critically, *permission to fail* if the idea does not work out. One finance director referred to this idea of a space for nurturing new ideas as a 'virtual laboratory' (author's personal experience; at the Institution of Water Officers Northern Ireland Conference, April 2009). Furthermore, many of these new ideas must be tried out *in collaboration* with other parties, customers or the supply chain, so part of the 'experiment' consists of these stakeholders learning to work together, and to trust each other, in such a space.

Much progress in step (4) needs much more effort and attention, as there are typically few incentives for engineers to write up such innovation stories. Speeding up the crossing of the knowledge gap requires two things: first, incentives and management discipline to write up innovation case studies quickly; and, second, organisations need to have systems for looking outside their own boundaries for new ideas. The latter is often done informally by an organisation's innovators. It is important that these stories of innovation also tell you about *how* the innovation was done, as well as *what* was the content, to help application by others (Ainger, 2013).

Finally, for sector change – raising the bar, (step (6)) – lessons from the banks' Equator Principles experience (see Box 11.5) show that some key 'enablers' are needed.

- To start the process there needs to be a strong common driver, and key people need to know and trust each other as individuals. In the Equator Principles case, most projects need multi-bank finance, so banks had to work together, and people already know each other, and who they trust.
- The small group starting the process needs to be sufficiently respected or dominant in the sector to create good momentum towards critical mass.
- The group needs a 'safe', well-facilitated space for frank discussions – a discreet location, working to the Chatham House rule (see Box 11.5), with an expert facilitator, who knows the sector.
- It helps if there is already available an authoritative draft of the new method or guidelines you are trying to agree on, in order to minimise partisan battles about wording and to avoid months of argument. When setting up the Equator Principles the group used the World Bank/IFC guidelines.

For infrastructure, such conditions now probably apply in the best 'partnered' client/supply chain programmes. Professional or sector bodies should also be able to provide 'safe space'.

11.4.2 Using performance measurement tools to start

As a starting point for change, if you are trying to get an organisation or project to accept some more radical goals, you can suggest first trying out some extra measurements as a 'what-if?' exercise: 'How would we do, if we decided to set targets?' Because the criteria for sustainability are wider than organisations are used to, just trying to define what they should be can generate a valuable debate as to what sustainability means, and start the learning process. The story in Box 11.3 illustrates starting a deeper change process by beginning with performance measurement.

A range of sustainability tools for setting targets and measuring performance are discussed in Part IV. They include those used to quantify and measure the sustainability of projects, such as BREEAM and CEQUAL (see Section 13.5), but there are many corporate social responsibility type tools too (see Section 13.8), for measuring organisational sustainability. However, be cautious about choosing which tool to suggest. Many (almost too many) 'pure' tools are available, and these can be cumbersome to fit in with business frameworks and strategies. You may do better to add sustainability goals into standard business performance measurement systems (see Section 3.3). Make sure that in doing so you still include the fundamental principles (A.1 – Environmental sustainability, within limits – and A.2 – Socio-economic sustainability – 'development', see Chapter 2), as embodied in frameworks such as the Five Capitals Model (Forum for the Future, 2011) and the Natural Step (see Appendix A). You can use these tools as a logical framework on which to construct key performance indicators, but convert the language to fit the organisation.

Because change is difficult, organisations can convince themselves that they are doing better than they are, or can declare 'success' too soon. Then asking: 'Why don't we just do a check on how far we have got?' may be useful. Specific things you could say might include the following.

- Where are you, really, on the S-curve (Figure 11.4)? Well done for having done a 'flagship' sustainability project, but wasn't that an exception to your current practice? Now how can you embed that innovation into all your future projects?
- Yes, you are doing well against the measurement standard that you set three years ago, but that was a starter, not matching the latest sustainability challenges. Should you now use a more searching set of targets? Don't stand still.

One more rigorous tool for assessing an organisation's progress on its global warming response is the Performance Acceleration for Climate Tool (PACT) framework (see the case study in Hampshire County Council (2008)). PACT measures an organisation's progress through six stages of competence as it responds to climate change (it could be adapted for more general sustainability use). The top two stages – the 'what' of progress – are more challenging than in previous tools. It also measures the capabilities for change – the 'how' of progress – along nine *developmental pathways*. These include three often ignored, but key, factors: having agents of change, working together and learning. These strongly reflect our own experience of the critical capabilities required in effective change processes.

Recently, PACT was used to reassess several leading companies that scored top marks in the Carbon Disclosure Project rankings:

> The findings reveal how these companies are striving to manage GHG emissions efficiently, but they currently lack the necessary capabilities for higher response levels to effectively mitigate climate change ... The PACT framework has helped participating companies realise how far they still need to go.
>
> (Mehta, 2012)

So, the use of a familiar type of tool – performance measurement – but with more challenging questions, may be an effective way for you to gain a reality check, or to restart a change programme.

11.4.3 Getting your timing right

When is the right time to act? The overall answer is, of course, now, urgently and at every opportunity. But choosing the *right moment to act*, in detail, is an important part of being effective in change. The combination of events that makes it the right time will vary, depending on whether you are working in a defined project or programme team, or more generally within your organisation or professional (or academic) body.

The ideal opportunity to act for change for sustainability is when you have *leverage*. This means, when you can see the right argument or evidence, with time to act, and when the project team or organisation also has some readiness and capability to act in new ways. The key link will be your personal credibility (see Section 12.3) – your power to influence your audience. Here are some typical opportunities, in infrastructure planning and delivery.

- Internal to your organisation – perhaps a first demonstration project has given good results, and helps make a business case, so you can now raise the idea of embedding the new ideas into standard practice.

- New external drivers – such as new government regulations, or new utility client demands, enable you to make a better business case.
- Starting work in new joint working teams, such as partnering, which require setting up new methods, which can often be more innovative that your standard ones.
- Utilities can influence supply chains – from a position of strength, set out new procurement qualification requirements, such as 'You must be able to account for your carbon'.
- Innovation in marketing – before a bid document comes out (which would have been framed in more conventional terms than it could have been), you suggest new ideas to the client, to include in the bid terms of reference.
- Professional encounters – you use conference or journal papers to set out new ideas. (It might incentivise engineers to do this if their professional institutions made a point of awarding higher 'training' points for papers on innovation.)
- Academic–practitioner interactions – you use academic 'free' researcher time, at Masters' or PhD level, to help answer 'what-if?' questions, applying new ideas to utility (see Box 11.4).
- Supply chains influencing their clients and partners – if you are working on projects, use the small windows at the start of each project stage throughout project delivery to raise new ideas (see Part II, particularly outline design (Section 7.3)).

The last of these is illustrated by the story in the Box 11.1. The right moment to suggest changing to refurbishment was, *only*, right at the start of the commission, which was the beginning of the outline design stage. 'Redefining the brief' was raised before the work content and plan had been decided. Other examples have been given earlier in the book.

To sum up, you should take action for change urgently, and at every opportunity. The occurrence of opportunities can be hard to predict, so making too-rigid a 'change project plan' may lead you to miss those that arise suddenly. A better approach is probably to adopt a 'planned opportunism' attitude. Have a general strategy and goals, recognising that this is a journey of change, made up of many actions, and then continuously look for opportunities to act at each 'right moment', when that 'leverage' combination appears. Some questions to help you think about opportunities are given in Box 11.6.

11.4.4 Choosing your argument and language

As you persuade a group to try out an innovation for the first time, or to spread it into wider practice, your language and arguments will matter. They need to match the organisation's culture and situation.

- If it sees sustainability as a risk and focuses on 'protecting the brand' and 'compliance', use the language of risk and reputation; but if they see sustainability more as a strategic opportunity, use that language. (Would it not incentivise some companies more, if you called not for zero waste but for 100% profit?)
- If it is very aware of competitors, watching their every move, use competitors' sustainability actions to spur change; but if its self-image is of being a leader, 'the

---

Box 11.6 Questions to help find opportunities to act

Current view of the world?
■ What is the organisation's current view of sustainability – is it on their 'radar screen'.
■ Does the organisation see itself as a leader or a follower – could it be interested in sustainability as a possible 'next big thing', because it's a leader?
■ What current accepted organisational change process, or 'language', might provide a 'hook' for sustainable development?

External drivers?
■ Are there any new regulatory or statutory pressures, risks, customer pressures or top-level commitments to corporate responsibility (however 'public-relations' driven) that can be used as a 'hook' on which to hang arguments for sustainability issues?

High level champions?
■ Are there senior 'strategists' who might become champions, for top-down support?

Culture, people and power?
■ Is there a culture in which you could talk at the office about your own human values?
■ Is the culture one of 'knowledge sharing'?
■ Are there internal communication and feedback channels, and data or knowledge networks, in which content is free to be decided by staff who might be interested in sustainability?
■ Which possibly peripheral group or individuals, not recognised as having much managerial power, actually do have some power, and might be keen on sustainability:
  – middle managers
  – knowledge managers
  – 'training'
  – 'communications'
  – IT, finance/accounting?
■ Are the values and interests of the youngest, brightest staff being listened to and looked after; could their time and enthusiasm be harnessed?

---

best in class', then appeal to that thinking. In this case, gaining awards or joining a league table on sustainability may be an effective suggestion.

As you push innovations up the S-curve, the critical point for changing language and argument is when you shift from 'first project' to embedding it in practice – i.e. step (5) in 11.4 above. As you are persuading people to 'buy' something new, it is helpful to consider a 'marketing' analysis. This is what Figure 11.5 does, showing the 'market segments' in your organisation, and a language and values analysis as interpreted in a useful book *Creating Contagious Commitment*, by Shapiro (2010). The left to right progress here mirrors the bottom to top 'change' axis in Figure 11.4, with progress up the S-curve.

To embed the innovation in the organisation's systems you must attain a critical mass of regular use, going well beyond the 16% of 'early adopters' (roughly your

Figure 11.5 Getting from 'first use' to embedded 'critical mass'. (Quoted in Shapiro (2003)

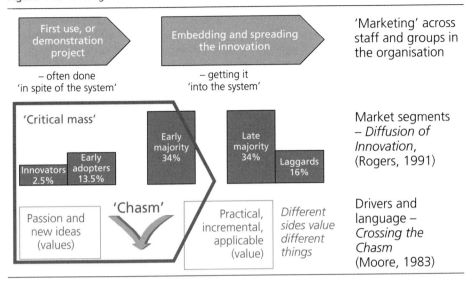

'innovators' in Figure 11.1, and steps (2) and (3) 11.4 above). You must reach, and convince, at least the 'early majority', perhaps 50%. This change corresponds roughly to the 'tipping point' in Figure 11.4, and step (4) above; it must involve some managers in that group.

To do this, you have to cross the 'chasm' in understanding; because the motivation and values, and thus the language needed to convince the 'majority' groups, are completely different from those that worked for the innovators. You must move from 'values'-driven action, done as an exception, to 'value'-producing action, serving the system (see Section 11.3). So your arguments must show 'do-ability', commercial value and relevance to customers. An argument such as 'reducing carbon can save money' meets those chasm-crossing requirements (see the story in Box 11.7). It also demolishes the myth that 'sustainability costs more' (see Section 11.3), and so is an effective piece of evidence to present. This getting 'across the chasm' to critical mass also represents the point at which your management style shifts from a bottom-up creative approach to needing the addition of some top-down discipline.

Choosing new or familiar language may also depend on your immediate purpose. If you are brainstorming opportunities or new challenges, or trying to invent new solutions, then deliberately choosing to use new, unfamiliar language, and 'big picture' ideas, can help to get people to be creative. Because they cannot relate these to their usual range of thinking, it forces them out of their usual assumptions. However, you will not want to take engineers so far out of their comfort zone that they just disengage. One example of an extreme reaction to using non-engineering language is given in Box 11.8. Actually, in this case, starting off from this very unfamiliar language produced

Box 11.7 Management discipline drives CO_2e reduction, and cost savings too

To embed CO_2e reduction goals in project investment decisions, a utility set a new threshold goal for choosing the 'preferred option' in feasibility studies: *any chosen solution must demonstrate that it had at least 50% lower CO_2e emissions than business-as-usual.* This was in addition to the usual requirement for cost savings below programme targets.

In one study, a preferred option that showed the best cost savings, but did not meet the CO_2e goal (even though the study team had tried hard to achieve it), was sent to the capital investment review team – a senior management body that had to approve all choices. Perhaps the study team thought the new CO_2e goal was only a 'nice to have' – but the senior management review team insisted that the CO_2e goal was not negotiable, and must be met, even though it might cost more money.

So the study team went away and brainstormed again: and came back with a solution that met the CO_2e goal – *and cost still less than the earlier 'lowest cost' option.* It was, of course, accepted.

Lesson: The project team had sincerely looked for cost reductions, within their current 'mindset'. The harder CO_2e goal pushed them out of this mindset, to think in a more innovative space, which delivered both lower CO_2e and lower costs. And it was necessary to apply top-down management discipline – 'we really mean that CO_2e goal' – to push the study team into it.

(Source: Author's personal communication)

Box 11.8 Using deliberately unfamiliar language

In our 2006 paper 'Widening Horizons for Engineers', we deliberately used eight very unfamiliar labels for sustainability issues, to derive some practical questions about how sustainable engineers are being (see Figure 1.5 and Section 1.5) and not to allow the pigeon-holing of answers into easy boxes such as sustainability = environment. One engineer's reaction was this:

It was very illuminating and confirmed my worst fears ... that the sustainability it deals with is in a kind of separate universe, eschewing links to established society, writing its own laws ... A political orientation, even a quasi-religion, not much to do with civil engineering providing society with infrastructure to sustain lives and livelihoods sustainably, i.e. within the resources available in nature.

Not that the example doesn't show that any effort at sustainability has to grapple with the same issues, but they seem more likely to be resolved by [engineer-led methodologies] ... and science ... than appeals to holistic financial accountability, etc. I do think that those in this kind of sustainability should make an effort to rejoin society.

(Source: Personal communication to author)

useful and practical questions that were wider than engineers usually consider; this individual's shocked reaction was purely to the unfamiliar language style used, rather than to the questions derived from them.

On the other hand, if you already are seeking to embed some new practice within operational procedures, maybe one that has already been demonstrated in a case study, then acceptance may be gained more easily if you use *familiar* language that is already well known, to help make the practice do-able and acceptable. For example, this could be: 'What are our targets and actions going to be on energy efficiency (including CO_2e), reducing waste, customer service and regulator consultation?'

These examples have concentrated on choosing the right *arguments* and *language*. When you want to convince your organisation's highly technical, knowledge leaders to change, who are used to having responsibility and making their own judgements, then you must also make a careful choice of the *methods* you use to spread knowledge.

11.4.5 Spreading innovation

Once a new innovation is available, and you are trying to spread it, it is tempting to think of its lack of general acceptance as a *problem* to be solved. The top-down approach is to have the solution developed by the innovators (usually not the group of technical people you are trying to spread it to), write a manual and then roll the innovation out through the ranks. However, if the recipient group is composed of technical leaders who are trusted to use their own technical judgement on projects, they are unlikely to accept being told what to do. This is particularly true when they have a strong attachment to their existing practices and there is no obvious crisis reason for change. So, you need to be cleverer than this, in how you get buy-in for the innovation.

One way to get buy-in is to use the *recipient group themselves* to investigate the innovation case study (advised by the innovators) and to write their own manual – a bottom-up approach.

> The key is to engage the members of the community you want to change in the process of discovery, making them the evangelists of their own conversion experience.
>
> (Ludema *et al.*, 2001)

This 'appreciative inquiry' approach is so called because it values and learns from what is *being done better already*, rather than by defining a 'problem' and solving it.

Accepting change is a learning process, and the investigation allows the technical group to *learn their way* into the change, catalysing them to be more innovative more quickly, and to feel that they thought of the ideas themselves. The group members also trust their own evaluation, and thus their capability to implement it. Ideally, the group will contain members from several internal sections and offices, each of whom will be a local champion for the method as it is rolled out. This all reflects the Lao Tse quotation given at the start of Part III.

Amusingly, Pascale and Sternin (2005) in an article in the *Harvard Business Review* describing this process, call it using 'positive deviance' – implying that, to a rigidly top-down manager, anybody doing something better 'without permission' is a deviant. This shows that the biggest challenge in adopting this way of spreading knowledge may be the very different management style you need. The manager leading the embedding of the innovation needs to play an open-ended and facilitative role, not a dominant one.

This practical bottom-up way of *doing and learning* for change together demonstrates a key tool that can support your effectiveness in individual action too. Now that we have set out the 'arena' for your individual action for change, we will address in Chapter 12 how to do the best you can in that role.

Understanding your arena for change: summary

By understanding your context for action you can increase your capability. Your individual actions each help to change 'the system', so it is a complex journey.

Understand and use the tensions that can 'pull' change, and choose your role. As change for sustainability is deep and contains many barriers, know the barriers and get around them.

Follow the innovation S-curve model, and persuade, not tell; plan to cross the gaps, garnering better 'evidence' at each stage. Challenge the assumption that 'sustainability costs more' at every opportunity.

Nurture cultures and systems that favour innovation – experimental mindsets, time to write things up, and a safe space for delicate discussions.

Start or restart change by measuring sustainability performance against 'stretch' targets.

Choose the right time to act; deploy the right argument, and choose your language to match; and use the recipients, not the innovators, to learn and spread the innovation.

Suggested Further Reading is listed after References.

REFERENCES

Ainger C (2012) Organisational change – for sustainable development. MSt in Sustainability Leadership, Cambridge Programme for Sustainability Leadership, Cambridge.

Ainger C (2013) Briefing: Speeding up 'innovation' by better first use reporting. *Proceedings of the ICE – Engineering Sustainability* **1(1)**: 8–10.

Ainger C and Pickford W (2007) Staff values as enablers for change – 'volunteering' for sustainability. *Organisations and People Journal* **14(4)**.

Ainger K (2003) Against the misery of power, the politics of happiness. *New Internationalist* **360**. Available at: http://newint.org/columns/essays/2003/09/01/reinventing-power (accessed 3 August 2013).

Aldersgate Group (n.d.) Available at: http://www.aldersgategroup.org.uk/home (accessed 3 August 2013).

Binney G and Williams C (1997) *Leaning into the Future*. Nicholas Brealey, Boston, MA.

BT (2003) *Just Values. Beyond the Business Case for Sustainable Development*. Available at: http://advancingsustainability.com/documents/Just_values.pdf (accessed 3 August 2013).

C40 cities Climate Leadership Group (n.d.) Available at: http://www.c40cities.org (accessed 3 August 2013).

Clements (2006) *Complexity, Responsibility, Intention and Evolution*. Talk to Bath RBP MSc.

Fenner RA, Ainger C, Guthrie P and Cruickshank HJ (2006) Widening horizons for civil engineers – addressing the complexity of sustainable development. *Proceedings of the ICE – Engineering Sustainability* **159(ES 4)**: 145–154.

Forum for the Future (2011) *The Five Capitals Model – A Framework for Sustainability*. Available at: http://www.forumforthefuture.org/sites/default/files/project/downloads/five-capitals-model.pdf (accessed 3 August 2013).

Gladwell M (2002) *The Tipping Point – How Little Things Can Make a Big Difference*. Abacus, London.

Hamel G (2000) Waking up IBM – how a gang of unlikely rebels transformed Big Blue. *Harvard Business Review* **July–August**.

Hampshire County Council (2008) *Case Study: The Hampshire Performance Acceleration Climate Tool – Hampshire PACT*. Available at: http://www3.hants.gov.uk/_hf000000607078__pact_case_study_template_v_2_2009-06-12__2_.pdf (accessed 3 August 2013).

Hawley C and Fenner RA (2012) The potential for thermal energy recovery from waste-water treatment works in Southern England. *IWA Journal of Water and Climate Change* **3(4)**: 287–299.

IEMA (Institute of Environmental Management & Assessment) (2006) *Change Management for Sustainable Development – A Work Book*. Best Practice Guide, Vol. 8. IEMA, Lincoln.

Kotter J (2012) *Leading Change*. Harvard Business Review Press, Cambridge, MA.

Lawson E and Price C (2003) The psychology of change management. *The McKinsey Quarterly* **2**. Available at: https://www.mckinseyquarterly.com/Organization/Change_Management/The_psychology_of_change_management_1316 (accessed 3 August 2013).

Ludema JD, Cooperrider DL and Barrett FJ (2001) Appreciative inquiry: the power of the unconditional positive question. In *Handbook of Action Research* (Reason R and Bradbury H (eds)). Sage, London.

Mehta N (2012) Corporate leaders on the road to rapid decarbonisation? – Towards more effective responses to climate change: Exploring the capabilities of leading FTSE companies. MSt in Sustainability Leadership, Cambridge Programme for Sustainability Leadership, Cambridge.

Meyerson DE (2001) Radical change – the quiet way. *Harvard Business Review*, October.

Moore GA (1991) *Crossing the Chasm: Marketing and Selling High-Tech Products to Mainstream Customers*. Harper Business Essentials (later editions of the book exist).

Pascale RT and Sternin G (2005) Your company's secret change agents. *Harvard Business Review* **May**: 124–132.

Prince of Wales Corporate Leaders Group (2013) Available at: http://www.cpsl.cam.ac.uk/Business-Platforms/The-Prince-of-Wales-Corporate-Leaders-Group-on-Climate-Change/UK-CLG.aspx (accessed 3 August 2013).

Reason P, Coleman G, Ballard D *et al.* (n.d.) *Insider Voices: Human Dimensions of Low Carbon Technology.* Lowcarbonworks, University of Bath. Available at: http://www.bath.ac.uk/management/news_events/pdf/lowcarbon_insider_voices.pdf (accessed 3 August 2013).

Rogers EM (1983) *Diffusion of Innovations.* New York: Free Press (later editions of the book exist).

Senge P, Kleiner A, Roberts C, Ross R, Roth G and Smith B (1999) *The Dance of Change.* Nicholas Brealey, Boston, MA.

Shapiro A (2010) *Creating Contagious Commitment: Applying the Tipping Point to Organizational Change.* Strategy Perspective, Hillsborough, NC.

SustainAbility (2008) *The Social Intrapreneur. A Field Guide for Corporate Changemakers,* 1st edn. Available at: http://www.echoinggreen.org/sites/default/files/The_Social_Intrapreneurs.pdf (accessed 3 August 2013).

SustainAbility (n.d.) *Developing Value.* Available at: http://www.sustainability.com/developing-value (accessed 3 August 2013).

The Equator Principles Association (2011) Available at: www.equator-principles.com (accessed 3 August 2013).

Wheatley MJ (2000) *Leadership and the New Science.* Berrett-Koehler, San Francisco, CA.

World Business Council for Sustainable Development (n.d.) Available at: http://www.wbcsd.org/home.aspx (accessed 3 August 2013).

FURTHER READING

Kotter J (2012) *Leading Change.* Harvard Business Review Press, Harvard, MA.

Shapiro A (2003) *Creating Contagious Commitment: Applying the Tipping Point to Organizational Change.* Strategy Perspective, Hillsborough, NC. Second edition (2010).

Strategy Perspective (n.d.) Learning through action workshops. Available at: http://www.4-perspective.com (accessed 3 August 2013).

Sustainable Infrastructure: Principles into Practice
ISBN 978-0-7277-5754-8

ICE Publishing: All rights reserved
http://dx.doi.org/10.1680/sipp.57548.245

Chapter 12
Individual action for change

> People are much more likely to act their way into a new way of thinking, than to
> think their way into a new way of acting.
>
> (Pascale and Sternin, 2000)

Having set the scene in Chapter 11, you as an individual are now at the heart of
the change process. The whole of this book aims to help you have the confidence and
capability to make a real and practical difference. Much of the advice in this chapter
is at the personal level, and addresses how you act and feel personally – what motivation
you have, what roles you take, what power you can use, what support you need, and how
some practical tools can help you.

12.1. Motivation and creativity

Most organisations need to embrace change, and this includes those people on the
'inside' of infrastructure delivery, as well as those in a regulatory or civil society role,
who aim to enable the introduction and adoption of sustainability best practices. In
each of these roles there are *innovators* and *managers*, and such people often have a
two-fold view of the world. They are loyal to and work hard internally for their
organisation, but often they can also see new challenges, such as sustainability in the
outside world, that they are passionate about. They recognise that these issues affect
their organisation's *purpose* in the world. So, they want to act to change their organisation, for it to adopt and work to that wider new purpose.

So there is one key advantage that change for sustainability has over, say, changing to a
new accounting system. Because it connects to such powerful human values, it can draw
on a deep well of such people's interest, passion and enthusiasm, as a source of creativity
and action for change. If you are reading this book you may share these attributes and
concerns:

> So, find your organisational purpose and social context, to create meaning and
> creativity for innovation.
>
> (Ainger, 2010)

From experience, such people are key actors for change, in all the stages of innovation
that we discussed throughout Chapter 11. It is usually an informal leadership role,
and is not much talked about in mainstream management texts, but a valuable handbook
to have is Penny Walker's *Change Management for Sustainable Development – A*

245

Workbook (2006). Hopefully, you can identify with this kind of person, or want to try out this kind of role. If so, this chapter is particularly for you.

12.2. Roles and influence

When you decide to try changing something, it is worth first considering which role (or roles – you can often play more than one) you are best placed to play, depending on your organisation, position and opportunity. Then, you can consider how much that allows you to influence things.

12.2.1 Thinking about your role

Several different 'positional' roles were identified in the tensions diagram in Figure 11.1. They include *infrastructure organisations* (public or private), *activists* who create the tension for change, *regulators* who raise the playing field, for the laggards, *entrepreneurs* who challenge practice by competition from the outside, and *consultants* (and all other members of the supply chain) who can assist in change. Any one of these has the key internal roles of the *innovator*, who invents new practice, and the *manager*, who embeds new practice in the system. In any of these organisations, if you are planning a change action, it will be useful to work out whether you are being an innovator or a manager, in each instance.

Furthermore, there are at least four different *types* of individual role you can play, sometimes more than one at once (adapted from Visser, 2005). It may help you to further apply the right actions and skills to identify which role you are playing.

■ The *expert* – using your sustainability expertise directly to advise or advocate.

Or, a group one.

■ The *project leader* – taking on the role of directly inspiring, planning and organising an innovative project.
■ The *facilitator* – bringing people and their own sustainability ideas together, and helping them to formulate and develop joint ideas.
■ The *catalyst* – connecting people to do things together (often across organisational boundaries), or adding some small ingredient of money, time or evidence to make things happen that otherwise would not, and letting others get on with it.

It may help to have a checklist of questions to ask, to help you decide on where and in what role to act. Some such questions are given in Box 12.1. They include reference to issues raised later in this chapter.

You may often start as an individual sustainability *expert*, because you have educated yourself in the issues ahead of your organisation. As support grows, your role may need to change too, to more of a group one. Just because you care intensely about the topic, it can be surprisingly hard to change from the individual expert role, where your ideas rule, to one where you must listen to and accept others' ideas:

Box 12.1 Questions to help you decide on where and in what role to act

---

- What opportunities will I have to act for change, and when?
- Which *power arena*(s) do I have to effect change in, and what *formal* power do I have in (them)?
- How much will I need to use *informal* power as well, and what are my *personal* power sources?
- What *constraints* do I see on my likely success?
- What *opportunities* are there to help me? Sector or competitor actions to emulate? A chance to do a 'what if question' or an 'experiment'?
- What *champions and supporters* do I have – formal and informal – and how will I discover and confirm them?

---

the more passionate I am about the outcome, the harder it is to move from individualist (expert, activist) to strategist (facilitator, catalyst) ... and start 'letting go'.

(Ainger, 2001)

Each of these different roles will make different demands on your *credibility* (see Section 12.3).

12.2.2 Using your individual authority or influence

In Section 11.1, we introduced the different aspects of 'power' – top-down authority, as in 'power over' people – and bottom-up credibility and influence, giving 'power to' do things with people. In any role, you will need to assess which form of power will be effective and how much power you actually have, and use it accordingly. In an infrastructure organisation your authority might include

- decision-making responsibility, such as approving outline designs to proceed for construction
- control of resources, people and projects – budgets and time allocations (via time sheets), costs, project management teams and controls
- human resources issues – rewards (thanks, public acknowledgement, awards, pay, bonus) and sanctions (blame, discipline, dismissals)
- more subtly, control of language, information and processes – what is in the strategy and business plan, what is measured as 'performance', what is allowed on meeting agendas (what can be talked about), having some control of the 'culture'.

If you have authority, you need to choose carefully when to use it. The story in Box 11.1 showed a very appropriate combination, in which top-down decision-making authority (the first point in the above list) was used to drive project teams to greater bottom-up creativity. One the other hand, sometimes the authority of a formal role is worth nothing, because you are acting in an informal situation, as the story in Box 12.2

Box 12.2 When notional authority did not work, and 'influencing' was king

---

Failure, in spite of job title and notional authority

The context was a company global strategy meeting. My formal role was UK Director of Strategy, with authority to look at the future, but I was in an audience of about 70 senior people, listening to an invited speaker on 'future trends'.

- The 'guru' futurist speaker gave a dissertation on trends, and said 'technology will save us'.
- A colleague and I asked about 'downsides', and disagreed on the balance of risk.
- The futurist dismissed our question summarily (ours was the only challenge to his views).
- I hated getting 'slapped down' in response, and felt I had 'failed'. I thought that I should have continued the challenge, but that was not possible in this power relationship.

Lesson

In this context, buried in an *informal* audience of 70 seniors, my *formal* authority as Director of Strategy had no value. The guru futurist speaker successfully used his *informal* power of influence, of being the expert up on the podium, to dismiss me.

But this led to some unexpected 'change' progress. Afterwards, in the coffee break, two senior people came up and said 'well done'; they agreed with me – so I gained two new potential sustainability allies. It was not entirely a failure after all.

(Source: Author's personal experience)

---

shows. The power of authority, or influence, always depends on the context. Note, however, that in the story in Box 12.2 some unexpected progress was still made.

As we said throughout Chapter 11, in some early stages of introducing innovation you will certainly need to use informal 'influencing for change'. 'Influencing' persuaded companies to take part in the CIRIA indicators project (see Box 11.3), even though it was not what some of them might have seen initially as useful. Some typical dilemmas about the reality of authority, and of opportunities, are illustrated in Box 12.3.

12.3. Influencing power

So you will likely spend much time engaged in 'influencing for change', often with others. You will be exercising *power to influence*; that is, power to be listened to and to be taken seriously. What sources of personal power might you have, how are they relevant to your role, and how do they depend on the surrounding culture?

12.3.1 Personal power

In Section 11.4, we suggested that *credibility* with your audience is a key part of your effectiveness. Box 12.4 lists a range of sources of *personal power* to influence, entirely independent of any formal authority that anyone might have.

Some of these power sources may be more available to senior, long-term employees. These include *expertise, loyalty and trust, group support, people and information access.*

Box 12.3 Some typical change agent dilemmas, self-questioning and feelings

---

I've got a Sustainability/Corporate Responsibility job title, but actually I have few resources and little formal authority.
– Can I try 'influencing' – who do I know?

My ideas are more radical than the organisation is ready for.
– Can I think of a less-threatening action, which might lead to deeper questions?

There are too many other competing change programmes.
– Can I hitch my sustainability issue onto one of those already going?

Our sustainability commitment has been reduced by the difficult financial results.
– Can I come up with an initially cheap idea that has a long-term business case?

This stuff is so complex – where do I start?
– Can I just test a few starting ideas out against these frameworks – try some out, learn from the results and develop from there?

Am I credible? How do I get my confidence up?
– Can I try out my idea with friends; then just go for it, and still learn even if I fail?

However, some may also be available to more junior, newer employees, particularly in relation to sustainability issues. These include *external legitimacy, scarcity* and *freedom.* Your chosen role (see Section 12.2) will also affect which power sources are most critical. You need to harness your skills so you can move from, for instance, *expertise,* to being an expert, from *trust* and *group support* to being a facilitator, or from *people access* and *information* to being a catalyst. You may have to fit your role, to your personal power to influence things, in any opportunity.

Another key source of power may lie in your access to a key resource needed to try out new ideas, often by just getting round the rules. In the consulting organisation example in

---

Box 12.4 Sources of personal power, or *credibility* – 'worth listening to'

- *Expertise* – you have valued, possibly rare, skills and experience.
- *Loyalty and trust* – known to be acting in the organisation's interests.
- *Group support* – support from many people.
- *People access* – access to people with authority.
- *Information* – access to special kinds of information.
- *External legitimacy* – knowledge of esoteric 'new' and 'future' issues – like sustainability.
- *Confidence* – from all the above.
- *Scarcity* – being a key resource that is, or will be, in short supply.
- *Freedom* – to leave, maybe (or not to join).

(Source: Palmer and Hardy (2000: Box 4.2) and Ainger (2001))

Box 12.5 Power for change – informal action by staff using time and enthusiasm

---

Box 11.2 described how a sustainability interest group (SIG) was set up. The SIG started to develop sustainability tools and client offerings; over three years, this developed into a formal, funded, Knowledge Community.

The most critical management control in such a consulting company is the allocation of time, through close time-sheet control and only to authorised budgets. At the start, there were no budgets or time codes for work on sustainability, but it was regular custom to work longer hours each day that the formal $7\frac{1}{2}$ hours that had to be entered on time sheets. So the group initially met after 4.30 pm, working in their own time, after completing timesheet time. In this way, they used another source of personal power – *time and enthusiasm* – to start work on sustainability.

By giving their own time, the SIG members became free of formal budget control; they controlled their own purpose and agenda from the start. This 'unpaid-for' start-up approach is particularly relevant for organisations that have a very strong focus on meeting 'time utilisation' targets.

(Source: Ainger and Pickford (2007))

Box 12.5, that key resource was staff time to develop new sustainability materials. In others it might be access to computing capability in order to model a new process, or access to specialist manufacturing to develop a pilot technology.

Finally, being able to use your personal power sources also depends on knowing and understanding your audience's *culture*. In some organisations, for *expertise* (on management issues such as 'change', not on technical ones) internal staff are usually overlooked and only external consultants are thought to have credibility; in others, it is exactly the other way around. In many organisations, it would be entirely acceptable for you to seek *group support* on sustainability issues, by sending out an email, or blogging on a company site, seeking others' interest, but in others this might be seen as 'misuse of the IT system for personal purposes' and a disciplinary offence. The culture may also be different in different departments in an organisation:

> This dynamic of more than one culture in an organization is common. You yourself might work for an organization where the head office is 'networked', the marketing division 'mercenary', and the poorly performing manufacturing section 'fragmented' – but within manufacturing there happens to be one highly 'communal' 'skunk works' team. Indeed, given the different work done by varied parts of an organization, their different managers, different customers, and different competitive environments, a uniform culture is hard to find.
>
> (Goffee and Jones, 1998)

So, know your audience's culture, in planning a response to an opportunity.

12.3.2 Skills and ethics

Here is one way of summing up all these skills: the ability to combine the right language, argument, role, understanding of culture and audience, and to make effective use of personal power by applying leverage to any opportunity for innovation.

> True innovators ... choose to make their mark from within. They are likely to have ... characteristics of innovation such as political savvy, tact, teamwork and patience. Furthermore, rather than getting their way via force of personality or charismatic zeal, these innovators learn how to bring projects to life through the deft manipulation of the latent intellectual and financial capital inside their organizations.
>
> (The IDEO team, in SustainAbility, 2008)

In the above quote, the usually negative word 'manipulation' is used to describe informally influencing people in your organisation in order to pursue a sustainability purpose. To be 'manipulative' does not necessarily imply 'devious' or 'underhand'. One dictionary definition (*Collins English Dictionary*, 2nd edn, 1986) states that manipulating involves doing what is: 'illicit – not allowed or approved by common custom, rule or standard', which actually describes well the innovation role, that is, challenging business-as-usual. However, it still raises important points about ethics and integrity, in terms of both *what* you take it upon yourself to advocate, and *how* you act.

As with any question of professional ethics, the best test is to be always ready to be asked to justify, transparently and in public, *what* it is you advocate. But regular self-questioning is good practice. Here's how one change agent felt about this:

> For some time I worried that I shouldn't be using my influence with large organisations to promote an 'agenda' (sustainability), but I have concluded that this is not only the only way I can work, but that the influence I have is a gift and privilege I must make the most of.
>
> The response given by a change agent in the
> Organisational Change Agents survey (Walker, 2007)

12.3.3 Respecting others

The other important aspect of ethics in acting for change is in *how* you act, in dealing with the people you work with and want to influence. Sustainability can be a deep personal and emotive issue, as well as an intellectual one, and a business driver – you can see this from the emotions apparent in the climate change debate. There are many individual routes to engagement with sustainability, depending on people's own history, personality and experience. A common feature may be some kind of 'insight' experience, which moves people from awareness and unease to action. The connection with children seems to be one of the most powerful ways in which people become emotionally engaged:

> We do not inherit the earth from our ancestors; we borrow it from our children.
>
> Anon.

So, do not make assumptions that others' motivation is the same as yours, or about their feelings about it. They do not need to have the same motivation as you to work with you. Be careful not to let your passion for the topic spill over into being a zealot (this will quickly destroy your credibility, too), and respect everyone's different journey – be careful with people, and with yourself.

12.4. Looking after yourself

Sustainability may well be an emotive issue for you too, and you may feel passionately about it. Unless you are a natural and instinctive leader, acting to initiate change may mean working outside your comfort zone some of the time. So, you could get your balance of passion and expertise wrong, perhaps by taking on too much yourself. You could also wrongly blame yourself too much, when progress is hard, and even find yourself getting burnt out.

12.4.1 Watching your own feelings

Looking after how you feel – and how this affects the way people see that you 'are', in yourself – affects your effectiveness too. If you can show emotional engagement, confidence, energy and enthusiasm, this will communicate itself to your audiences, and may help to motivate and engage them more too.

People need validation and reassurance about their abilities, to initiate empowerment and encourage their determination to act. A survey by Penny Walker (2007) investigated how change agents feel, and where they seek support. You may sometimes be supported, empowered and enthusiastic:

> certain staff already have a natural skill or education in sustainability issues. We have tried to harness that by giving them the time and encouragement to develop their expertise and then feed back into the organisation as a whole.

But at other times:

> they feel negative – fearful, angry, despairing – about the amount of change that is needed and their effectiveness in bringing about these changes.

So, be careful to think about and acknowledge your feelings, to nurture your credibility and confidence, and to protect yourself. Professor Judi Marshall (Lancaster University Management School) advocates: 'know when to persist, and when to desist'. Sometimes, the right decision may just be: 'Anyone else feel like packing it in and going to the pub?'.

12.4.2 Finding support

Little of the change management literature pays much attention to the roles and skills associated with your emotional competence – courage, letting go, inspiration and finding support. These are all the things that you need when stepping out of your comfort zone to initiate and push for change.

Almost all the 'tempered radicals' [internal change agents] we spoke to emphasised the importance of maintaining strong ties with individuals, communities or groups outside of their organization.

(Myerson and Scully, 1995)

Penny Walker's survey subjects found this too:

My role in sustainability is a bit lonely. I have been promoted at work partly because of it, but the promotion has resulted in me having almost no time to further it at the moment. Very frustrating! ... Having more contact with others in a similar position would be motivating and invigorating.

The response given by a change agent in the
Organisational Change Agents, survey (Walker, 2007)

Sources of support can be essentially from others who share the same roles, commitment and experiences, as well as friends, for support and sense-making. The survey found that the following were popular.

- Conversations and sharing experience with peers/competitors in other organisations.
- Talking with friends who work in the same field.
- Formal networking with other people concerned with sustainable development (e.g. through organised networks, professional bodies).
- Attendance at training courses and learning programmes, where at least part of the purpose is to share experience and get tips on making sustainable development a reality.
- Signing up to e-newsgroups, email groups or web-based discussion forums that have a sustainable development focus.
- Informal networking (e.g. 'green drinks', socialising, occasional coffees) with other people concerned with sustainable development.

In extremis, if you feel you might despair, you may find Macey and Johnstone's (2012) book *Active Hope: How to Face the Mess We're in Without Going Crazy* helpful.

One strong motivation for writing Part III of this book is the idea that more knowledge about acting for change can help you *to look after yourself*, as you act, and therefore be more effective too. Doing some analysis of realistic potential change expectations beforehand can help you decide not to 'beat yourself up' by trying to do the impossible ones. You may even expect to fail, without blaming yourself, on those impossible challenges where being true to your values pushes you to speak up anyway. The personal practices and tools discussed below may further help in this.

12.5. Tools for effectiveness

Chapter 13 summarises assessment tools that can be applied to projects, and Section 11.4 presented tools for measuring progress towards sustainability. This section suggests *personal* tools that you can use to assist your thinking, action and sense-making for change. Here are some simple practices to use, and some tools to help.

Box 12.6 Good approaches and communication for sustainability – not 'doom and gloom'

Futerra's 10 principles

1 *Big picture* – make connections, demonstrate long-term thinking, blow myths.
2 *Technically correct* – be trustworthy, provide transparency, give real facts.
3 *Be cool* – sexy, mainstream and non-patronising, avoid worthy and be brave – stand out!
4 *Belong* – invitation to a massive world-wide change, positive conformity; join a success.
5 *Only stories work* – empathy and emotions are a powerful tool, stories hold people's attention.
6 *Optimism* – sustainable development is achievable; avoid too much guilt.
7 *'Glory button'* – feel good about yourself, give yourself a pat on the back – sustainable development makes you a great person and we love you for it.
8 *Change is for all* – breaking stereotypes, inclusive language and images, mass ownership.
9 *We need more heroes* – icons to emulate; 'be like me'
10 *Personal circle* – relate to everyday life, give a familiar context for big ideas.

(Source: www.futerracom.org)

Rob Hoskin's Transition Town movement philosophy
Change our lifestyles – together in our communities – and have fun!

(Source: Hopkins (2008))

- Remain *curious rather than certain* – be open to, and learn from, unexpected outcomes (see Box 12.2).
- Build *time for reflection* into your work (and life?). Introduce conscious cycles of action and reflection – use *action inquiry* (see more below), write a journal, diary or notebook, to help this and allow re-engagement. Try out *freefall writing* – 'What do I really think, and feel about this?'; record interviews and meetings, and listen to them again later.
- As part of reflection, *seek feedback, interrupt your habits* – find friends who are willing to act as 'enemies'.
- When stuck on what to do, try a *two-on-one interview* (see more below) – 'How do I know what I think until I hear what I say?'.
- For more persuasive conversations and arguments – try *four parts of speech* (see more below), and remember some principles of good sustainability communication (Shone and Marx, 2009). Some from Futerra and Transition Towns are given in Box 12.6.

More detail on three of those tools is given below.

12.5.1 Action inquiry – learning by doing

Action inquiry is a practice of 'learning by doing' (Fisher *et al.*, 2000; Reason and Bradbury, 2001). It turns any action into a *learning cycle*, as shown in Figure 12.1. It is like the well-known management mantra: 'Plan, Act, Measure, Review', but with much more emphasis on not just reviewing progress, but also making sense of and

Figure 12.1 Action inquiry – a cycle of action plus learning

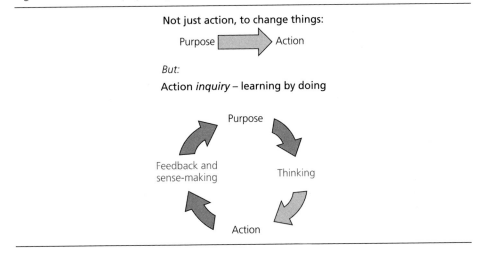

learning from whatever happens, including (especially) unexpected outcomes, and failure as well as success.

Rather than just treating each action as a linear 'I have a purpose, then I act', you add *inquiry* into your *action*. This is achieved by thoughtful thinking beforehand, by actively seeking feedback from your audience, and by giving time and attention to sense-making afterwards.

- Thinking – *use any information on the organisation and tools, ideas, research* – to plan your action, perhaps using some of the ideas in this chapter.
- Feedback – *gain real evidence*, preferably not just your own impression, on the results.
- Sense-making – *review and make sense of the results* of the action, using all this information. Ask 'What happened, and why?' Think about how you feel. What is the learning for next time?

An example of using this approach is one of the author's early (1999) international actions to advocate sustainability in his company (Box 12.7). The story also shows the *personal power* (see Box 12.4) that he used.

The example in Box 12.7 shows action inquiry by one individual. However, the approach can also very effectively be used to investigate innovation for sustainability by a cooperative group. In this case the group would meet regularly and successively during a project or investigation. Each meeting would contain new thinking inputs for planning future action. Between meetings all members would try out new *action*, and the first session in each following meeting would be to tell each other what happened, for *feedback* and *sense-making*. So, the group learning cycles would follow each other, making more learning progress at each meeting. Such 'cooperative inquiry' is particularly

---

Box 12.7 Using action inquiry at an international knowledge conference

---

Purpose: to informally introduce sustainability onto the company's 'radar screen'; test for the sustainability awareness of international staff, and start an interest group.

Thinking: used the annual Knowledge Management (KM) meeting in the USA; used *personal power to influence*; got a speech slot through *credibility* as a KM founder, and *political access* to the organisers; attached the sustainability arguments to recognised company language.

Action: the speech was short, at lunchtime, with no slides; asked for *feedback* written on extra paper napkins at each place setting (the jokey idea encouraged the audience to respond).

Feedback and sense-making: feedback was very positive – 'You were inspiring'; 'It needs to be said'. This reinforced his confidence in speaking; gave 'firm' (because written) evidence on staff interest, which he then used to argue for the relevance and importance of sustainability to the company, and initiated it as a KM topic every year since.

(Source: Author's personal experience)

---

effective where the learning is by a crossing-boundaries group, who have to learn new collective behaviour – to know, understand and trust each others' position, and build a team. It would be an ideal process to use for the 'spreading innovation' group advocated in Section 11.4.

12.5.2 Two-on-one interviews

If you are stuck, or unsure about what to do next, or how to address an opportunity – try using the 'How do I know what I think until I hear what I say?' approach. This just formalises what we often do in an accidental conversation in the corridor or by the coffee machine.

Ask two colleagues, who you think can help, to join you for a 'corridor meeting', with a maximum time of 15 minutes. Use the first 5 minutes (maximum) to give them a very brief (1 minute) explanation of what you are stuck on, and then have them push you hard to frame your uncertainty into a 'How do I do ... something?' question. Then use this question. Spend the remaining minutes having them question you, to help you answer it, connecting each of their successive questions to your last answer. Listen to what you say in response, particularly your unplanned answers to the unexpected questions. Then go back to your desk and capture all that you said and have your colleagues help you to remember.

Usually, you will find that you have at least defined your question; and often you have made progress and got unstuck, toward the start of an answer. This gives you something to work on in more detail. If you can record it, that is even better. Go back to this conversation a day or two later, and listen again. Your mind may have been working on the issue subconsciously, and new insights may emerge.

12.5.3 Good conversations – 'four parts of speech'

This simple process helps you to have clearer, more persuasive conversations, and to reinforce your arguments (Fisher *et al.*, 2000). It makes it clearer why you are having the conversation, better explains what you are advocating, and seeks direct feedback (as for action inquiry, above). The process consists of using 'four parts of speech' – they can be in any order.

- Frame – you explicitly say what your purpose is, that is, where you are 'coming from'.
- Advocate – you assert an option, perception, feeling, opinion or proposal.
- Illustrate – you add a story to illustrate your point, i.e. to 'put meat on the bone' of the advocacy.
- Inquire – you explicitly question your audience, to learn their reaction.

For example, where you have an opportunity to speak at an internal meeting, contribute your answer to a question about the relevance of sustainability to your consultancy, or to more general questions, such as 'How can we differentiate ourselves from competitors?' or 'What new issues do we need to look out for?'. What you want to advocate is: 'We need to take sustainability more seriously'. But, to improve your persuasiveness, you use the 'four parts of speech' around this, in this order.

- To get attention, and business relevance, you start by expressing a possible threat to the consultancy's appeal to clients. You *frame*: 'I've heard that public clients will soon be making expertise in sustainability a part of their pre-qualification criteria'.
- To add to the urgency of your suggestion, you refer to a competitor. You *illustrate*: 'XXX [a competitor] already have their own sustainability methodology, and may be ahead of us.'
- Then you *advocate*: 'We need to take sustainability more seriously.'
- And, finally, you *inquire*, to ask for reactions, but also to suggest an action: 'Why don't we try out using some sustainability choice criteria, in our next feasibility study, as a "what-if"?'

Often, people leave out the *framing* and the *illustrating*. Adding the framing helps to stop a sceptical audience dismissing your advocacy from the start by making some wrong, self-excusing, assumption about why you are saying it: 'It's just his personal bee in the bonnet', perhaps. Adding the illustration allows you to really drive home the meaning of what you advocate – bringing in a *story* is a powerful way of capturing people's attention. We used it by providing the detail in Section 11.4, to illustrate the meaning of the innovation curve in Figure 11.3.

Framing can also be useful with public audiences. Box 12.8 shows an example of using it at the start of a public speech, to 'insulate' your company's reputation from the content, when planning to say something that might be seen as too radical. The last two bullet points use the 'court jester' analogue to ask that the audience to give the speaker a license to challenge authority, and be radical. If the audience likes what the speaker says, the company gains some credit; if the audience does not like it, the framing has distanced it from company policy.

Box 12.8 Gaining permission to be radical

Where am I coming from? My position?
- Engineering experience, science interest, management experience – 'big picture'.
- 30+ years as a water and environment contractor, consultant, strategy advisor.
- Speak from company base but have 'course jester' license – allowed to ask the awkward questions – 'the emperor has no clothes?'.
- I ask your permission to do the same here.

(Source: author's personal experience)

12.6. Summary – 'facilitating the emergence of change'

The 'change' ideas described in Part III are not a 'magic formula' for success. We have provided a menu of different perspectives and approaches that we have found helpful, hoping that some of them may be useful to you. Look them over, see what appeals to you and suits your way of working, and try some out. To summarise, here are some lessons learnt from experience of acting for change.

Be an 'insider–outsider' by searching externally for new sustainability evidence and practice, and bringing it inside your organisation. To be credible, to be listened to by others, always show commitment and loyalty, and add value in your 'everyday' work. Use a balance, and often a combination, of

- external regulatory, market and competitor pressure, and internal action
- top-down support and 'championing', and peoples' 'bottom-up enthusiasm
- planned and opportunistic action
- informal 'influencing' power, and formal 'authority'
- action, and personal learning and development
- committing yourself, and protecting yourself.

An experienced construction practitioner's summary of these skills is given in Box 12.9.

Box 12.9 'The change agent's CV'

- Demonstrate authenticity and integrity (do not be not a power holder, or power hungry).
- Your own belief system is the crux.
- Have 'stickability' (this can be a lonely and thankless role).
- Succeed by stealth (this does not mean deceit, which is a killer for credibility).
- Be a known face (external 'management consultants' are at a disadvantage).
- Be politically astute, understand how power is held, and works, in organisations.
- Most essentially, have an effective sense to know *when* to apply the right tactics.

(Source: Guest Speaker to the University of Cambridge MPhil
Engineering for Sustainable Development lecture, 2002–2003)

Finally, do not get trapped in a requirement that any alternative to business as usual must be 'proved to be perfect' before you can try it. Make liberal use of the 'What if?' and 'Why not?' questions, and try just acting as well as talking. That takes you back to the quotation with which we started the chapter.

Individual action for change: summary

Every change starts with someone saying 'this can't go on', and deciding to act. Is that you?

A passion for sustainability motivates creativity for the innovation we need, but combine it with technical expertise and integrity, to be a force to be reckoned with.

You do have personal power to influence for change – more than you think – even if you have little formal 'authority'. Think about your role, and match it to each opportunity for change; use authority or influence to fit the opportunity.

Use your skills for change ethically, and respect others' positions; look after your own feelings too, and find support.

To be most effective, use personal tools, including learning by doing, and develop good communication skills.

Suggested Further Reading is listed after References.

REFERENCES

Ainger C (2001) Walking the tightrope: learning to be a change agent for sustainability. MSc Dissertation, University of Bath School of Management, Ch. 6.

Ainger C (2010) 'Chasing the buzz' – building business through culture based innovation. Presented at CEDA Twilight Event, Australia.

Ainger C and Pickford W (2007) Staff values as enablers for change – 'volunteering' for sustainability. *Organisations and People* **November**.

Fisher D, Rooke D and Torbert B (2000) *Personal and Organizational Transformations: Through Action Inquiry*. Harthill Group, Lydney.

Goffee R and Jones G (1998) *The Character of a Corporation. How your Company's Culture Can Make or Break Your Business*. HarperCollins, London, Ch. 2.

Hopkins R (2008) *The Transition Handbook*. Green Books, Cambridge.

Macey J and Johnstone C (2012) *Active Hope: How to Face the Mess We're in Without Going Crazy*. New World Library, Novato, CA.

Myerson DE and Scully MA (1995) Tempered radicalism and the politics of ambivalence and change. *Organisation Science* **6(5)**: 585–600.

Palmer I and Hardy C (2000) Chapter 4: Managing Power: Everything or nothing? In *Thinking About Management: Implications of Organizational Debates for Practice*. Sage Publications, London.

Pascale RT and Sternin G (2005) Your company's secret change agents. *Harvard Business Review* **May**: 124–132.

Reason P and Bradbury H (2001) The action research cycle. In *Handbook of Action Research: Participative Inquiry and Practice*. Sage, London.

Shone D and Marx S (2009) *The Psychology of Climate Change Communication: A Guide for Scientists, Journalists, Educators, Political Aides, and the Interested Public*. Centre for Research on Environmental Decisions, Colombia University, New York. Available at: http://www.csc.noaa.gov/digitalcoast/_/pdf/CRED_Psychology_Climate_Change_ Communication.pdf (accessed 13 August 2013).

SustainAbility (2008) *The Social Intrapreneur. A Field Guide for Corporate Changemakers*, 1st edn. Available at: http://www.echoinggreen.org/sites/default/files/The_Social_ Intrapreneurs.pdf (accessed 3 August 2013).

Visser W (2005) University of Nottingham, ICCSR: presentation at Sustainability Learning Network Workshop, Cambridge Programme for Sustainability Leadership, 8–10th June 2005.

Walker P (2006) *Change Management for Sustainable Development – A Workbook*. Practitioner Best Practice Series, Vol. 8. The Institute of Environmental Management and Assessment, Lincoln.

Walker P (2007) Supporting the change agents: keeping ourselves effective on the journey of change. *Greener Management International* **54**.

FURTHER READING

CRED (Center for Research and Environmental Decisions) (n.d.) Available at: http:// www.cred.columbia.edu/guide (accessed 13 August 2013).

Marshall J, Coleman G and Reason P (eds) (2011) *Leadership for Sustainability – An Action Research Approach*. Greenleaf, Sheffield.

Transition Network (n.d.) Available at: http://www.transitionnetwork.org (accessed 13 August 2013).

Part IV

Tools

This part provides more detailed sources and references on tools.

Chapter 13 describes some key tools that are specifically aimed at 'enabling' and measuring sustainability. These tools can be useful in helping you apply the practices suggested in Part II, in planning, delivering and operating infrastructure projects.

Then, Chapter 14 provides some closing thoughts as End Words.

The expectations of life depend upon diligence;
the mechanic that would perfect his work must first sharpen his tools.

(Confucius)

Sustainable Infrastructure: Principles into Practice
ISBN 978-0-7277-5754-8

ICE Publishing: All rights reserved
http://dx.doi.org/10.1680/sipp.57548.263

Chapter 13
Tools for sustainability practice

There are many tools for sustainability practice available, and we aim to help you be aware of the more commonly used ones. We do not provide a detailed explanation for each of them, as this information can be found elsewhere, and sources are signposted in the following pages. We do comment, from experience, on their applicability, their key issues and their limitations. We try to provide a balanced view, and hope this will help you to select the right tools.

The tools we present here match the range of our four absolute principles presented in Chapter 2, as summarised in Table 13.1.

Table 13.1 How the tools described relate to the absolute principles

Tool	A1 Environmental sustainability – within limits	A2 Socio-economic sustainability – 'development'	A3 Intergenerational stewardship	A4 Complex Systems
Environmental impact assessment (EIA) and strategic environmental assessment (SEA)	✓		✓	✓
Life-cycle assessment (LCA)	✓		✓	
Carbon footprinting	✓		✓	
Environmental management systems (EMS)	✓			
Building rating systems (BRS)	✓	✓		✓
Ecosystem services and green infrastructure valuation	✓	✓		
Whole-life cost accounting (WLC)	✓	✓	✓	
Corporate social responsibility (CSR) and sustainability reporting		✓		
Backcasting, forecasting and scenario planning			✓	
Multi-criteria decision-making (MCDA)		✓		✓
Systems dynamics				✓

One reason why some useful tools are not taken up by infrastructure organisations is that some were designed for other aspects of engineering or have been seen to be cumbersome or difficult to apply to infrastructure. One way past this is for you to first review a potential tool. If its key principles and applicability fit your need, make sure you keep them, but customise the detail and language to fit more easily into your organisation's standard processes. Some of these tools can be effective when used in combination, too.

13.1. Environmental impact assessments and strategic environmental assessment
13.1.1 Summary

Environmental impact assessment (EIA) is a familiar tool, having been in use for over 40 years. They have been required in the UK for large infrastructure projects since the 1980s (European Directive 85/337/CEE). Each assessment is carried out by the proponents of a project, and is used by decision-making bodies such as local planning authorities to understand the environmental affects and future consequences of a development. In the UK, the EIA Regulations are implemented under Town and Country Planning legislation. Certain specific types of infrastructure (e.g. ports and harbours) is not covered by this, and is subject to separate legislation.

An EIA emphasises a systematic analysis of information, to evaluate the importance and significance of predicted effects. The output information provides a focus for public scrutiny of the project, and encourages the developer to modify designs and prepare mitigating measures. The EIA moves beyond the traditional feasibility study, which focuses on technical aspects of project design (Will it work?) and financial considerations (Will it make money and not cost too much?), usually the first interests of the project developer. It widens the assessment to consider both the environmental issues (Will it damage or enhance the environment?) and the social concerns (Will it benefit or disrupt the neighbourhood, region or country?) of the project neighbour community. More recently, such assessments might ask: 'Will the project contribute to sustainable development?'

In the UK, an EIA is required for infrastructure projects listed in Schedule 1, such as chemical works, waste incinerators, power plants and major road schemes. Other projects, such as quarries, overhead transmission lines, waste water treatment plants and a wide range of others identified under Schedule 2, must be subject to an EIA if the nature, size or location may create a significant environmental impact. The deciding criteria are based on thresholds that relate to these issues.

An EIA essentially asks five questions.

- What are the existing characteristics of the environment in the area to be used by the proposed development?
- What is the nature of the development?
- What effects will the development have on the existing environment?
- What measures can be taken to mitigate any of its adverse affects?
- What would happen if it did not proceed?

In addition to an EIA, a social impact assessment (SIA) may also be carried out. These assessments have been relatively underfunded and neglected, and their status and influence have grown more slowly. This has been due to ambiguity over their legal status, a wide diversity of possible methodologies and inadequate data availability, as well as a lack of expertise. Even when there is no specific requirement to consider social impacts, there may be a range of regulations that apply to social issues, such as employment conditions, ambient noise levels, protection of heritage sites, residential zoning requirements and sanitation standards (Box 13.1).

Box 13.1 Examples of social aspects and impacts

Social aspects	Social impacts
Employment and employee benefits	People's way of life
Provision of training	Their culture
Tax payments	Their community
Procurement of goods and services	Their political systems
Improved infrastructure	Their health and well-being
Loss of access to land/other resources	Their personal and property rights
Physical displacement of people	Their fears and aspirations
Causes	**Effects**

Project-based EIA's can have difficulty in evaluating indirect impacts, they cannot address alternatives that have already been eliminated earlier in the planning process, and they do not address cumulative impacts from multiple schemes. They have also been criticised as being too narrowly focused and reactive, resulting in fiddling with detail rather than addressing fundamental issues. Strategic environmental assessments (SEA) have therefore been developed, focusing on policies, plans and programmes. They consider a greater scale and longer time interval than project EIAs. Impact predictions are subject to much greater uncertainty, but the time for data gathering is often longer and the degree of detail required is much less than that for project evaluations. SEAs take the impact assessment 'upstream' into infrastructure planning (see Chapters 3 and 4) rather than outline design (see Chapter 7), and broaden its scope in space and time.

Preparation of an EIA is required by international agencies such as the World Bank, and is enshrined in international agreements such as Principle 17 of the Rio Declaration. They are required by project financiers, as more than 80% of them have adopted the Equator Principles (see Box 11.5) for commercial projects. These principles include assessing potential investments with EIAs and SIAs.

13.1.2 Application

A number of key stages are essential in an EIA. The first principle is to consider main alternatives to the development. The 'do nothing' option is always a possibility. It may sometimes have worse environmental consequences than proceeding with the scheme, and should always be assessed. Other alternatives include other sites, and solutions outside the operational control of the proponent (e.g. alternative disposal options for waste), as well as different types of processes and changes to the physical appearance and site layout of buildings.

Early EIA's attempted to study the effect of 'everything on everything', a too unwieldy process. Scoping is now used to determine where the key issues are, and to focus on these. So, for an airport expansion, noise will be a central issue, whereas eutrophication of watercourses may not be. Issues of less significance will need only brief treatment. Developers can seek a formal scoping opinion from planning officers on what should be included in the final Environmental Statement (ES).

A baseline survey, sometimes referred to as an 'environmental inventory', is then carried out to establish the pre-existing conditions in the environment. This should include work by an environmental science team, and also take into account local knowledge drawn from the affected community. This can be highly informative about the baseline conditions. The developer who failed to find out why his site was locally referred to as 'Spring Meadow' must have rued his inaction when the new houses were subject to groundwater flooding only 6 months after completion.

In the UK, the person undertaking the EIA is required to consult with a set of statutory consultees, such as the Environment Agency, English Heritage and Natural England. They must provide information already in their possession, and usually in the public domain, but are not required to conduct any extra research. The EIA regulations do not require any wider public consultations but it is good practice to consult directly with the community and other local organisations. Local groups may have conducted their own monitoring and have data that are valuable to the assessment.

The output of the EIA is the ES, which includes a description of the physical characteristics of the development, and the land-use requirements during construction and operation. It should also contain a description of the main characteristics of the production processes, such as the quantity and nature of materials used, and an estimate of any wastes or emissions. Any indirect, secondary or cumulative impacts should also be established; temporary and permanent effects should be distinguished, and positive and negative impacts clearly identified, together with any knowledge of data gaps or missing information (including lack of knowledge or technical know-how). A non-technical summary must be published separately for circulation to local residents or other interested parties. The main ES should suggest mitigation measures, which might also include monitoring arrangements throughout the lifetime of the scheme. It should also say clearly how the ES was prepared, including what consultations took place and when, what impact prediction methods were used, and when the process began and finished. The EIA process is shown in Figure 13.1.

13.1.3 Key issues
Environmental impacts can significantly influence public opinion about a development. Some are obvious to the public, such as visual intrusion, traffic congestion, severance of

Figure 13.1 The environmental impact assessment process

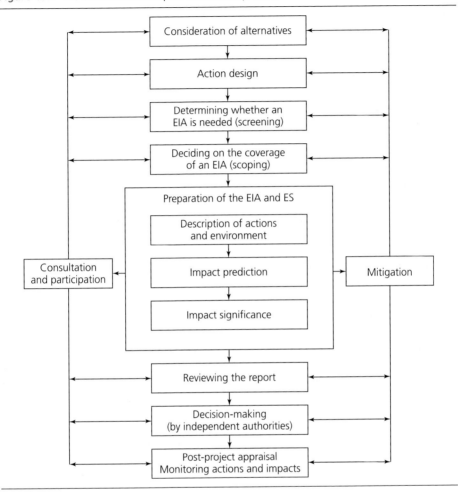

land holdings, loss of agricultural land, noise and changes in property values. Others, such as hidden pollution, job losses and other economic gains or losses, ecological, impacts and effects on scarce resources, require specialists to assess them. Issues that usually attract opposition are

- impact on the visual quality of the landscape
- pollution and disturbance to the ecology of an area
- the land take of houses and agricultural land
- the effect of new infrastructure on human activities.

Positive aspects of EIAs are that they anticipate and integrate impacts across all types of pollution *and* economic development. In general, EIAs have not led to huge increases in project costs or created long delays in planning. Many projects have been usefully amended at the design stage as a result of the assessment.

13.1.4 Criticisms and drawbacks

The ES output of an EIA can be seen as just about producing a 'box-ticking' report. Some have commented that it is often a sales document for the applicant, and have increasingly called for an independent commission of EIAs, to take them out of the hands of those with a vested interest (Friends of the Earth, 2005). The best result of an EIA is the process that a developer is forced to go through, identifying areas of concern and mitigating responses. The procedure should be iterative, with a clear feedback into the design itself.

The format and contents of an ES can often be inadequate, either in terms of quality or because key parts are missing. Frequent defects include failures to produce a non-technical summary, to adequately consider human health, to include proper consideration of alternatives, and to admit uncertainty. Non-technical summaries can be vague and generalised. EIAs have also been criticised for having too narrow a spatial and temporal scope, and their outcome can be complex and confusing, leaving local communities unsure of how a development affects them. They can also be done too late in the overall process.

The quality of ESs has varied substantially, with some not even making it clear whether a consultation process has taken place. Other criticisms are: a lack of data in a quantitative form, no detail about the timing of field monitoring surveys, and inadequate impact predictions. They can also fail to judge the significance of each impact. A good final step is to conduct a post-project appraisal, to learn whether the expected affects did actually accrue, but this is rarely done in practice.

UK regulations have been criticised for requiring EIAs for some types of development and not others. This uncertainty can be a source of major dispute between communities and local authorities. Some of the criteria determining the need for an EIA can be a matter of opinion, and different authorities may reach different views.

Finally, in the past many EIAs were done only after main design decisions on the project had been taken. As a result, mitigating the impacts highlighted by the EIA almost always involved add-ons, because it was too late for fundamental redesign. These add-ons usually added cost, and so the, often false, perception has developed that 'protecting the environment always costs more'. Fighting this perception, with evidence, is now very important (see Sections 7.3 and 11.3).

13.1.5 Sources of further information

Carrol B and Turpin T (2009) *Environmental Impact Assessment Handbook: A Practical Guide for Planners, Developers and Communities*. Thomas Telford, London.

Glasson J, Therivel R and Chadwick C (2012) *Introduction to Environmental Impact Assessment*. Routledge, New York.

Hardisty PE (2012) *Environmental and Economic Sustainability (Environmental and Ecological Risk Assessment)*. CRC Press, Boca Raton, FL.

Morris P and Thervel R (2009) *Methods of Environmental Impact Assessment*. Routledge, New York.

Therivel R (2010) *Strategic Environmental Assessment in Action*, 2nd edn. Routledge, New York.

Tromans S (2012) *Environmental Impact Assessment*, 2nd edn. Bloomsbury Professional, London.

13.2. Life-cycle assessment
13.2.1 Summary

Life-cycle assessment (LCA) is based on a 'cradle-to-grave' approach to assessing the environmental impacts of a product, project or service throughout its life cycle. This is achieved by considering the primary resources consumed and the emissions and wastes generated during the extraction and processing of raw materials, manufacturing, transportation and distribution, use, reuse and maintenance, recycling and final disposal. The LCA has been used most frequently in the manufacturing and process industries, as it enables the environmental performance of different products to be compared on a common basis, and the identification of the life-cycle stages that are the most damaging in terms of pollution emissions (e.g. the greenhouse gas emissions released at each stage calculated in kilogrammes of carbon dioxide equivalents (CO_2e)). The LCA methodology is standardised by a series of ISO standards (ISO 14040: 2006). For a concise summary of the method, see Azapagic and Perdan (2011).

13.2.2 Application

The method reflects a systems approach and evaluates environmental impacts across multiple criteria. It can be used for strategic decision-making, policy assessment and the management of environmental risks through the supply chain, through appropriate materials selection and consideration of end of use disposal options. It can be applied not only to the manufacturing of products but also to an entire building or infrastructure asset over its life. LCA can also be used to evaluate service requirements such as waste management practices, transportation services and energy provision. For example, the principles of using LCA have been applied in bridge analysis (Horvath, 2009; Kendall et al., 2008) and in comparing alternative technologies for manufacturing Portland cement (Huntzinger and Eatmon, 2009), as well as in urban water provision (in California) (Stokes and Horvath, 2010), in buildings (Diakaki and Kolokatas, 2009) and in heavy construction activities (Ries et al., 2010). The method also underpins well-established sets of tools such as the Building Research Establishment's Environmental Assessment Method (BREEAM) for buildings and Arup's SPeAR. It is also the basis of the *Green Guide to Specification* (BRE, 2009), which provides advice on how to make the best environmental choice when selecting construction materials and components by presenting Ecopoint scores, which are converted into final ratings on the scale A+, A, B, C, D and E. BRE's Envest2 software tool combines LCA with whole-life costs to allow both environmental and financial trade-offs to be made explicit in the design process (BRE, n.d.). However, there is a real and pressing need for an easy to use tool for civil engineers that takes into account both the LCA of the materials used and the environmental impacts these have on infrastructure projects (Ghumra et al., 2009).

13.2.3 Key issues

The outcome of the analysis is largely determined by the goal and scope of the study and how the *systems boundaries* are set. Boundaries should reflect the purpose of the analysis, such as whether it is aimed at reducing environmental impact or, more specifically, seeking to establish the best source for a material or the best disposal route for waste. This involves deciding whether the analysis should only go back to the supplier or back to primary mineral extraction. Sometimes the calculation stops at the completion of the manufacture of a product and does not include the use or final disposal stages. This is sometimes referred to as considering impacts from 'cradle to (factory) gate'.

The subject of the analysis is expressed as a *functional unit*. This represents a quantitative measure of the output of products or services that the system delivers, and allows comparison between alternative designs or processes on a common basis. Examples might be: 200 000 m^2 of office development with a 60-year life, the transport of two executives with luggage between London and Paris (in which alternative transport modes can be compared: road, train, air, ferry etc.), or the water-treatment infrastructure required to provide a family of four with their domestic water consumption needs for a year.

The output of the analysis is typically in the form of a series of *impact categories*, such as global warming potential, non-renewable resource depletion, ozone depletion potential, acidification potential, eutrophication potential, human toxicity potential, aquatic toxicity potential and so on. These impacts are usually reported relative to a reference substance. So, for greenhouse gases everything is translated into an equivalent amount of carbon dioxide which will have the same effect. The values can then be recorded in absolute terms; or sometimes it is more revealing to normalise them against some reference value so that the relative magnitude of each impact can be easily compared. This might be to express the impact as a ratio to the total national impact in a year of that impact category, or relative to the average quantity produced annually by one person. A further consideration is to make sure that where multiple activities generate a given impact this is *allocated fairly* across each contributing process. The ISO 14040 series provides a procedure by which this can be done.

The approach follows a series of steps including scoping the issues, mapping the life cycle, usually in a flow diagram of various process (or project) stages, compiling inventories of materials and energy consumed at each stage, allocating the environmental burdens in appropriate units for each impact category, analysing the results (sometimes including applying significance weightings to some categories) and interpreting the results.

In practice, many of these procedures are handled using commercial LCA software, which includes databases that translate quantities of raw material, such as steel and concrete, into amounts of associated carbon dioxide, etc. per unit mass. Several such software packages include Simapro and GaBi. A database that is often used is the Inventory of Carbon and Energy (ICE) produced by the University of Bath (Hammond and Jones, 2008a, 2008b), which can applied to a wide range of building materials.

13.2.4   Criticisms and drawbacks

LCA has been subject to a range of criticisms, including that it can take an excessive amount of time and effort to assemble the detailed amounts of data needed to drive the calculations. This has meant that it has been used mainly in product development in manufacturing (which creates millions of identical outputs), and not extensively in one-off infrastructure projects. The accuracy and availability of data is also a significant issue. Some have even argued that LCA provides no consistent results, as the answers it yields are wholly dependent on the choice of systems boundaries adopted in each calculation. The method also cannot easily represent social impacts or problems associated with activities that generate nuisance such as noise, dust and vibration.

LCA is based on a listing of input and output streams at a point in time and does not deal well with waste material (e.g. to landfill) where emissions may arise over long periods of time. It is useful for designers of static components and assemblies but is less easily applied to processes and services. Furthermore, as the central object of LCA is a functional unit, the actual total amount of pollution created is not always dealt with directly. Other major problem areas around the methodological limitations of LCA are issues such as spatial variation and local environmental uniqueness, as well as data quality and data availability (Reap *et al.*, 2008a, 2008b). Nevertheless, the discipline of conducting an LCA can reveal much useful performance information that can lead to suitable mitigations being considered as part of a dynamic and iterative design process, which may be of more value than the absolute results the tool generates.

13.2.5   Sources of further information

Crawford R (2011) *Life Cycle Analysis in the Built Environment.* Taylor & Francis, London.

Forum for the Future/The Natural Step (2007) *Streamlined Life Cycle Analysis (SLCA).* Available at: http://www.forumforthefuture.org/sites/default/files/project/downloads/slca-2-pager-intronov-2007.pdf (accessed 20 August 2013).

Strauss A, Frangopol D and Bergmeister K (2012) Life-Cycle and Sustainability of Civil Infrastructure Systems. *Proceedings of the Third International Symposium on Life-Cycle Civil Engineering (IALCCE'12), Vienna.* CRC Press, Boca Raton, FL.

The International Journal of Life Cycle Assessment (n.d.) Available at: http://www.springerlink.com/content/112849 (accessed 20 August 2013).

13.3.   Carbon footprinting

13.3.1   Summary

With a strong emphasis on tackling climate change, governments are increasingly requiring organisations to reduce their carbon emissions and minimise their energy requirements. Monitoring how this is achieved can be done by calculating a carbon footprint for an organisation or activity. This provides a measure of the mass of carbon (or carbon equivalent) emission associated with the activity in question. Essentially this is done through a simplified LCA, which provides a single impact category of global climate change/global warming.

One helpful definition states that 'the carbon footprint is a measure of the exclusive total amount of carbon dioxide emissions that is directly and indirectly caused by an activity or is accumulated over the life stages of a product' (Wiedmann and Minx, 2008). In this approach only carbon dioxide is included (e.g. methane is excluded), and so the description differentiates between a true *carbon* footprint and a *climate* footprint, which would be based on other greenhouse gases.

13.3.2 Application

The calculation is usually based on developing a process map for the activity over what is considered to be the relevant life-cycle stages, which most often are taken to cover extraction of raw materials, manufacture, transport to site, on-site construction of assets, off-site disposal of waste from construction, operational use and end-of-life disposal and recycling. The relevant significance of these stages in the carbon and embodied energy contained in a building's structural elements is shown in Figure 13.2

Figure 13.2 Assessing life-cycle embodied energy of building structural elements (Vukotic *et al.*, 2010)

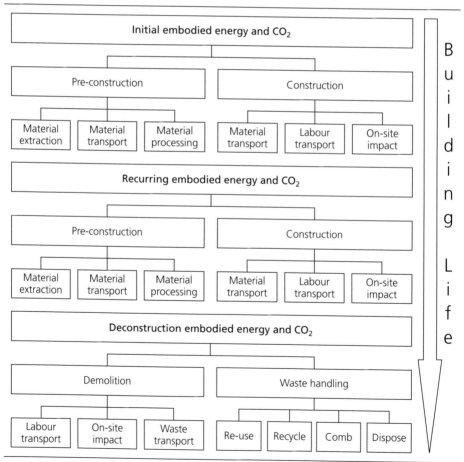

(Vukotic *et al.*, 2010). Material selection, material sourcing and waste handling at end of life have been shown to be the phases offering the most potential for reductions in embodied energy and carbon.

As with an LCA, the exercise requires the collection of a lot of data. For example, the activity data needed consists of the types and quantity of all inputs and outputs, including material inputs (often usefully sourced from records of bills of quantities if dealing with physical infrastructure), product outputs and waste. The type, source and quantity of all energy used can be determined from bills for electricity and other fuels. If there are any direct greenhouse gas emissions (e.g. from methane release in sewage treatment) these need to be assessed, as do the vehicles and average distances travelled in any transport component of a scheme. Similar to the concept of a functional unit in an LCA these quantities should be expressed per unit of finished product or service. The carbon footprint can then be calculated as follows:

Carbon footprint of a given activity

$$= \text{activity data (mass/volume/kW h/km)} \times \text{emission factor } (CO_2e \text{ per unit})$$

Figure 13.3 shows the tonnes of CO_2e emissions for 1000 m of 700 mm diameter ductile iron pipeline laid 2 m below ground, broken down to reveal the contribution from each life-cycle stage, and Figure 13.4 provides an example of how cumulative emissions are calculated for a pipeline and pumping station over a 40-year asset life.

To demonstrate the importance of the contribution of infrastructure assets to global warming it is worth briefly reviewing the results of a recent study (Griffiths-Sattenspiel and Wilson, 2009) which reported on the energy and carbon emissions embedded in the US's water supplies. It estimated that US water-related energy use is at least 521 million MW h/year – equivalent to 13% of the nation's electricity consumption – and that the carbon footprint currently associated with moving, treating and heating water in the US is at least 290 million tonne/year. The CO_2 embedded in the nation's water represents 5% of all US carbon emissions, and is equivalent to the emissions of over 62 coal-fired power plants.

13.3.3 Key issues
Should the carbon footprint reflect all life-cycle impacts of goods and services used in a project or activity? If the answer to this is 'yes', then where should the boundary be drawn? Some argue the measure should include all emissions, including those that do not stem from fossil fuels. Key questions to ask are the following.

- Have all raw materials been traced back to origin?
- Were any by-products created during manufacturing?
- Have all waste streams and emissions been accounted for?
- Has the transport of waste been accounted for?
- Have multiple distribution stages been accounted for, including all transport links and storage conditions?
- Is energy consumed during the use phase?

Figure 13.3 Embodied carbon in a buried pipeline and pumping station. (© MWH based on UKWIR (2012) guidelines worked example)

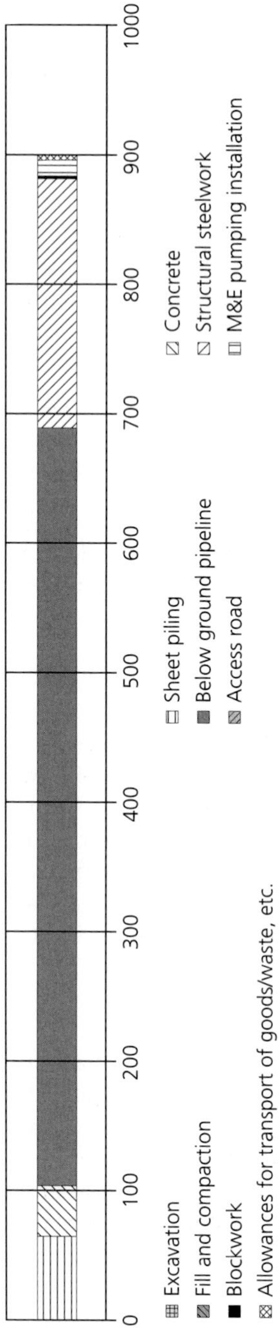
Carbon emissions (tonne CO_2e)
0
100
200
300
400
500
600
700
800
900
1000
Excavation
Fill and compaction
Blockwork
Allowances for transport of goods/waste, etc.
Sheet piling
Below ground pipeline
Access road
Concrete
Structural steelwork
M&E pumping installation

Figure 13.4 Cumulative total emissions for a pipeline and pumping station. (Source: MWH based on UKWIR (2012) guidelines worked example)

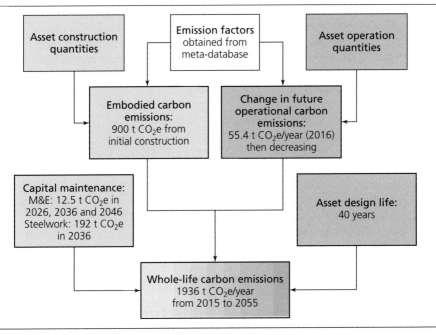

13.3.4 Criticisms

There are differing methodologies in existence for calculating carbon, and many consultancies now offer carbon footprinting services. This has led to problems of standardisation, and so any results considered need to be interpreted against the particular methodology that was used to calculate them. Results can sometimes be difficult to compare, and need to be approached with some caution unless the calculation procedures behind the outputs are fully understood. Definitions range from restricting footprints to direct CO_2 emissions, through to full life-cycle greenhouse gas emissions measured in kg CO_2e, (including methane and nitrogen oxides), and the units of measurement are sometimes not clear.

By focusing on carbon footprints and embodied energy derived from the use of fossil fuels other causes of climate change from non-fuel-related greenhouse gases such as methane may not be considered. Methane from animals and release from ground sources is 22 times more powerful a greenhouse gas than carbon dioxide. It is also sometimes difficult to attribute emission to a single activity. For example, is the carbon used in travelling by a commuter attributable to his or her individual lifestyle or to the productivity of their workplace?

The focus on carbon can dominate policy decisions at the expense of proper consideration of other parameters such as biodiversity, social fairness and poverty reduction, although authors such as Nicholas Stern (2010) argue strongly that these issues are critically interlinked. Some experts have claimed that inaccurate calculation of greenhouse gas emissions, in particular carbon dioxide, using so-called 'carbon footprint calculators' is undermining the method of using carbon offsetting to combat climate change. As a result, critics of the system say that those who rely on them may be deceiving themselves that their resulting offset purchases are having their intended effect (Nash, 2008). Nevertheless, reducing carbon footprints can save money by lowering operating costs and can be a motivating factor for employees, as well as meeting the environmental demands of investors.

13.3.5 Sources of further information

BSI (2011) PAS 2050:2011 *Specification for the Assessment of the Life Cycle Greenhouse Gas Emissions of Goods and Services*. Carbon Trust/Defra/BSI, London.

Franchetti MJ and Apul D (2012) *Carbon Footprint Analysis: Concepts, Methods, Implementation, and Case Studies*. CRC Press, Boca Raton, FL.

Hoffman J (2007) *Carbon Strategies: How Leading Companies Are Reducing Their Climate Change Footprint*. University of Michigan Press, Ann Arbor, MI.

The Carbon Trust (2013) Available at: http://www.carbontrust.com (accessed 20 August 2013).

UKWIR (UK Water Industry Research) (2010) *Workbook for Estimating Operational GHG Emissions*, Version 4. Ref. 10/CL/01/12. UKWIR, London.

13.4. Environmental management systems
13.4.1 Summary

The UK Department for the Environment, Food, and Rural Affairs (Defra) states:

> a formal EMS can provide organisations with a practical tool to help them understand and describe their organisations impacts on the environment, manage these in a credible way and evaluate and improve their performance in a verifiable way. [And it can be used] to help drive performance through the supply chain, and support and encourage suppliers to attain more transparent and higher levels of financial, environmental and sustainability performance.
>
> (Defra, 2008)

In other words, an environmental management system (EMS) can provide, for an infrastructure organisation and its supply chain

- a systematic way of managing their impact on the environment
- a framework for continual improvements of their systems
- encouragement to all to acknowledge and address environmental issues.

Effectively applied, an EMS can help integrate environmental considerations with overall operations and help ensure regulatory compliance. By exposing inefficiencies and waste an EMS can help effectively reduce the use of resources, and so reduce costs. There are three national or international standards that an organisation can adopt.

- ISO 14001 is the international standard for environmental management systems, and provides systematic guidance on how to identify, evaluate, manage and improve the environmental impacts of an organisation's activities, products and services.
- EMAS (the EU Eco Management and Audit Scheme) is a voluntary EU-wide environmental registration scheme which requires organisations to produce a public statement about their performance against targets and objectives (and incorporates ISO 14001).
- BS 8555 is a British Standard which breaks down the preceding standards into implementation stages that are especially relevant for small companies.

13.4.2 Application

To set up an EMS, start with a baseline assessment of where the organisation or project currently stands in terms of environmental management. The following should be compiled.

- A site plan.
- An environmental history of the business (from any environmental records that may be available).
- The environmental impacts of products and services that the company or business is responsible for – both good and bad.
- Current risks.

Examine all processes and procedures carefully, for their strengths, weaknesses, opportunities and threats posed to good environmental management. Once environmental impacts have been identified, assess their significance, to develop ways in which they can be managed. The process is shown in Figure 13.5.

Figure 13.5 The environmental management system (EMS) process

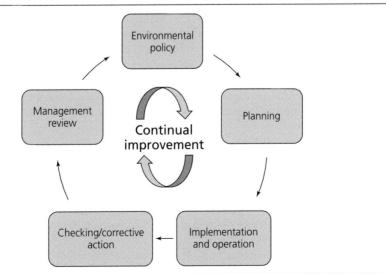

Two international studies throw light on the various motives for adopting an EMS.

- Internal motivations such as employee commitment and environmental concern produced better overall performance than external ones, such as in response to regulatory pressures (Darnall *et al.*, 2008).
- Motives ranged from 'better compliance and improved performance', to 'indicating good environmental practices to competitors and regulators'. The relative importance of these drivers was dependent on the size of the manufacturing facility (Johnstone and Labonne, 2009).

A UK study (Griffith and Bhutto, 2008) found that a number of prominent UK principal contractors have introduced effective EMSs – integrated standards-based systems for managing a construction project's environment, quality, and safety. However, they also found problems, including:

a lack of management system awareness, passive environmental standpoints, litigious project-participant relationships, and cost driven, rather than environmentally empathetic, cultures.

13.4.3 Key issues
Senior management needs to be committed to the development, implementation and maintenance of the EMS. If implementation is left to the specialist coordinator or champion alone it stands a good chance of being ineffective. Also, systems can be made too complicated. If they are confusing and difficult to manage, the links between the standard and how the organisation demonstrates its compliance can be hazy. Pragmatic decisions need to be taken on the detail to be included; not every single step in a process or procedure needs to be catalogued, only those that interact directly with the environment. Including as many people as possible in scoping the EMS is a useful way of being both comprehensive and relevant.

Adequate human, financial and physical resources need to be available, and key responsibilities need to be set. This requires allowing staff time to integrate maintaining the system into their normal routines. Setting measurable targets makes it clear when objectives are achieved, and makes performance monitoring easier. Finally, an EMS requires a process of continual improvement and redevelopment as priorities change, new information becomes available, and more stringent targets are adopted.

13.4.4 Criticisms and drawbacks
An EMS does not address the social aspects of an operation. Measuring change in social interactions is not easily achieved and often not well recognised or rewarded by outside stakeholders (Carruthers and Vanclay, 2007). Beyond widening the scope of an EMS in this way, many organisations lack adequate environmental staff to competently develop a rigorous EMS. Developing one will also cost time and money.

Compared with the manufacturing industry, the multiple stages of project delivery and the presence of a multilayered subcontracting system in construction – with many individual relationships and responsibilities – makes the application of an EMS more complex and harder to do (Tse, 2001). For example, the architect and the builder are responsible for separate activities and have different contractual relationships with the clients.

Also, much documentation, checking, control forms and other paper work are needed to meet the requirement of, for example, ISO 14001. This may be difficult for organisations with limited resources. Nevertheless, for infrastructure, a well-designed EMA it is a good tool to implement, at the construction and in-use stages, the environmental policies, objectives and targets set earlier in the delivery process. It should contain predetermined indicators, measurable goals and a means of determining if the performance level has been reached.

13.4.5 Sources of further information

Griffith A (2010) *Integrated Management Systems for Construction: Quality, Environment and Safety*. Prentice Hall, Englewood Cliffs, NJ.

National Construction College (2008) Environmental Management Systems at Site Level One Day Course: NCC010. Construction Skills.

Sheldon C and IEMA (2007) BS 8555: 2003 Environmental management systems. Guide to the phased implementation of an environmental management system including the use of environmental performance evaluation. British Standards Institution, London.

Sheldon C and Yoxon M (2006) *Environmental Management Systems: A Step by Step Guide to Implementation and Maintenance*, 3rd edn. Earthscan, London.

13.5. Building rating systems and civil engineering awards schemes

13.5.1 Summary

Building rating systems (BRSs) are holistic, multi-dimensional, criteria-based assessment tools, often third-party verified, and tied to a green building certification scheme. They emerged in the early 1990s with the introduction of BREEAM in the UK, LEED (Leadership in Energy and Environmental Design) in the USA, and more recently the International Green Building Tool (Bernier *et al.*, 2010). Often employing multi-criteria decision-making methods to quantify value judgements, BRSs score different project alternatives, and facilitate the selection of a preferred way forward (Kiker *et al.*, 2005).

BRSs have many benefits, including 'providing compelling incentive to order parties, owners, architects, and users of buildings to promote development and diffusion of sustainable construction practices' (World Green Building Council, 2009). Their use has raised the profile of green buildings internationally, encouraged building owners and the construction industry to strive for higher levels of sustainability, and, with growing performance data, facilitated the adoption of green building requirements into leading regulations and planning mechanisms. Experience from the application of such systems can inform policy decisions aimed at market transformation in buildings, including adoption of minimum efficiency standards, creation of labels and quality marks, and selection of stretch incentives for leading parties.

Most building environmental assessment tools are based on some form of life-cycle assessment database. They fall into two categories. The first are assessment tools that provide quantitative performance indicators for design alternatives. Alternatively, rating tools determine the performance level of a building using a classification system (based on stars or descriptors) (Ding, 2008). The various tools can be divided as follows.

- *Knowledge-based tools* comprise manuals and information sources that serve as reference materials for designers.
- *Performance-based tools* use life-cycle impact assessment and simulation tools to calculate aspects such as energy consumption, lighting and indoor air quality.
- *Building rating tools* are design checklists and credit rating calculators developed to assist designers in identifying design criteria and documenting proposed design performance.

Five BRSs are compared in Table 13.2.

The use of BRSs should encourage greater dialogue and teamwork within project teams, and provide detailed performance summaries to communicate design points to stakeholders and investors. By declaring a set of environmental issues and assigning priorities to them, they effectively define an 'industry standard' of what constitutes a green building, while taking into account practicality and cost concerns. Finally, the use of an established methodology fosters innovation within the manufacturing and supply sectors, encouraging the creation of new environmentally friendly products.

Related to these tools are evidence-based award schemes such as CEEQUAL, launched in the UK in 2003, to drive high environmental and social performance in civil engineering, and infrastructure projects. The scheme rewards project and contract teams in which clients, designers and contractors go beyond the legal, environmental and social minima to achieve distinctive environmental and social performance in their work. In addition to its use as a rating system to assess performance, it also is designed to significantly influence project teams, because it encourages them to consider the issues in the question set *at the most appropriate time*. The CEEQUAL scheme is available in three forms: for the UK and Ireland, for international projects, and for term contracts. Using a copy of the applicable CEEQUAL manual, and an Online Assessment Tool for capturing scores and evidence, an assessor assesses and records the scores for each question relevant to the project/contract, together with details of evidence justifying the score. After ratification, award certificates are presented to all project partners at suitable events. While CEEQUAL assesses infrastructure projects across the whole sector, it does not specifically take into account the life-cycle assessment of the project and, in particular, the materials used (Ghumra *et al.*, 2009).

13.5.2 Application
BRSs are applied in three distinct stages.

- *Classification.* The various inputs (e.g. resources and raw materials used) and outputs (e.g. waste produced) are assigned to different impact categories (e.g. global warming or non-renewable resource depletion) based on their expected type of impact on the environment.
- *Characterisation.* Relative contributions of each input and output to its assigned impact category are assessed and the contributions are aggregated within the impact categories.
- *Valuation.* Also known as 'weighting', the seriousness of each category is assigned a value in relation to the other categories.

There are several benefits to employing a BRS. For instance, clients may request their use as they provide verification of meeting an accepted market standard for a green building, as well as acting as an auditing body for the design team. Designers and owners alike can translate a successful certification into increased bottom-line returns, raising the value of the structure while creating a market niche for the construction team. Setting targets for specific certification levels can, however, lead to 'points hunting' – a phenomenon where significant amounts of money and resources are expended to find ways to achieve various credits and ensure the design team's interpretations of any specific credit intent. Although this circumstance can be detrimental to the project, it indicates that the design team is fully exploring alternatives and searching for truly environmentally protective building options.

Building rating tools and assessment methods are most useful during the design stage, when measures can be incorporated to minimise environmental damages. A more effective way of achieving sustainability in a project is to consider and to incorporate environmental issues at a stage even before design is conceptualised (Ding, 2008), as we have argued in Part II of this book. However, BRSs are less useful for selecting the optimum project options because they are used to evaluate building design against a set of pre-designed environmental criteria. Environmental issues are only considered at the design stage of projects, while different development options or locations of development are decided at the feasibility stage.

13.5.3 Key issues

Some assessment tools such as BREEAM (UK), BEPAC (Canada), LEED (USA) and HK-BEAM (Hong Kong) do not include financial aspects in the evaluation framework and therefore do not reflect the fundamental principle of all projects – to generate a financial return. So, while a project may be very environmentally friendly, it may be very expensive to build. This is why the Green Building Challenge model includes economic issues in the evaluations.

Most building rating systems have been developed for local use and do not allow for national or regional variations. Therefore, for example, the points achieved for water-saving devices can remain the same whether the area in question is water rich or water scarce. However, regional and social and cultural variations are complex and the boundaries difficult to define (Ding, 2008). These variations include differences in climatic conditions, income level, building materials and techniques, building stocks and appreciation of historic value (Kohler, 1999).

Table 13.2 Comparison of five building rating systems*

	BREEAM	CASBEE	GBTool	Green Globes US	LEED
Applicability					
New construction	✓	✓	✓	✓	✓
Major renovations	✓	✓	✓	Under development	✓
Tenant build out				Under development	✓
Operation and maintenance	✓	✓	Under development	Under development	✓
System age	1990	2001	1996	2004	1998
Technical content					
Optimise site potential	15%	15%	15%	11.5%	15%
Optimise energy use	25%	20%	25%	36%	15%
Protect and conserve water	5%	2%		10%	
Use environmentally preferable products	10%	13%		10%	
Enhance indoor environmental quality	15%	20%	15%	20%	12.5%
Optimise operations and maintenance practices	15%	15%	15%		
Other	15%	15%	30%	12.5%	7%
Verification					
Level of detail of check	Detailed assessment of documentary evidence	Depends on the assessment tools used Document review is required	n/a	Review of documentation and site inspection	Administrative and credit audit

Third party	✓	✓		✓	✓
Assessor qualifications	Trained and licensed by BRE	Trained and must pass an assessor examination Must be a first-class architect to qualify	n/a	Under development	Trained and must pass an assessor examination
Communicability					
Results	Pass, Good, Very Good, Excellent, Outstanding	'Spider web' diagram, histograms and BEE graph	Range of detailed and broad histograms	One to four globes (1 = 35–54%, 2 = 55–69%, 3 = 70–84%, 4 = +85%)	Certified (40%), Silver (50%), Gold (60%), Platinum (80%)
Product	Certificate	Certificate and website published results	n/a	Plaque, report and case study	Award letter, certificate and plaque

* Adapted from Pacific NorthWest National Laboratory (2006)

The more comprehensive the method used the more detailed is the information that needs to be assembled and analysed. A balance needs to be struck between completeness of coverage and simplicity of use, and this remains one of the challenges in developing effective and efficient BRSs. Assessment includes quantitative data (such as water consumption, annual energy use, greenhouse gas emissions) and qualitative data (such as impact on ecological value), which are difficult to evaluate using market-based approaches. Some environmental issues can, therefore, only be evaluated based on points being awarded for the presence or absence of desirable features.

Most methods use their own points scoring system without reference to a single consistent measurement scale, which makes comparing assessment results of similar building types across countries difficult. For a more detailed discussion of issues involved in preparing a building assessment see Fenner and Rice (2008).

13.5.4 Criticisms and drawbacks

Some consider that BRSs have not fostered the creation of sustainable buildings, as they currently only seek to minimise 'un-sustainability'. Such systems explicitly seek to minimise environmental impacts, but can fail adequately to take into account social and economic issues. Other critics also point to the following shortcomings.

- *They are not universally applicable.* Currently, assessment is only being encouraged in the narrow sector of stand-alone building construction. Assessment methods will need to be implemented in a much wider array of infrastructure categories to foster true environmental protection. CEEQUAL is one response to this, for civil engineering projects.
- *They require constant updating.* A rigorous revision schedule is necessary to maintain accuracy of the assessment, as well as maintain the potency and attraction of the certification.
- *Effective application requires an integrated approach.* An integrated design strategy greatly benefits the application of a rating scheme; current schemes do little to foster this type of approach.
- *Environmental impact projections are based on assumptions.* Assessment schemes rely on designers to estimate the amount of energy/resources/carbon consumed by building users. These estimations largely ignore behavioural issues, which can greatly affect a building's overall performance.
- *Buildings can have many lives with different uses.* As the operational life of a building is typically far greater than that of its occupants, a building will have several 'service lives' during its 'design life'. Current rating schemes usually examine only the building as it is first commissioned. A BREEAM or LEED 'in-use' option (see Box 10.1) addresses this concern.

Another criticism is of forcing unrelated system elements and different types of criteria into a single rating 'answer', using questionable weighting systems. This unduly simplifies a complex situation, and may add little value. Advice on this issue is given, for all decision-making, in Section 7.6, and those comments apply to BRSs too. One approach is to change weights based on regional differences.

In CASBEE (Comprehensive Assessment System for Building Environmental Efficiency) the weighting coefficients are determined by questionnaire survey to obtain opinions from the designers, building owners and operators, and other stakeholders. At present there is neither a consensus-based approach nor a satisfactory method to guide weightings (Ding, 2008). In response to this important challenge, a growing number of BRSs are modified on a project-by-project basis, to reflect the objectives and circumstances of the development.

The tendency to 'points chase' is addressed by some systems through mandatory compliance issues or minimum scoring thresholds, but underlying problems remain. The practice of reducing a complex, multi-dimensional problem to a single rating presents many issues, including a loss of visibility of underlying drivers.

However, with their evolution, BRSs have increasingly addressed the criticisms against them and generally provide a logical, structured, decision-support tool for complex problems. On the whole, applied with care, they can provide significant value in driving sustainable improvements to the construction industry and the built environment.

13.5.5   Sources of further information
Barlow S (2011) *Guide to BREEAM*. RIBA, London.
Cinquemani V and Prior J (2011) *Integrating BREEAM Throughout the Design Process: A Guide to Achieving Higher BREEAM and Code for Sustainable Homes Ratings*. FB 28. IHS BRE Press, London.
Kubba S (2012) *Handbook of Green Building Design and Construction: LEED, BREEAM, and Green Globes: LEEDS, BREEAM, and Green Globes*. Butterworth-Heinemann, London.
Reeder L (2010) *Guide to Green Building Rating Systems: Understanding LEED, Green Globes, Energy Star, the National Green Building Standard, and More*. John Wiley, New York.

13.6.   Ecosystems services and green infrastructure valuation
13.6.1   Summary
Recognising the additional services that can be provided by some types of infrastructure, by valuing the wider non-use services they provide, can change the balance in asset investment decision-making. Similarly, understanding the value of natural environments which may be lost as a result of constructing new infrastructure can also radically change how such projects are viewed. Therefore, the case for assessing ecosystem services, and then valuing them, is that it will contribute towards better decision-making on infrastructure projects, by ensuring that policy and project appraisals fully take into account the costs and benefits to the natural environment. Otherwise, projects may actually generate wider benefits, and/or cause environmental losses, that go unmeasured. Taking both into account can radically affect the project 'balance sheet' and change decision-making.

Ecosystems such as wetlands, forests or estuaries can be characterised by the physical, chemical and biological processes that occur within them. These 'ecosystem functions' can provide services such as provision of wildlife habitat, carbon sequestration and water and nutrient recycling, as well as maintaining wilderness qualities and natural landscape features, which have high amenity value and are valued in many recreational pursuits. These can be described as either 'provisioning services', which produce products such as food, materials, biochemicals and genetic resources from ecosystems, or 'regulating services', such as air quality maintenance, climate and water regulation, erosion control and protections from natural hazards such as storms and floods, as well as removal of pollutants from waste. There are also 'cultural services', which reflect aesthetic pleasure and heritage value in important landscapes of outstanding natural beauty, as well as 'support services', which have no direct on people but which are necessary maintaining healthy ecosystems such as soil formation and retention, nutrient cycling, primary biomass production, production of atmospheric oxygen and provision of habitat.

Many of these services are 'public goods' or 'global commons' (Global Commons Trust, 2013), and they have often been considered 'externalities', outside project costs: as free resources to exploit or damage without consequence. So at the heart of this difficulty is a failure to recognise and credit the economic value of an ecosystem.

The policy and project appraisal system that can take account of such externalities is cost–benefit analysis. However, the analysis is usually too narrowly framed and fails to include ecosystem services and other externalities because hitherto it has been felt to be difficult to assess them in monetary terms. In economics this can be considered a market failure and has led to calls, for example, to include a carbon price in all cost appraisal calculations (Stern, 2006).

It is an advance to be able now to include ecosystem services, but we note here that cost–benefit analysis itself has a number of key issues and criticisms, which are discussed later in this chapter.

The total economic value of an ecosystem can be broken down into several components, as shown in Figure 13.6.

Figure 13.6 A total economic value framework. (Source: Defra (2007). © Crown Copyright 2013, reproduced courtesy of Defra)

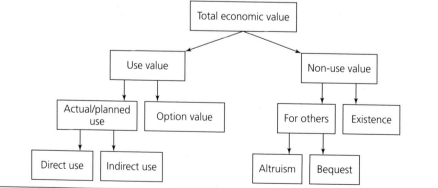

The *use* value is derived from either the *actual* use of a good or service (such as hunting, fishing, bird watching or hiking, either directly by visiting an area, or indirectly by experiencing it through film or television) or the *option* value that people place on the ability to enjoy something in the future. *Non-use* values may simply reflect the knowledge that future generations will have the option to derive benefits (*bequest* value) or simply that people place value on something just by knowing it exists, even though they will never see or use it (*existence* value).

For example, guidance in Defra's *An Introductory Guide to Valuing Ecosystem Services* (Defra, 2007) summarises the key steps required, and gives advice on the different valuation methods available.

- Establish an environmental baseline.
- From this, do a qualitative assessment of the potential impacts of the proposal on ecosystems services.
- Quantify these on specific services, assessing the effects on human welfare.
- Then, you can compute the economic value in the changes of these services.

Many techniques exist to determine these economic 'externality' values. These are summarised in Figure 13.7. Revealed preference methods rely on how individuals choose goods with environmental attributes and are based on actual market transactions. Stated preference methods use carefully structured questionnaires to elicit individual preferences for a given change in a natural resource or environmental attribute. A brief overview of each technique is given in Table 13.3.

Benefits transfer is a process where the economic values that have been generated for one site are applied to another context that has similarities. Willingness-to-pay information is often estimated from a meta-analysis of previous studies, and this can be based on a database of more than 1900 valuation studies (10% of which are from the UK) provided by the Environmental Reference Inventory (EVRI, 2013).

Another useful tool is 'The Green Infrastructure Valuation Toolkit', which provides a detailed methodology on how to derive the economic value of green infrastructure. An example of the steps this involves for valuing an urban tree is given in Table 13.4 (Green Infrastructure Valuation Network, 2010). It involves describing the asset, identifying its function, calculating its benefit and allocating a value. The method assesses green infrastructure across 11 categories, shown in Table 13.5, which indicates where the most benefit of some typical infrastructure features may occur.

13.6.2 Application
Some examples demonstrate the decision-making impact of the approach. Contingent valuation was used to study how much water to allocate Los Angeles from Mono Lake, as reduced flows would affect food supplies for nesting and migratory birds. The study found that Californians were prepared to pay an average of $13 a month for extra water. This priced the total environmental benefits from the lake as 50 times more that the $26 million cost of finding the water from an alternative source; it

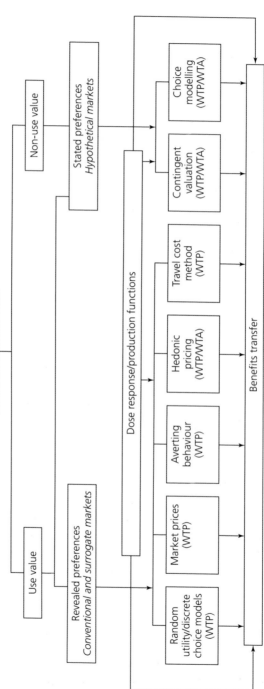

Figure 13.7 Techniques for monetary evaluation. (Source: Defra (2007) © Crown Copyright 2013, reproduced courtesy of Defra)

Table 13.3 Methods for estimating the values of natural capital and ecosystem services

Method	Overview
Market price method	Many ecosystem goods or services are bought and sold in commercial markets. Although externalities exist, market prices provide a starting point in estimating the value of related natural capital and ecosystem services
Productivity method	Economic values may be estimated for intermediate ecosystem goods or services that contribute to the production of commercially marketed final goods
Hedonic pricing method	Economic values may be estimated for ecosystem goods or services that directly affect prices of some other marketed good or service, often local property prices
Travel cost method	Based on the assumption that the value of a recreational site is reflected in how much people are willing to pay to visit the site, economic values associated with ecosystems or parcels of land that are used for recreation
Damage cost avoided, replacement cost and substitute cost methods	When an ecosystem is protected from economic or other disruptive activities, damage costs are avoided, as are the costs of replacing ecosystem goods and services or of providing substitute goods and services
Contingent valuation method	Economic values for virtually anything may be estimated contingent upon certain hypothetical scenarios. For example, people may be asked how much they are 'willing to pay' for the protection of an ecosystem or certain of its natural capital stocks or ecosystem services
Benefit transfer method	Estimates economic values by 'transferring' (or extrapolating) existing estimates obtained from studies already completed in other areas

howed that the population cared about fish and birds, as well as cheap water. As a result, the California Water Resources Control Board did reduce Los Angeles water rights from the lake by half.

The travel cost approach was applied to the Hells Canyon and Snake River environmentally sensitive area on the Oregon–Idaho border in the US. This was the site for a proposed dam and hydropower scheme; the net economic value of using this site was $80 000 higher than the 'next best' site, which was in a less environmentally sensitive

Table 13.4 Valuing the multiple benefits from green infrastructure assets (Green Infrastructure Valuation Network (2010))

Green infrastructure asset/intervention	→ Function	→ Benefit	→ Value
	Shelter from wind	→ Lower heating costs	→ £ Reduced building heating (M) £ Avoided CO_2 (M)
	Evapo-transpiration	→ Climate change adaptation and mitigation	→ °C reduced temperatures (Qt)
	Carbon sequestration	→ Less CO_2 in atmosphere	→ £ Market value of CO_2 stored (M)
	Particulate filtering	→ Health and well-being	→ £ Reduced pollution control (M) Reduction pulmonary diseases (Ql)

M, in monetary terms; Qt, in quantitative terms; Ql, in qualitative terms

area. A survey showed that the recreational value of Hells Canyon was about $900 000; so, even if this was ten times less, it would be still be greater than the additional economic value from using that dam site. Based on this non-market valuation study, Congress voted to prohibit the development of Hells Canyon.

A study used hedonic pricing, in a rapidly urbanising rural area at Southold, Long Island, to estimate the value of preserving a parcel of open space, by calculating the effect on adjacent property prices. For a 10 hectare plot, surrounded by 15 'average' properties, this was found to be $411 000.

In the UK, a number of studies have calculated the monetary value of the wider benefits from infrastructure projects. An example from the London Borough of Bexley is given in Box 13.2. In Liverpool, an initiative to introduce green roofs and 3300 trees around the University was costed using these techniques. The benefits included climate-change adaptation, flood alleviation, property-value enhancement, a general employment uplift, savings in energy costs and carbon emissions, carbon sequestration and pollution reduction. The total net present value (NPV) of the benefits gained was calculated to be £45.6 million, set against the £29.7 million in costs, showing a net gain of £15.9 million (Green Infrastructure Valuation Network, 2010).

13.6.3 Key issues
The services that ecosystems provide are both wide ranging and critical, we simply would not exist without them. The substitute technologies for these natural services are either prohibitively expensive or non-existent, and our overall understanding of them (the

Table 13.5 Benefits for green infrastructure features (Green Infrastructure Valuation Network (2010))

Green infrastructure features (non-exhaustive list)	Climate change adaptation and mitigation 1	Flood alleviation and management 2	Place and communities 3	Health and well-being 4	Land and property values 5	Investment 6	Labour productivity 7	Tourism 8	Recreation and leisure 9	Biodiversity 10	Land management 11
Allotments	□	□	▒	■	□	□	■	□	□	■	■
Open and running water – canal, river, stream, marsh, wetland, pond	■	■	□	▒	▒	▒	▒	▒	▒	■	□
Riverbank	■	■	■	▒	■	▒	▒	▒	▒	■	■
Path – footpath, cycle path, bridleway	□	□	■	■	▒	□	■	▒	■	□	□
SUDS – swales, ditches, filter drain, infiltration trenches	■	■	□	□	▒	▒	□	□	□	▒	□
Green roofs and walls	■	■	▒	▒	▒	▒	▒	□	□	▒	□
Trees	■	■	■	■	■	■	▒	■	▒	■	▒
Verges and hedges	▒	□	▒	□	▒	▒	▒	□	□	■	□
Woodland	■	■	▒	■	■	▒	▒	▒	▒	■	■
Grassland – meadow, rough, heath	■	■	▒	▒	■	■	■	■	■	■	■
Grassland – lawn	▒	■	■	■	▒	▒	▒	▒	■	□	□
Playing fields	▒	▒	▒	■	▒	▒	▒	□	■	□	▒

■ Shows the biggest impact
▒ Shows that the space or asset will provide some benefit
□ Shows that the benefit is small or not relevant

Box 13.2 London Borough of Bexley redevelopment of 12.5 ha of derelict land, construction of the new link road

Preparation: understanding physical characteristics and beneficiaries

- 15 km of drainage dykes, 156 ha of existing marshland.
- Main direct beneficiaries are local residents.
- The number of recreational users is predicted to be 237 600 (based on a likely ten visits each year, 50% of which are assumed additional to the existing baseline figure).

Assessment: identifying potential benefit areas and applying relevant tools
Marshes are likely to have a positive impact on:

- climate change adaptation
- flood alleviation
- general quality of place.

The green corridor element (paths and cycleways) will have benefits on:

- tourism
- transport
- public health and well-being.

Benefits were deemed to last for ten years.

In each case, the valuations were discounted to give a present value (PV) figure.

Reporting: articulating a strong return-on-investment case
The benefits were calculated as follows:

Climate change adaptation and mitigation
The marshes and other areas of greenspace exhibit a significant urban cooling effect – impact on 2000 to 2500 households within 300–450 m of the marshes.

Water management and flood alleviation
Energy costs and carbon emissions relating to water treatment will be reduced through improvement of the natural drainage system on the marshes. (Value of £0.6 million and £0.3 million, at PV.)

Health and well-being
The calculation of reduction in mortality rates from increased take-up of moderate exercise (walking and cycling) was estimated to be £7.4 million (PV) for walking and £1.5 million (PV) for cycling.

Land and property values
Residential land and property uplift within a 450 m radius was estimated to be £9.5 million (PV).

Investment
For employment, by 2016 the link road might provide an additional 2200 jobs, and the green links 650, but together the increase is predicted to be a net 8700. Adjusted for the relative importance of the green infrastructure, the estimation of site employment capacity and employment based GVA assessment was £31 million (PV).

Labour productivity
Reduced absenteeism was calculated to be worth between £0.1 million and £0.5 million (PV).

Recreation and leisure
Based on a 'willingness to pay' measure, the benefits were estimated to be £1.64 million (PV).

Biodiversity
Enhancement will bring increased qualitative biodiversity benefit, especially the promotion of rare and specially protected species such as the water vole.

Land management
Direct management of the land was estimated to generate employment for three people, calculated at a benefit value of £0.6 million (PV).

Benefit value £55.8 million; cost £10.5 million; gain £45.3 million.

contributions of individual species, threshold effects, synergies, etc.) is poor. Moreover, even taking into account the difficulty of valuation, ecosystem services have extraordinarily high value. A 1997 study reported in the journal *Nature* (Costanza *et al.*, 1997) estimated their aggregate value as being $16–54 trillion per year (global GNP is around $18 trillion). Reflect for a moment on the 'paradox of value' (Simpson, 2011): Why is water, which is so essential to life, so cheap, while diamonds, which have such limited, and largely ornamental uses, so expensive? The answer is that water is (generally) abundant with respect to the uses to which it is put, while diamonds are (generally) scarce relative to the demand for them. The exceptions prove the rule: someone dying of thirst – experiencing an extreme scarcity of water – would surely trade all the diamonds he had for a drink. Now consider a corollary, a 'paradox of valuation'. If something is more or less valuable depending on whether it is more or less scarce, then the only way to place a value on a good that does not have an evident market price is to identify some circumstances under which it is more scarce, and others under which it is less so. Now this will be relatively easy to do if the public good whose value we are trying to establish is relatively local; that is, it has a discernible effect on those who are close enough to its source to enjoy its benefits, but others who are more distant receive only negligible benefits. If the benefits afforded by such a public good are localised, however, it begs the question as to whether the good in question ought to be considered 'public' at all. Moreover, even if it were possible to identify *some* variation in the intensity with which ecosystem services are provided across the landscape by different sources, it becomes increasingly difficult to disentangle the effect of any particular wetland, forest, meadow, etc., from that of the multitude of others that would influence production if the benefits of each do not dissipate relatively quickly in the distance from source to receptor areas.

Evaluating ecosystem services, and using a cost–benefit analysis in general, have the great attraction of giving you a single number that tells you 'the answer', and it is in terms of money, which is a very important choice criterion.

There is broad agreement that is it is both right and useful to *quantify* project impacts on ecosystem services, such as 'tonnes of CO_2e emissions', or 'hectares of lost biodiversity' (and also on social impacts and benefits, such as 'numbers of people displaced by a dam', or 'number of accident fatalities saved by a safer road'). These quantities themselves can then directly influence decisions and reflect the relative scale and importance of the benefits and impacts being attributed to a project. The continuing debate is about whether such 'priceless' items as a stable climate, or 'my favourite view', or a human's attachment to their ancestral village, can or *should* then be *priced*, and included in a cost–benefit analysis calculation.

13.6.4   Criticisms and drawbacks

The philosophical concern gets mixed up with questions about the validity of the whole-life cost accounting process (see Section 13.7) that estimates a net present (net) value (NPV) as the cost–benefit analysis result. The detailed concerns have been well articulated by infrastructure engineers (Jowitt, 2010); three in particular come to a head in pricing and accounting for CO_2e emissions.

■ The highly esoteric *methods* used to calculate it (as described above). There is an active ongoing debate among environmental economists on these issues, and between 2008 and 2011 the UK Government abandoned the use of the 'shadow price of carbon' (SPC) as methodologically inappropriate (DECC, 2009). The methodological approach underlying the updated traded carbon values used for appraisal has been recently revised. A market-based approach using futures prices has now been used for estimating carbon values.

■ The way that discounting the future value of money in NPV calculations produces a result that seems against common sense: once you price it into money terms, each further future year's tonne of CO_2e emissions counts *less* in your NPV, just as its actual global warming impact is becoming *more and more* severe.

■ A concern that the cost–benefit analysis result *alone* should not determine a decision. This is because putting a price on any ecosystem service implicitly assumes that there is no boundary limit (see Section 2.3) that must not be exceeded (that would require a price that is infinite). In practice, for the global scale, irreversible, climate change and ecosystem impacts discussed in Table 2.3 there *are* boundaries. Cost–benefit analysis is fundamentally inconsistent with that. If everything in nature gets a price tag it can be bought, bartered and sold like any other commodity (McCauley, 2006).

Recent UK water sector work on including CO_2e in investment appraisal contains a useful summary of the implications for infrastructure (UKWIR, 2012).

A further criticism involves the key concerns listed above, which need to be acknowledged and taken account of when using the approach. They do apply particularly to ecosystem services that are of global scale and irreversible – all CO_2e emissions, because they cause global warming, and, in some cases, internationally critical biodiversity resources. For these, you should always *set limit targets* (see Section 2.3), not just allocate prices and rely on cost–benefit analysis.

However, for project impacts on many regional or more local ecosystems, such as those studies reported above, the approach is a valuable addition to decision-making, and can be used, with care. Two practical drawbacks must still be taken account of.

- Difficulty in estimating an ecosystem service price. Market prices can only be used where a market exists for specific ecosystem services (usually for products such as timber, fish, etc.). Cost-based approaches can potentially overestimate actual value, while hedonic pricing is very data intensive and limited mainly to services related to property. The travel cost method is based on observed behaviour and is generally limited to recreational benefits and difficulties that arise when trips are made to multiple destinations. Contingent valuation is able to capture both use and non-use values, but is subject to bias in responses, and is resource intensive. Developing experience should gradually improve both the data and the methods available.
- Remember that cost–benefit analysis produces just a net result; it does not recognise that the costs and benefits will likely *be borne by different people*, nor does it identify who these people are. You should always also find out who the affected are, and take their needs into account (see Section 5.3).

13.6.5 Sources of further information

Defra (Department for Environment, Food and Rural Affairs) (2007) *An Introductory Guide to Valuing Ecosystem Services.* PB12852. Available at: https://www.gov.uk/government/publications/an-introductory-guide-to-valuing-ecosystem-services (accessed 20 August 2013).

Grant G (2012) *Ecosystem Services Come to Town: Greening Cities by Working with Nature.* Wiley-Blackwell, London.

Green Infrastructure Valuation Network (2010) *Building Natural Value for Sustainable Economic Development. The Green Infrastructure Valuation Toolkit User Guide.* Available at: http://www.greeninfrastructurenw.co.uk/resources/Green_Infrastructure_Valuation_Toolkit_UserGuide.pdf (accessed 20 August 2013).

Hanley N and Barbier EB (2009) *Pricing Nature: Cost–Benefit Analysis and Environmental Policy.* Edward Elgar, Cheltenham.

Matlock MD and Morgan RA (2011) *Ecological Engineering Design: Restoring and Conserving Ecosystem Services.* John Wiley, New York.

Simpson DR (2011) *The Ecosystem Service Framework – A Critical Approach.* UN EP Ecosystem Services Economics Working Paper Series Paper No. 5.

Wise S, Braden J, Ghalayini D *et al.* (2010) Integrating valuation methods to recognise green infrastructure multiple benefits. In *Low Impact Development: Redefining Water in the City* (Struck S and Lichten K (eds)). American Society of Civil Engineers, Reston, VA.

13.7. Whole-life cost accounting
13.7.1 Summary

Techniques need to be used that look at project costs over the whole of a project's life-time, while at the same time building in the environmental externalities, which would

Figure 13.8 The relationship between whole-life-cost savings and time of implementation (Flanagan *et al.*, 1989)

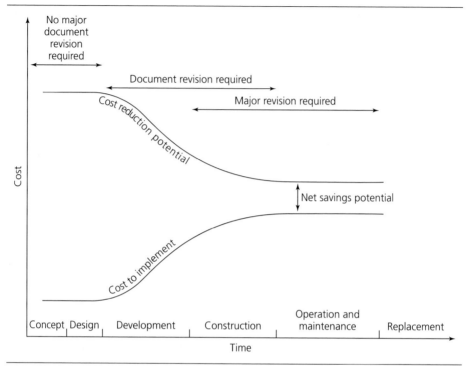

hitherto have not been accounted for in investment decision-making and project appraisal. Section 13.6 showed that such an expanded calculation can change significantly the balance of a decision when viewed simply on cost grounds (see Box 13.2).

Whole-life costing is used to make choices between a range of project alternatives, and as Figure 13.8 demonstrates is used most effectively during early design stages, when the most number of options are open to consideration. This mirrors the argument in this book which suggests that decisions concerning sustainability can create the most scope for innovation at early project stages.

The basic premise of the analysis is that a sum of money in hand today is worth more than the same sum of money at a later date, because of the interest that could be earned in the interim. The time value of money, therefore, must be taken into account. Kishk *et al.* (2003) provide a good summary of the mathematical modelling of whole-life costs; refer to that for details. However, almost all models are based on the concept of net present value, which is the sum of money that must be invested today to meet all future financial requirements, as they arise throughout the life of the project.

Crucially, in the context of sustainability, time, like other goods and services, has its price. The price is incorporated by *discounting* the values of goods and services that lack earliness; that is, those delivered in the more-or-less distant future. This is accomplished by dividing a presumed willingness to pay for a future product by 1 + (interest rate), once for every extra year into the future that its use is delayed. This leads to the familiar *present value formula*:

$$PV = FV/(1 + I)^t$$

where PV is the present (discounted) value, FV is the future value, I is the interest rate, and t is the number of years of delay to consumption. (Rearranged this equation gives the formula for compound interest, that is, it converts the present sum of money into its future values.)

13.7.2   Application

An appraisal of the application of life-cycle costing within the construction industry was carried out recently in the UK. It pointed to a lack of understanding of the technique, and the absence of a standardised methodology, as a key factor limiting wider implementation (Olubodun *et al.*, 2010). More specifically, a whole-life cost approach to sewerage and potable water services has been proposed (Savic *et al.*, 2008). The heart of this approach is gauging the impacts that each unit of investment has on many different levels of service (at different points in time) and assessing the whole-life costs (both capital and operational) of interventions associated with that investment. Whole-life costing has been applied to bridges (Ryall, 2011) and to municipal infrastructure such as a trunk sewer (Table 13.6).

13.7.3   Key issues

In many infrastructure projects, the acquisition and construction costs are the ones that receive the most attention, but they are really just 'the tip of the iceberg' (Figure 13.9). The other 'down the line' costs are also very significant, but are often ignored, as developer-led projects have no real interest in subsequent operational issues or maintenance issues. A figure often quoted is 1 : 5 : 200 as the ratio of capital expenditure to maintenance and building operating costs to business operating costs. Therefore, considering whole-life costs is the most informed way to make decisions, as a key part of sustainability, reflecting all the life-cycle discussions in earlier sections in this chapter. It even earns BREEAM points. A summary of its application to costing sustainable urban drainage systems is given in Box 13.3.

13.7.4   Criticisms and drawbacks

One of the reasons why whole-life costing has not always appeared attractive in the construction industry is that, once a project has been completed, the responsibility for running and operating it is often handed over to someone else. The growing use of new models through private finance initiatives and build, design and operate contracts is beginning to change this. Another difficulty is the uncertainty inherent in forecasting a long way ahead with respect to future maintenance costs, discount and interest rates and changes in anticipated operational use.

Table 13.6 Maintenance/renewal plan for a trunk sewer (from Rahman and Vanier, 2004)

Steps	Examples and detailed activities	A (10 years)	B (20 years)	C	Action
1 Problem statement	Concrete trunk sewer (length 100 m and diameter 600 mm) with breakage and seepage problems				Appropriate inspection and condition evaluation
2 Select analysis period	20, 40, 60, 80 or 100 years				80 years
3 Propose alternatives	(A) Chemical grouting, spot repairs and joint sealing for every 10 years (B) Joint sealing and sliplining for every 20 years (C) Complete replacement of trunk sewer @ 80 years				Verify alternatives
4 Choose economic cost model	PV method, uniform annualised cost method, benefit–cost ratio or rate of return method (discount = 4% and $1 US = $1.30 CDN)				PV method
5 Prepare cost breakdown	Option and renewal cycle	A (10 years)	B (20 years)	C	Prepare detailed cost breakdown for every item
(1) Initial cost	(1) Design*	3100	6000	10 920	
	(1) Construction cost (CC)	52 000	100 000	182 000	
	Contingency cost (8% CC)	4160	8000	14 560	
	Construction cost subtotal	56 160	108 000	196 560	
	Admin. And legal costs (+20%)	11 232	21 600	33 312	
	Total construction costs	67 392	129 600	235 872	
(2) Maintainance and operational cost	(2) Preventive maintenance (1% CC/year)	5200	10 000	45 500†	
	(2) Major repair (50% CC each cycle)	26 000	50 000	–	
	(2) Emergency costs	10 000	10 000	10 000†	
	(2) User costs (social, delay, service, etc.) (10% CC)	2500	10 000	18 200†	
(3) Salvage value	(3) Salvage value (2% CC)	1040	2000	3640	

6 Cost estimate	Initial costs in present dollar	70 512	135 600	246 792	Cost calculation in present value (PV)
	Maintenance cost in PV	80 280	53 182	31 558	
	User cost in PV	10 132	7 597	10 349	
	Salvage value in PV	45	87	158	
7 Determine LCC	Lifecycle cost	$160 880	$196 292	$288 541	
8 Evaluate results	Analytical approaches (sensitivity analysis and Monte Carlo simulation), budget and risk evaluation				Choose lowest LCC
9 Decision (choose preferred action)	Alternative A, B or C				Alternative A

Note: All costs are assumed for model calculation only and based on constant dollars. In some cases, future costs such as preventive maintenance and salvage value are estimated as a percentage of the construction costs whenever this correlation is applicable

* Costs considered every 25 years and † lump sum amount

Box 13.3 Whole-life costs for sustainable urban drainage

The capital costs of sustainable urban drainage systems (SUDSs) are likely to be lower than for conventional systems but, because of their higher operational maintenance requirements an understanding of the whole-life costs of SUDSs is important. The breakdown of costs into financial and economic costs is shown in Figure i.

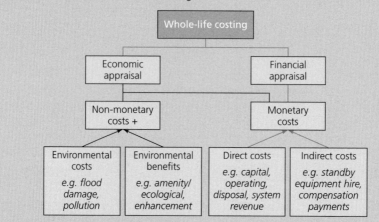

Figure i Two aspects of cost appraisal

It is appropriate to use the present value (PV) for SUDS schemes, defined as the value of a stream of cost or benefits when discounted back to the present time. The PV approach is summarised in Figure ii.

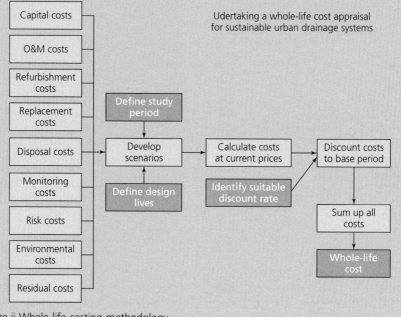

Figure ii Whole-life costing methodology

Box 13.3 Continued

Data required

Design life – minimum length of time for which a scheme is required to perform its intended function.

SUDS capital costs should include

- planning and site investigation costs
- design and project management/site supervision costs
- clearance and land preparation work
- material costs
- construction (labour and equipment) costs
- planting and post-construction landscaping costs
- cost of land-take.

SUDS maintenance costs should include

- monitoring
- regular planned maintenance
- intermittent irregular maintenance
- unplanned maintenance and rehabilitation.

Risk costs – flooding or water quality impacts from CSO spills on conventional systems have been the responsibility of the adopting authorities (water utilities). In SUDS schemes, the risks are likely to be public or societal, and not borne by the SUDS owner or operator.

Environmental costs and benefits – quantify these with monetary values

- amenity and recreation opportunities
- biodiversity and ecological enhancement
- aquifer and base flow augmentation
- water quality improvements
- net flood risk reduction.

Disposal costs – material needing disposal includes

- vegetation (e.g. aquatic planting and grass turfing)
- granular fill
- permeable surface blockwork
- sediment
- geotextiles.

Residual costs – the residual value of the land should be accounted for based on its net present worth following the nominal operational lifetime.

Discount rate and discount period – in the public sector, this is set by the Treasury (which was 2.2% in March 2013 for periods over ten years) which puts a higher weight on future costs, with the aim of encouraging longer term, more sustainable development.

(Source: HR Wallingford (2004). © HR Wallingford Ltd)

Figure 13.9 Acquisition and construction costs are just the tip of an iceberg. (Adapted from Lockie (2008))

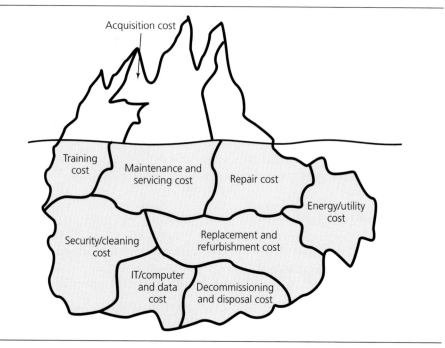

13.7.5 Sources of further information

Capelhorn P (2012) *Whole Life Costing: A New Approach*. Routledge, London.

Ellingham I and Fawcett W (2006) *New Generation Whole Life Costing: Property and Construction Decision Making Under Uncertainty*, 1st edn. Taylor and Francis, Oxford.

Flanagan R, Jewell C and Norman G (2005) *Whole Life Appraisal for Construction*. Blackwell Science, Oxford.

Office of Government and Commerce (2007) *Whole-life Costing and Cost Management: Achieving Excellence in Construction Procurement Guide*. Available at: http://spce.ac.uk/wp-content/uploads/1807.pdf (accessed 20 August 2013).

13.8. Corporate social responsibility and sustainability reporting

13.8.1 Summary

Corporate social responsibility (CSR) sets out that businesses manage the economic, social and environmental impacts of their operations to maximise the benefits and minimise the downsides and should make a positive contribution in the following three areas of influence (Nelson, 2002).

- Core business activities by ensuring responsible implementation.
- Poverty focused social investment and philanthropy programmes.
- Institution building and public policy dialogues by involvement with stakeholders.

The treatment of stakeholders in an ethically and socially responsible manner is at the core of CSR. Stakeholder management has been seen as the tool to connect strategy to social and ethical issues. It is also now emphasised that it is not just about what a company does with its profits, but how it earns them too. This is clear from the Harvey Kennedy School of Government's definition of corporate responsibility (CSR is often now referred to as 'corporate responsibility', as it is about more than just the 'social 'side):

> Corporate Responsibility goes beyond philanthropy and compliance to address the manner in which companies manage their economic, social and environmental impacts and their stakeholder relationships in all their spheres of influence: the workplace, the marketplace, the supply chain, the community and the public policy realm.

As a support to CSR, non-financial reporting on sustainability has expanded over the last 20 years. The Global Reporting Initiative (GRI), set up in 1997 jointly by UNEP and CERES (see Appendix A) produces one of the world's most prevalent standards for sustainability reporting. GRI is a network-based organisation that produces a comprehensive sustainability reporting framework which is widely used around the world, and provides sector guidance for organisations in the construction sector.

13.8.2 Application
Criteria for assessing the relationship that a company has with its internal and external stakeholders are as follows (Labuschagne *et al.*, 2005).

- *Information provision.* The quantity and quality of information shared with stakeholders are measured. Information can either be shared openly with all stakeholders or shared with targeted specific groups.
- *Stakeholder influence.* Stakeholder participation has only really succeeded if the stakeholders' opinion is known throughout the company. To achieve this, structures must be set up to distribute the information. The degree to which the company actually incorporates stakeholder opinions in operational decision-making should therefore be evaluated.

Internal human resources focuses on the company's social responsibility towards its workforce, which includes employment stability, employment practices, health and safety, and capacity development through offering training for its staff.

With regard to the impact of the company on the wider community, the following issues are considered (Labuschagne *et al.*, 2005).

- *Human capital*, which relates to an individual's ability to work in order to generate an income. This encompasses aspects such as health, psychological well-being, education, training skills levels and health.
- *Productive capital* reflects the assets and infrastructure that an individual needs in order to maintain a productive life, and the strain on these assets and services by

business initiatives should be measured. This would include impacts on housing, service infrastructure (water, electricity wastewater), mobility and public transport, and the quantity, quality and burden on transport networks.

■ *Community capital* takes into account the effect of a project on the social and institutional relationships and networks of trust – reciprocity. Issues here relate to aesthetics, noise and odour levels, cultural and social issues, as well as induced or increased crime, economic welfare and impacts on poverty, and social cohesion through demographic changes. In short, this reflects how people feel about their surroundings and way of life, and is the core of social sustainability. Finally, a company performance at a regional or national level can be considered through measuring the contribution that is made to GDP, taxes and trading.

While this definition of CSR does not cover the environment, sustainability reporting assesses the organisation's overall economic, environmental and social dimensions, and communicates its efforts and sustainability progress to its stakeholders (Lozano and Huisingh, 2011). Guidelines used include the ISO 14000 series (especially ISO 14031) and EMAS, the Social Accountability 8000 standard, and the GRI Sustainability Guidelines (GRI, 2006).

13.8.3 Key issues

Sustainability reporting is increasingly seen as a widespread element of an organisation's contribution to sustainability, and the number of companies participating in this activity is increasing, particularly in Europe and Japan. It has been suggested that 'money spent on social responsibility' can conflict with the company goal to maximise profit.

However, not all responsibilities to other stakeholder groups necessarily conflict with shareholder interest, and expenditure on CSR does not have to have a negative effect on the bottom line. In fact, it can have an extremely positive impact. So, the idea that 'ethics is an idea we can't afford' is replaced by 'ethics pays', as regulation and legislation are not always effective or adequate in preventing immoral behaviour and the law often lags behind technological development.

By taking a proactive role to collect, analyse and report those steps taken by the organisation to reduce potential business risk, companies can remain in control of the message they want delivered, as well as promote transparency and accountability. In essence, the advantage to an organisation is in enhanced reputation, which some have described as 'Doing well by doing good'.

13.8.4 Criticisms and drawbacks

Despite an increase in the number of companies producing sustainability reports, it is still insignificant compared with the number of businesses in the world today. Moreover, the quality of the disclosures has yet to translate into meaningful and comprehensive documents, as many reports fall short of GRI guidelines (Lozano and Huisingh, 2011).

Critics of CSR, as well as proponents, debate a number of concerns related to the approach. These include the relationship of CSR to the fundamental purpose and nature of business; and questionable motives for engaging in CSR, including concerns about insincerity and hypocrisy ('greenwashing'). Some organisations that commit to CSR and sustainability continue nevertheless to be involved in poor business practices, and use CSR as a distraction from their core activities.

As this form of reporting is voluntary and lacks formal regulation, the information provided is not standardised, so it is difficult to compare performance, especially across geographic regions.

13.8.5 Sources of further information

Asbury S and Ball R (2009) *Do the Right Thing: The Practical, Jargon-free Guide to Corporate Social Responsibility*. IOSH Services Ltd, Sudbury.

Benn S and Bolton D (2011) *Key Concepts in Corporate Social Responsibility*. Sage, London.

Tapscott D, Eccles RG and Krzus MP (2010) *One Report: Integrated Reporting for a Sustainable Strategy*. John Wiley, New York.

White G (2009) *Sustainability Reporting: Doing Well by Doing Good*. Kendall Hunt, Dubuque, IA.

13.9. Backcasting, forecasting and scenario planning
13.9.1 Summary

Defining a desirable outcome or future is at the heart of the 'backcasting' technique (Dreborg, 1996). This has been described as: 'an attempt to envision an acceptable future system state, taking into account the status of as many important defining constraints and criteria as possible, including the requirement to meet "needs"'. This system state is then used as a reference for tracing pathways back to the present, for placing milestones along those pathways, and for identifying short-term challenges and obstacles that will have to be overcome on the way (Figure 13.10). Backcasting thus provides a way of 'connecting the future to the present' (Weaver *et al.*, 2000).

The backcasting method is advocated by The Natural Step and Forum for the Future as a way of determining a route towards sustainability (Holmberg, 1998). The chief virtue of the technique is that it helps focus on a development trajectory that avoids merely extrapolating forward from within present constraints. In contrast, forecasting can be limited by using models calibrated against past and present circumstances, which do not deal well with likely complex future situations and require many assumptions, rendering predictions often simplistic, reductionist and even absurd.

The Natural Step (n.d.) provides a helpful description of backcasting:

> Strategic sustainable development relies on 'backcasting from *sustainability principles*' – which are based in science, and represent something we can all agree on: if these principles are violated, our global society is un-sustainable. To achieve a sustainable society, we know we have to *not violate* those principles – we don't know exactly what that society will look like, but we can define success on a principle level. In that way, backcasting from principles is more like chess –

Figure 13.10 Forecasting, foresighting and backcasting compared. (Source: Adapted from The Natural Step (n.d.) and Darton (2004))

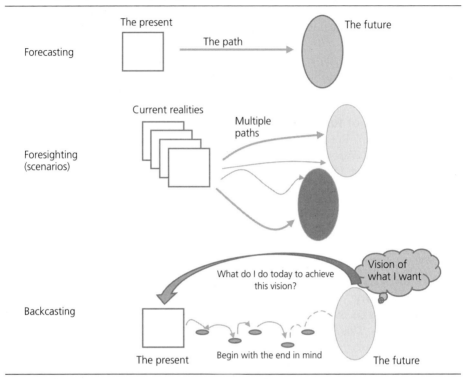

we don't know exactly what success will look like, but we know the principles of checkmate – and we go about playing the game in a strategic ways, always keeping that vision of future success in mind.

The key to successful backcasting is that future scenarios that underpin it are desired outcomes, which have emerged from a collective envisioning process carried out by a diverse range of community and professional stakeholders. Analysing from a long-term perspective has a major advantage: if stakeholders are challenged to reason from a distant future, they are less likely to focus on their own direct interests (Mulder, 2006).

While backcasting comes from the future back to us, forecasting and scenario planning go in the other direction: from here forward. Backcasting is different from scenario planning in that ordinary scenarios are coherent forecasts of several paths into the future, while backcasting is trying to identify paths that finish at a specified end situation.

Forecasting and scenario planning are more familiar techniques to many. Indeed, *forecasting* is rooted in the application of predictive modelling tools, which are developed and calibrated on the basis of objective science, and appear attractive as they can often appear to promise precise answers. The approach can be effective in circumstances where the problem is relatively simple, and the answer is required only over the short term.

Forecasting becomes less reliable over longer time horizons (greater than, say, 10 years), and where systems are complex and subject to unpredictable perturbations (e.g. the structural integrity of tall buildings being intentionally struck by aircraft) and uncertainties of the kind discussed in Chapter 2 arising from incomplete knowledge, inherent unpredictabilities and multiple knowledge frames. One way of addressing these issues is to incorporate a sensitivity analysis in the model forecasting, by varying the ranges of the input parameters in a systematic way. However, the models themselves are often limited by our current understanding of the physical world. Also, while models might calculate the physical magnitude of some future state, they do not interpret the value or significance of the result in any contextual manner. There are clear dangers in merely extrapolating forward while assuming all other constraints and conditions remain the same as in the present. Thus forecasting is only useful where the rules connecting the present and future are not in doubt.

A way of dealing with these limitations is to adopt a *foresighting (scenario planning)* approach, which does not set out to predict the future but seeks to set out a number of different ways in which the future might happen. Typically this might be carried out over time horizons ranging from 20 to 80 years, in which trends in society are examined against a range of possible technological developments. This requires the development of a series of self-consistent and recognisable scenarios, in which separate and distinct views of possible futures are developed, which are relevant to our development decision-making now. They are not extrapolations of present trends, and do not carry any notional likelihood of any particular alternative future occurring.

Scenario development has been described as 'a thinking tool that focuses on the most important and uncertain strategy factors, and challenges us to consider various alternative outcomes' (Darton, 2004). Whereas forecasting uses our analytical capabilities, scenarios demand our creativity and imagination. We cannot know what the future is but we can plan for different eventualities. The idea is that the scenarios are used to challenge the strategy, as we develop it, to see how it performs in various possible future worlds. Good scenarios should go beyond just a straight extrapolation of the present situation, and should be more than just a set of independent assumptions, but should be plausible circumstances far from just business-as-usual descriptions.

13.9.2 Application

A major distinguishing characteristic of backcasting is:

> a concern, not with what futures are likely to happen, but with how desirable futures can be attained. It is thus explicitly normative, involving working backwards from a particular desirable end point to the present in order to determine the physical suitability of that future and what policy measures would be required to reach that point.
>
> (Robinson, 1990)

Backcasting techniques have been widely used for assessing and planning large-scale future infrastructure projects. For example, the Halcrow Group (2008) has used visioning and backcasting for transport in India at both the city scale (Delhi) and at the national level. Others have applied the technique to examine how to decrease energy use in buildings by 50% by 2050 (Svenfelt *et al.*, 2010). An example questioning how hemp might become more used as a building material is given in Box 13.4.

Box 13.4 A backcasting example (from the Transition Network)

If, by 2020, 50% of any new buildings in a community are required to include 50% local materials, backcasting is very useful in identifying what would need to be done and by when, in order for this to be a possibility. If construction-grade hemp is to be a key part of that, backcasting allows the following to be considered.

■ By when would the infrastructure for locally growing hemp need to be in place?
■ At what stage would it be necessary to begin training local builders to help in construction?
■ When would the first trials on farmland need to take place?

(Source: Adapted from Hopkins (2010))

The UK Government has developed a number of 'Foresight' initiatives for such issues as land-use futures (2010), sustainable energy management (2008), intelligent infrastructure systems (2006) and flooding and coastal defence (2004) (Foresight, n.d.). These serve to inform policy, through evidence-based, peer-reviewed, strategic insights. For example, the Foresight project on future flooding examined four scenarios, based on different combinations of socio-economic futures and climate-change scenarios (Foresight, 2004), and concluded either more needed to be invested in sustainable approaches to flood and coastal management, or society would have to learn to live with more increased flooding.

The Engineer of 2020 report (National Academy of Engineering, 2004) used a series of diverse scenarios to test the demands that might be placed on engineering over the next couple of decades. Specific scenarios considered were: (1) The Next Scientific Revolution, (2) The Biotechnology Revolution in a Societal Context, (3) The Natural World Interrupts the Technology Cycle, and (4) Global Conflict or Globalization? (National Academy of Engineering, 2004).

13.9.3 Key issues
Backcasting is well suited to problems having the following features (Dreborg, 1996).

■ When a problem is complex.
■ When there is a need for major change and marginal changes will be insufficient to solve a problem.
■ When dominant trends are part of the problem.
■ When a problem is largely a matter of externalities.
■ When the timescales are long enough to give considerable scope for deliberate choice.

The steps in a backcasting process involve analysing needs, identifying options for improvement, creating a common future vision with stakeholders, developing pathways that could lead to this vision and developing consensus on these pathways (Mulder, 2006).

The important steps in scenario planning are, first, to define the questions and issues to be addressed, and identify the external forces in operation, and the pressures they exert. Scenarios are then developed that reflect alternative views of the future, and the options available are considered within these scenarios. This may lead to the identification of innovations, new services, projects and opportunities. The potential implications and impacts of scenarios on the identified options are discussed, and this can lead to an action plan from which to move forward.

13.9.4   Criticism and drawbacks

One problem with backcasting is that, to be effective in stimulating creativity, it requires the setting of 'stretch goals' to challenge how future solutions can be envisioned. This may create a tendency to stimulate unrealistic flights of fancy, and so needs careful moderation and facilitation between participants.

Similarly, scenario planning has been criticised as being subjective and heuristic, with it not being known whether useful scenarios have been developed. There are few academically validated analyses, as the technique has emerged largely from business practice, with companies such as Shell being in the forefront of its use. While the process is good for reframing perceptions and addressing issues of uncertainty, it is often misused when decision-makers want to adopt one specific scenario, rather than engaging with a series of alternative futures.

A further pitfall is to interpret scenarios too literally, which can restrict the analysis to static options rather than to delineating boundaries for possible futures, which can be adjusted as things unfold and become known over time. In discussing scenarios, the dynamic of the group can introduce problems such as political derailing, agenda control, myopia and limited imagination, and so the approach is subject to organisational limitations.

13.9.5   Sources of further information

Dreborg KH (1996) Essence of backcasting. *Futures* **28(9)**: 813–828.

Lindgren M and Bandhold H (2009) *Scenario Planning – The Link Between Future and Strategy*. Palgrave Macmillan, New York.

Miller FP, Vandome AF and Brewster J (2009) *Futures Techniques: Futurology, Futurist, Delphi Method, Causal layered Analysis, Scenario Planning, Future History, Backcasting, Back-view Mirror Analysis, Cross-impact Analysis, Future Workshop*. VDM, Saarbrücken.

Schwartz P (1998) *The Art of the Long View: Planning for the Future in an Uncertain World*. John Wiley, Chichester.

Van der Heijden K (2004) *Scenarios. The Art of Strategic Conversation*, 2nd edn. John Wiley, Chichester.

13.10. Multi-criteria decision-making

13.10.1 Summary

A key principle, argued throughout this book, is for civil engineers to adopt a wider range of criteria when judging the performance and outcomes of a project, and when making decisions at each stage of the project life cycle. These criteria should reflect a range of economic, social and environmental considerations. However, in moving beyond straightforward engineering and financial variables, the problem arises of how to handle such different and seemingly intractable information, in a consistent, objective and robust manner. If delivering sustainable infrastructure means gathering and considering more information, then how can that extra information be handled effectively?

Multi-criteria decision-analysis (MCDA) can be used to evaluate alternatives based on a large number of attributes that are expressed in incommensurate units. Usually, performance across different criteria is not converted to a single scale such as money, and no single criterion is selected as being most important. In general, MCDA is a technique designed to manage decision processes typically characterised by many assessment criteria, alternatives and actions.

Techniques are either qualitative or quantitative, depending on the information input they are able to handle. Qualitative techniques include the regime method, frequency analysis or the analytical hierarchy process, whereas quantitative techniques include weighted summation and concordance discordance analysis.

13.10.2 Application

For renewable energy planning, where an integrated appraisal of such complex projects is needed, multi-criteria techniques based on outranking methods such as Novel Approach to Imprecise Assessment and Decision Environments (NAIADE) and the regime method have been found to be the most suitable (Polatides *et al.*, 2006). These allow a general ordering of the alternatives, while allowing individual pairs of options to remain un-compared, when there is insufficient to distinguish between them. In contrast, any additive method such as multi-attribute utility theory (MAUT) or the analytical hierarchy process (AHP), which generates a single score for each alternative, requires that all options be directly comparable with each other, even when such comparisons are questionable because of a lack of suitable data. The methods are compared in Table 13.7.

In construction practice, the Electre III method has been used to choose a sustainable demolition waste management strategy for a case study of 25 buildings at a military camp in the city of Lyon, France (Roussat *et al.*, 2009). Nine alternatives for demolition waste management were compared using eight criteria, taking into account energy consumption, depletion of abiotic resources, global warming, dispersion of dangerous substances in the environment, economic activity, employment, and quality of life of the local population. The recommended solution was a selective deconstruction of each building, with local material recovery in road engineering of inert wastes, local energy recovery of wood wastes, and specific treatments for hazardous wastes.

A good example of adopting the AHP technique was in the development of an index for prioritising a range of water management objectives based on the estimation of weights

Table 13.7 Comparison of several MCDA methods*

Method	Important elements	Strengths	Weaknesses
Multi-attribute utility theory (MAUT)	■ Expression of overall performance of an alternative in a single, non-monetary number representing the utility of that alternative ■ Criteria weights are often obtained by directly surveying stakeholders	■ Easier to compare alternatives, the overall scores of which are expressed as single numbers ■ Choice of an alternative can be transparent if the highest scoring alternative is chosen ■ Theoretically sound – based on utilitarian philosophy ■ Many people prefer to express net utility in non-monetary terms	■ Maximisation of utility may not be important to decision-makers ■ Criteria weights obtained through less rigorous stakeholder surveys may not reflect stakeholders' true preferences ■ Rigorous stakeholder preference elicitations are expensive
Analytical hierarchy process (AHP)	■ Criteria weights and scores are based on pairwise comparisons of criteria and alternatives. respectively	■ Surveying pairwise comparisons is easy to implement	■ The weights obtained from pairwise comparisons are strongly criticised as not reflecting people's true preferences ■ Mathematical procedures can yield illogical and inconsistent results. For example, s rankings developed through AHP are sometimes not transitive
Outranking	One option outranks another if: 1 it outperforms the other on enough criteria of sufficient importance (as reflected by the sum of criteria weights), and 2 it is not outperformed by the other in the sense of recording a significantly inferior performance on any one criterion. Allows options to be classified as incomparable	■ Does not require the reduction of all criteria to a single unit ■ Explicit consideration of the possibility that very poor performance on a single criterion may eliminate an alternative from consideration, even if that criterion's performance is compensated for by very good performance on other criteria	■ Does not always take into account whether overperformance on one criterion can make up for underperformance on another ■ The algorithms used in outranking are often relatively complex and not well understood by decision-makers

* Source: Linkov and Steevens (2008). Reproduced with kind permission of Springer Science + Business Media

for a range of criteria that reflected residents' priorities (Cheung and Lee, 2009). Using such an approach, it was possible to draw a wider range of stakeholders into the decision-making process.

13.10.3 Key issues

In many infrastructure situations, goals and preferences are not easily agreed upon. Therefore, outranking methods which can reflect stakeholder preferences are appropriate and can be more useful than other methods. In selecting a suitable MCDA approach, the following should be considered (Kain *et al.*, 2007).

- It should be applicable to the infrastructure problem being considered.
- It should be able to manage a broad array of sustainability criteria.
- It should be able to integrate different types of knowledge.
- It should be able to handle different kinds of data (including both qualitative and quantitative).
- It should allow the broad participation of stakeholders.

A critical issue in MCDA is how different criteria are weighted. Some methods use ordinal weights, where criteria are ordered from the most important to the least important, but no attempt is made to quantify by how much one criterion is more or less important relative to another. Other approaches use a cardinal system, allowing weights to be allocated from insignificant to extreme importance. In many cases the systematic process of considering one criterion at a time can prove more useful than any aggregation of multiple criteria at the end, and as such the approach provides a good way of structuring all the facts of a problem (or a required decision). Simply testing a problem against different criteria can open up creative discussion among participants, and lead to the formulation and agreement around novel, unforeseen and creative solutions.

13.10.4 Criticisms and drawbacks

There are two main drawbacks to the use of these techniques. First, the apparent objectivity can imply an unwarranted degree of precision, especially where a single aggregated score is reported implying mathematical accuracy. Second, the process by which measures and weights are determined can be both difficult conceptually and, where there are several decision-makers, extremely contentious.

The techniques are therefore good for eliminating lower ranked and less favoured options, but should not be relied upon, on their own, to determine a final solution. A range of possible solutions still needs to be interpreted and discussed contextually, and with a wide group of stakeholders, before a final decision is reached. MCDA techniques are therefore best used as screening tools to refine a shortlist of preferred decisions. Hence in many situations a more informal approach is best, where a decision is reached through informed discussion of the relative merits of each option, its attributes and the interpretation of these attributes against jointly agreed criteria.

In addition, some methods are conducted using sophisticated computer software and there is a danger in this that transparency with different stakeholder groups can be lost.

13.10.5 Sources of further information

Erghott M, Figueira JR and Greco S (2010) *Trends in Multiple Criteria Decision Analysis*. Springer, New York.

Journal of Multi-criteria Decision Analysis (n.d.) John Wiley, New York. Edited by Theodor J. Stewart.

Linkov I and Moberg E (2011) *Multi-criteria Decision Analysis: Environmental Applications and Case Studies*. CRC Press, Boca Raton, FL.

Tryantaphyllou E (2010) *Multi-criteria Decision Making Methods: A Comparative Study*. Springer, New York.

13.11. System dynamics
13.11.1 Summary

We conclude this chapter with a brief introduction to systems dynamics and the power of systems thinking. This relates back strongly to where we started in Chapter 1, when we introduced the need to consider thinking about projects over wider horizons. By doing so, we begin to address the challenge embodied in our fourth fundamental principle A4 – Complex systems.

It is commonplace to think of problems in a linear way, in which the outcome of an event or action is assumed not to affect the input. This approach can be traced back to our mechanical image of the world based on linear causality. So, events cause events, which is not very helpful in understanding the causes of underlying performance (Box 13.5).

Box 13.5 Event-oriented thinking

Initial problem: No health improvement in a rural African village.

- Water supplies are contaminated *(the event that is the cause of the problem)*.
- Failure of simple treatment technology *(another problem)*.
- Lack of maintenance of hand pumps *(the cause of the new problem)*.
- Difficulty in getting spare parts *(another problem)*.
- Community cannot afford to replace parts *(the cause of the new problem)*, etc.

There is always another event which caused the one you thought was the cause.

Solution: The solution here is outside this causality chain. Health and hygiene education are needed so that the community can recognise the need for clean water.

13.11.2 Application

Systems dynamics was first developed in the 1950s by Jay Forrester, who believed that the biggest impediment to progress comes not from the engineering side of industrial problems but from the management side. He reasoned that social systems are much harder to understand and control than physical systems, and this is at the heart of our approach to delivering infrastructure, which must operate within a socio-technical system. In the 1960s, using a systems analysis he showed how an urban policy of building low income housing *creates* a poverty trap that contributes to the stagnation of a city, while tearing down low income housing creates jobs and raises living standards. It is this non-intuitive insight that a systems approach can help provide. His work led directly to the WORLD 3 model and the publication of the Club of Rome's Report: *The Limits to Growth* (Meadows *et al.*, 1972). This was a landmark work in understanding the Earth's finite limits by which all engineering activity must be constrained (Principle A1 – Environmental sustainability – within limits).

A *system* is an interdependent group of items forming a unified pattern, or a configuration of ordered relationships. The focus of any analysis is away from studying the isolated objects in the system, towards understanding the relationships *between* objects. The internal structure of the system is often more important than external events in generating a problem. One of the goals of systems dynamics is to expand the boundaries of our mental models, and to lengthen the time horizon we consider so we can see patterns of behaviour created by the underlying feedback structure, not only the most recent events (Sterman, 2002).

Key aspects of applying a systems approach to a problem are as follows.

- Understanding how the behaviour of the system arises from the interaction of its components over time (i.e. dynamic complexity).
- Discovering and representing feedback processes (both positive and negative) that underlie observed patterns of system behaviour.
- Developing stock and flow relationships.
- Recognising delays and understanding their impact.
- Identifying non-linearities.
- Recognising boundaries of mental models and challenging those boundaries.

A key advantage of this approach is the ability to use a wide range of qualitative and quantitative data, which may be connected through a series of feedback loops, which represent a closed path of action and information. Positive feedbacks are self-reinforcing: so a change in A creates a change in the same direction in B, and leads to exponential growth structures. Negative feedbacks are self-balancing: so a change in A produces a change in B *in the opposite direction,* and can generate goal-seeking behaviour around asymptotes, such as temperature regulation in a thermostatic control. An example of a mechanical feedback system is shown in Figure 13.11.

These feedback structures can be mapped onto qualitative 'causal loop diagrams', which effectively represent the system under consideration. The act of producing such a diagram forces a system mindset to be adopted, and the results can be powerful in understanding the wider implications of a set of actions. For example, Figure 13.12 shows how the effectiveness of policies for reducing pollution from vehicle exhausts can be understood.

These diagrams can help establish effective leverage or intervention points, where a small shift in one thing can produce big changes in everything. These points are often found by intuition but they are often pushed in the *wrong* direction (Meadows, 1999; Sterman, 2002).

Figure 13.11 A mechanical feedback loop in a steam governor. (Source: Capra (1997))

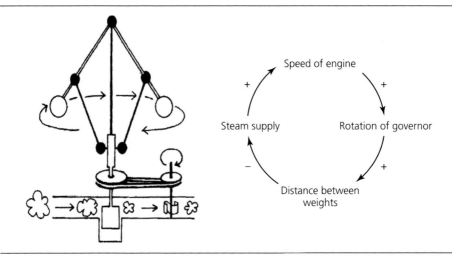

To generate true non-intuitive responses from the system, a quantitative model, based on mathematical formulation, is required, and such models are based on representing the 'stocks' and 'flows' involved. A stock is an accumulation of something (e.g. materials, people, capital, orders, money), while a flow is the movement from one stock to another. An example of this kind of detailed analysis has been carried out to represent the uptake of household water treatment systems in developing countries (Figure 13.13). Once such a model is built and validated, usually as a computer simulation, to deduce complex responses; it can be useful to test the consequences of certain policy interventions or decisions, and is good at answering 'What if we did this?' questions.

13.11.3 Key issues

A critical issue in developing a systems analysis is to decide where to stop – by clearly identifying boundaries to the problem at the outset. These can be determined in relation to the intended purpose for the analysis. For example, if the object is to optimise the allocation of resources across a construction activity, the site boundary may be sufficient. On the other hand, if the supply chain impacts are the concern, then the boundary may be set at the point of extraction of the primary materials. A further consideration is to account for time delays in the system, as these can often be overlooked and can have profound effects. Infrastructure is the underpinning support system for cities, but the urban system will take on different boundaries when tackling crime (as defined across political boundaries) and when tackling water supply (as defined by natural hydrological boundaries, but also legal boundaries in terms of the rights to abstract, and social boundaries in terms of demand and usage patterns and values placed on distant environmental features potentially impacted by expanding water supply schemes).

In dealing with systems, a distinction must be made between *simple* systems, which can be fully described from a single viewpoint perspective (e.g. Newtonian mechanics), *complicated* systems (which are rich in detail) and *complex* systems (which can only be fully

Figure 13.12 Strategies for reducing exhaust pollution from vehicle exhausts. (Source: Di Muro (2006) ©Johnathan Di Muro)

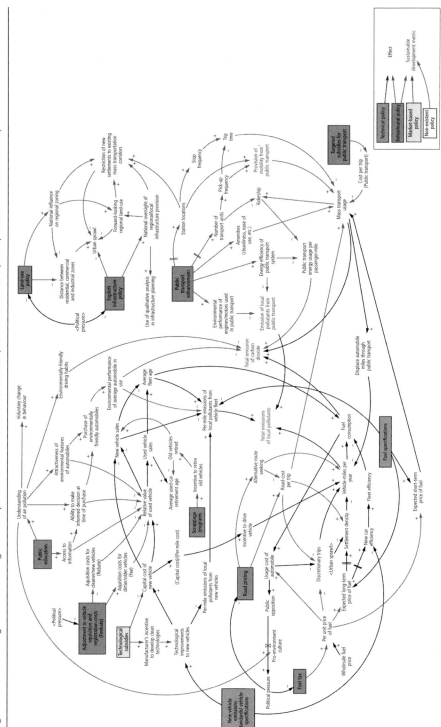

Figure 13.13 Outline structure of a systems dynamics model for the uptake of household water systems in Southern India (Ngai and Fenner, 2008)

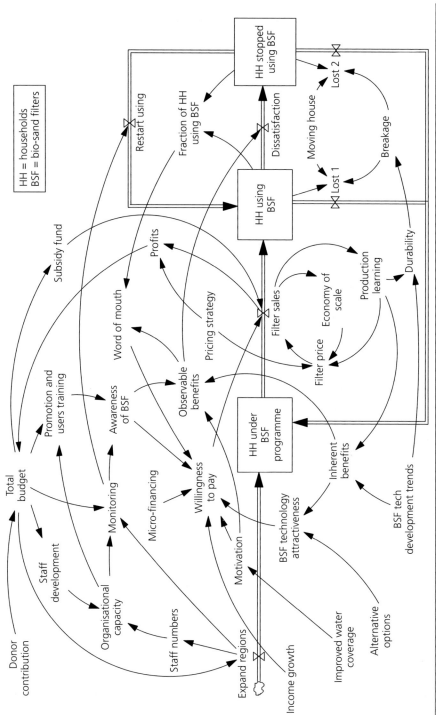

HH = households
BSF = bio-sand filters

explained from multiple irreducible viewpoints, such as law, society and physics, and where structure is the key to their understanding). These complex systems give rise to the 'wicked' problems and unexpected consequences we discussed in Chapters 1 and 2, which are often ill-defined and may contain conflicting objectives. Engineers have excelled at optimising the technical aspects of these systems, but often in isolation from these other legitimate interpretations of the problem.

In complex systems, physical assets, technological solutions and choices between alternatives can quickly get locked in (Allenby, 2012). Tightly coupled networks are more resistant to change than loosely coupled networks. Allenby suggests the following:

> Characteristics of complex systems explain, for example, why changes to pollution control regulations are more easily accomplished than changes to product design or manufacturing process regulations. In the former case, technology is only loosely coupled to underlying product and manufacturing networks, and thus can be changed with only minimal implications for other aspects of the product and the manufacturing networks; lock in exists, but is not significant. In the latter case, however the intervention is to tightly coupled networks based on complex product design and manufacturing processes which means that shifts in state propagate throughout the system. Lock in of changes to such systems occurs more easily, and is more irreversible, in the sense that changing them once implemented is much more difficult.
>
> (Allenby, 2012)

The infrastructure we build can lock us in to systems which are no longer sustainable in the way they deliver services. Much of the piped disposal of wastewater in sewerage systems is an example of this, with the dependence on water as a transport medium and high energy costs in pumping and end-of-pipe treatment. Alternative, decentralised systems are hard to introduce, however, to replace ageing systems, which were often developed in the nineteenth century. Systems such as the carbon-based energy system, which has benefited from a long period of increasing returns, may become 'locked-in', preventing the development and take-up of alternative technologies, such as low carbon, renewable energy sources (Foxton, 2006).

We are often faced with unintended consequences emerging from complex systems. Traffic congestion, deaths and injuries from car accidents, air pollution, and even global warming are unintended consequences of the invention and large-scale adoption of the automobile. Such effects can arise through ignorance and lack of knowledge, mistakes in the way problems are formulated, the influence of short-term versus long-term thinking, adoption of basic values that prohibit certain actions although the long-term consequences may be unfavourable, and adopting early policy choices and solutions before the problem occurs (which may not actually happen).

The way in which systems are structured also affects how they respond to changes in their environment, that is, their resilience. Many critical infrastructure systems are 'scale-free networks', which are based on a few super-connected hubs (e.g. the internet). These are quite resilient to random failure, because often, if a link goes down, the network can simply route around it; but they can be quite sensitive to targeted attack (Allenby, 2012). Systems whose underlying structure consists of random connections behave quite differently.

Resilience is the fourth dimension of sustainability, and is a function of system structure; and often may be the inverse of efficiency and optimisation. Optimisation assumes that change will be incremental and linear. But this does not always work because the world is configured and reconfigured by extreme events rather than by average, day-to-day events and incremental change. Efficiency leads to the elimination of redundancies, retaining only those elements that are immediately beneficial. The concepts of optimisation and efficiency diminish the importance of unquantifiable or unmarketed values (e.g. ecosystem services), and reduce time horizons below those at which important changes occur (e.g. climate change). Optimisation is a large part of the sustainability problem – *not* the solution – because in a non-equilibrium view there is no optimal state for a dynamic system. Therefore, embracing change is the essence of resilience. Practicing multi-functionality, modularity and redundancy through the creation of decentralised systems is a way of increasing resilience. When a minor urban function or service is provided by a single entity or infrastructure, it is more vulnerable to failure from a disturbance or extreme event. If the same function is provided by a distributed or modular system (or subsystems) it has an inherent insurance against failure.

13.11.4 Criticisms and drawbacks

One of the criticisms of conclusions drawn from systems dynamics models is to interpret the results literally as a precise forecast of future events. Thus the original limits-to-growth predictions were found lacking, when they did not occur exactly at the times they had been expected. This misunderstands the power of systems dynamics, which is not designed to be a mere forecasting tool, but to expose trends and pathways in future responses and behaviours, and to show how these can be modified by different ways in which the system can be influenced.

A further drawback is that models can quickly become convoluted, sometimes incorporating unnecessary elements. A key rule is that models should not contain more detail than is required to meet the modelling objectives. Modelling work must be done in close proximity to problem owners, who are able to see that their mental models are closely represented in the computer simulation.

Some have suggested that systems dynamics modelling takes a 'hard' deterministic view of human behaviour, where in reality system structure only partly explains human actions (Lane, 2000). More recently, new techniques have emerged in 'agent-based modelling'. This is a computational method that enables a researcher to create, analyse and experiment with models composed of agents that interact within an environment. This allows the simulation of the actions and interactions of autonomous agents (either individual or collective entities, such as organisations or groups) in order to assess their effects on the system as a whole. Agent-based models consist of dynamically interacting rule-based agents. The systems within which they interact can create real-world-like complexity, and they can be a further powerful tool in understanding the interactions in socio-technical systems (Gilbert, 2007).

13.11.5 Sources of further information

Gopalakrishnan K and Peeta S (eds) (2010) *Sustainable and Resilient Critical Infrastructure Systems: Simulation, Modeling, and Intelligent Engineering.* Springer-Verlag, Berlin.

Hjorth P and Bagheri A (2006) Navigating towards sustainable development: a system dynamics approach. *Future* **38(1)**: 74–92.

Meadows D (1999) Leverage Points: Places to Intervene in the System. The Sustainability Institute, Hartland, VT. Available at: http://www.sustainer.org/pubs/Leverage_Points. pdf (accessed 20 August 2013).

Sterman J (2002) All models are wrong. *System Dynamics Review* **18(4)**: 501–531.

Zigurds K, Mutule A, Meruryev Y and Oleinikova I (2011) *Dynamic Management of Sustainable Development: Methods for Large Technical Systems.* Springer-Verlag, Berlin.

REFERENCES

Allenby BR (2012) *The Theory and Practice of Sustainable Engineering.* Prentice Hall, Englewood Cliffs, NJ.

Azapagic A (2011) Life cycle thinking and life cycle assessment (LCA). In *Sustainable Development in Practice – Case Studies for Engineers and Scientists* (Azapagic A and Perdan S (eds)), 2nd edn. John Wiley, New York, Ch. 3.

Azapagic A and Perdan S (eds) (2011) *Sustainable Development in Practice – Case Studies for Engineers and Scientists*, 2nd edn. John Wiley, New York.

Bernier P, Fenner RA and Ainger C (2010) Assessing the sustainability merits of retrofitting existing homes. *Proceedings of the ICE – Engineering Sustainability* **63(4)**: 187–201.

BRE (Building Research Establishment) (n.d.) Envest 2 and IMPACT. Available at: http://www.bre.co.uk/page.jsp?id = 2181 (accessed 20 August 2013).

BRE (2009) *Green Guide to Specification.* Available at: http://www.thegreenguide.org.uk (accessed 20 August 2013).

BS 8555:2003 (2003) Environmental management systems. Guide to the phased implementation of an environmental management system including the use of environmental performance evaluation. British Standards Institution.

BS EN ISO 14001:2004 (2004) Environmental management systems. Requirements with guidance for use. British Standards Institution.

Bullock H (2008) What price carbon? *Green Futures*, 27 June. Available at: http://www.forumforthefuture.org/greenfutures/articles/what-price-carbon (accessed 20 August 2013).

Capra F (1997) *The Web of Life.* Flamingo Books, New York.

Carruthers G and Vanclay F (2007) Enhancing the social content of environmental managements systems in Australian agriculture. *International Journal of Agricultural Resources, Governance and Ecology* **6(3)**: 326–340.

CEEQUAL (2013) Available at: http://www.ceequal.com (accessed 20 August 2013).

Cheung E-S and Lee KS (2009) Prioritisation of water management for sustainability using hydrologic simulation models and multi criteria decision making. *Journal of Environmental Management* **90**: 1502–1511.

Costanza R, D'Arge R, De Groot R *et al.* (1997) The value of the world's ecosystems services and natural capital. *Nature* **387**: 253–260.

Darnall N, Henriques I and Sadorsky P (2008) Do environmental management systems improve business performance in an international setting? *Journal of International Management* **14(4)**: 364–376.

Darton R (2004) Scenario building and uncertainties: options for energy sources. In *Sustainable Development in Practice – Case studies for Engineers and Scientists* (Azapagic A, Perdan S and Clift R (eds)), 1st edn. John Wiley, New York, Ch. 9.

DECC (Department of Energy and Climate Change) (2009) *Carbon Valuation in UK Policy Appraisal: A Revised Approach.* See also peer reviewers' comments. Available at: https://www.gov.uk/carbon-valuation (accessed 20 August 2013).

Defra (Department for Environment, Food and Rural Affairs) (2007) *An Introductory Guide to Valuing Ecosystem Services.* PB12852. Available at: https://www.gov.uk/government/publications/an-introductory-guide-to-valuing-ecosystem-services (accessed 20 August 2013).

Defra (2008) *Defra Position Statement on Environmental Management Systems.* Available at: http://archive.defra.gov.uk/environment/business/scp/documents/position-statement.pdf (accessed 20 August 2013).

Di Muro J (2006) Analysis of the effects of US and EU environmental policy on mobile source pollution. MPhil Thesis, Centre for Sustainable Development, University of Cambridge.

Diakaki C and Kolokatas D (2009) Life cycle assessment of buildings. In *A Handbook of Sustainable Building Design and Engineering: An Integrated Approach to Energy, Health and Operational Performance* (Mumovic D and Santamouris M (eds)). Earthscan, London, Ch. 8.

Ding GKC (2008) Sustainable construction – the role of environmental assessment tools. *Journal of Environmental Management* **86**: 451–464.

Dreborg KH (1996) Essence of backcasting. *Futures* **28(9)**: 813–828.

EVRI (Environmental Valuation Reference Inventory) (2013) Available at: http://www.evri.ca (accessed 20 August 2013).

European Commission (1985) Environmental Assessment. European Directive 85/337/CEE. Available at: http://ec.europa.eu/environment/eia/full-legal-text/85337.htm (accessed 20 August 2013).

Fenner RA and Rice T (2008) A comparative analysis of two building rating systems. *Proceedings of the ICE – Engineering Sustainability* **161(1)**: 155–163.

Flanagan R, Norman G, Meadows J and Robinson G (1989) *Life Cycle Costing – Theory and Practice.* BSP Professional Books, London.

Foresight (n.d.) Published projects. Office of Science and Technology. Available at: http://www.bis.gov.uk/foresight/our-work/projects/published-projects (accessed 20 August 2013).

Foresight (2004) *Foresight Project: Flood and Coastal Defence.* Office of Science and Technology. Available at: http://www.bis.gov.uk/foresight/our-work/projects/published-projects/flood-and-coastal-defence (accessed 20 August 2013).

Forum for the Future/The Natural Step (2007) *Streamlined Life Cycle Analysis (SLCA).* Available at: http://www.forumforthefuture.org/sites/default/files/project/downloads/slca-2-pager-intronov-2007.pdf (accessed 20 August 2013).

Foxton T (2006) Technological lock-in and the role of innovation. In *Handbook of Sustainable Development* (Atkinson G, Dietz S and Neumayer E (eds)). Edward Elgar, Cheltenham, Ch. 22.

Friends of the Earth (2005) *Environmental Impact Assessment. A Campaigners Guide.* Friends of the Earth, London. Available at: http://www.foe.co.uk/resource/guides/ environmental_impact_asses1.pdf (accessed 20 August 2013).

Gabi Software (2013) Available at: http://www.gabi-software.com (accessed 20 August 2013).

Ghumra S, Watkins M, Phillips P, Glass J, Frost MW and Anderson J (2009) Developing an LCA-based tool for infrastructure projects. In *Proceedings of 25th Annual Conference of Association of Researchers in Construction Management ARCOM*, Nottingham (Dainty ARJ (ed.)), pp. 1003–1010.

Gilbert N (2007) *Agent Based Models.* Sage, London.

Global Commons Trust (2013) Available at: http://globalcommonstrust.org (accessed 20 August 2013).

Global Reporting Initiative (2013) Available at: https://www.globalreporting.org/reporting/ sectorguidance/sector-guidance/construction-and-real-estate (accessed 20 August 2013).

Green Infrastructure Valuation Network (2010) *Building Natural Value for Sustainable Economic Development. The Green Infrastructure Valuation Toolkit User Guide.* Available at: http://www.greeninfrastructurenw.co.uk/resources/Green_Infrastructure_Valuation_ Toolkit_UserGuide.pdf (accessed 20 August 2013).

GRI (Global Reporting Initiative) (2006) *Sustainability Reporting Guidelines Version 3.0.* GRI, Amsterdam.

Griffith A and Bhutto K (2008) Improving environmental performance through integrated management systems (IMS) in the UK. *Management of Environmental Quality: An International Journal* **19(5)**: 565–578.

Griffiths-Sattenspiel B and Wilson W (2009) *The Carbon Footprint of Water.* The River Network, Portland, OR. Available at: http://www.rivernetwork.org/resource-library/ carbon-footprint-water (accessed 20 August 2013).

Hacking T and Guthrie P (2008) A framework for clarifying the meaning of triple bottom-line, integrated, and sustainability assessment. *Environmental Impact Assessment Review* **28(1)**: 73–89.

Halcrow Group (2008) *Breaking the Trend: Visioning and Backcasting for Transport in India and Delhi.* Asian Development Bank, Manila.

Hammond G and Jones CI (2008a) *Inventory of Carbon and Energy (ICE)*, version 1.6a. University of Bath. Available at: http://web.mit.edu/2.813/www/readings/ICE.pdf (accessed 20 August 2013).

Hammond GP and Jones CI (2008b) Embodied energy and carbon in construction materials. *Proceedings of the ICE – Energy* **161(2)**: 87–98.

Harvard University, John F. Kennedy School of Government. Corporate Social Responsibility Initiative. Available at: http://www.hks.harvard.edu/m-rcbg/CSRI/ init_define.html.

Holmberg J (1998) Backcasting: a natural step in operationalising sustainable development. *Greener Management International* **23(Autumn)**: 30–51.

Hopkins R (2010) *Ingredients of Transition: Backcasting.* Transition Network. Available at: http://www.transitionnetwork.org/blogs/rob-hopkins/2010-11-15/ingredients-transition- backcasting (accessed 20 August 2013).

Horvath A (2009) Principles of using life-cycle assessment in bridge analysis. *Proceedings of US–Japan Workshop on Life Cycle Assessment of Sustainable Infrastructure Materials,* Sapporo.

HR Wallingford (2004) *Whole Life Costing for Sustainable Drainage* SR 627. HR Wallingford, Wallingford.

Huntzinger DN and Eatmon TD (2009) A life-cycle assessment of Portland cement manufacturing: comparing the traditional process with alternative technologies. *Journal of Cleaner Production* **17(7)**: 668–675.

ISO (International Organization for Standardization) (2006) ISO 14040: 2006 Environmental management – Life cycle assessment – Principles and framework. ISO, Geneva.

Johnstone N and Labonne J (2009) Why do manufacturing facilities introduce environmental management systems? Improving and/or signalling performance. *Ecological Economics* **68(3)**: 719–730.

Jowitt PW (2010) Now is the time. Abbreviated version of the 2009 ICE Presidential Address. *Proceedings of the ICE – Civil Engineering* **163(1)**: pages 3–8; discussion **163(3)**.

Kain J-J, Karrman E and Soderberg H (2007) Multi-criteria decision aids for sustainable water management. *Proceedings of the ICE – Engineering Sustainability* **160(2)**: 87–93.

Kendall A, Keoleian GA and Helfand GE (2008) Integrated life-cycle assessment and life-cycle cost analysis model for concrete bridge deck applications. *Journal of Infrastructure Systems, ASCE* **14(3)**: 214–222.

Kiker GA, Bridges TS, Varghese A, Seager TP and Linkov I (2005) Application of multicriteria decision analysis in environmental decision making. *Integrated Environmental Assessment and Management* **1(2)**: 95–108.

Kishk M, Al-Hajj A, Pollock R, Aouad G, Bakis N and Sun M (2003) Whole life costing on construction, a state of the art review. *RICS Foundation Research Paper* **(4)18**.

Kohler N (1999) The relevance of Green Building Challenge: an observer's perspective. *Building Research and Information* **27(4/5)**: 309–320.

Labuschagne C, Brent AC and van Erck RPG (2005) Assessing the sustainability performance of industries. *Journal of Cleaner Production* **13**: 373–385.

Lane D (2000) Should systems dynamics be described as a hard or deterministic systems approach? *Systems Research and Behavioral Science Systems Research* **17**: 3–22.

Linkov I and Steevens J (2008) Multi-criteria decision analysis. In *Cyanobacterial Harmful Algal Blooms: State of the Science and Research Needs* (Hudnell KH (ed.)). Advances in Experimental Medicine and Biology, Vol. 6192008. Springer-Verlag, Berlin, pp. 817–832.

Lockie S (2008) Making sense of whole life costing. *ACES Spring Conference, Birmingham.* ACES paper No. 08.5/7. Available at: http://www.aces.org.uk/publications/ASSET-08-05-07-Lockie.pdf?PHPSESSID = d4a18663c1728c000372039067c70081 (accessed 20 August 2013).

Lozano R and Huisingh D (2011) Inter-linking issues and dimensions in sustainability reporting. *Journal of Cleaner Production* **19**: 99–107.

McCauley DJ (2006) Selling out on nature. *Nature* **443**: 27–28.

Meadows DH, Meadows D, Randers J and Behrens III WW (1972) *The Limits to Growth.* Universe Books, New York.

Meadows D (1999) Leverage Points: Places to Intervene in the System. The Sustainability Institute, Hartland, VT. Available at: http://www.sustainer.org/pubs/Leverage_Points.pdf (accessed 20 August 2013).

Mulder K (2006) *Sustainable Development for Engineers: A Handbook and Resource Guide*. Greenleaf Publishing, Sheffield.

Nash J (2008) A Critical Look at Carbon Footprint Calculators. Available at: http://www.articlesbase.com/environment-articles/a-critical-look-at-carbon-footprint-calculators-539557.html (accessed 20 August 2013).

National Academy of Engineering (2004) *The Engineer of 2020: Visions of Engineering in the New Century*. Available at: http://www.nap.edu/catalog/10999.html (accessed 20 August 2013).

Nelson J (2002) Corporate social responsibility: passing fad or fundamental to a more sustainable future. *Sustainable Development* 7: 37–39.

Ngai T and Fenner RA (2008) A systems approach to characterise the dissemination process of household water treatment systems in developing countries. IWA World Water Congress, Vienna.

Olubodun F, Kangwa J, Oladapo A and Thompson J (2010) An appraisal of the level of application of life cycle costing within the construction industry in the UK. *Structural Survey* 28(4): 254–265.

Pacific NorthWest National Laboratory (2006) *Sustainable Building Rating Systems. Summary for the US Department of Energy*. PNNL-1585. Pacific NorthWest National Laboratory, Richland, WA.

Polatides H, Haralambopoulos DA, Munda G and Vreeker R (2006) Selecting an appropriate multi-criteria decision analysis technique for renewable energy planning. *Energy Sources, Part B* 1: 181–193.

Rahman S and Vanier DJ (2004) *Life Cycle Cost Analysis as a Decision Support Tool for Managing Municipal Infrastructure*. NRCC-46774 Institute for Research in Construction, National Research Council Canada.

Reap R, Roman F, Duncan S and Bras B (2008a) A survey of unresolved problems in life cycle assessment. Part 1: Goal and scope and inventory analysis. *The International Journal of Life Cycle Assessment* 13(4): 290–300.

Reap R, Roman F, Duncan S and Bras B (2008b) A survey of unresolved problems in life cycle assessment. Part 2: Impact assessment and interpretation. *The International Journal of Life Cycle Assessment* 13(5): 374–388.

Ries R, Velayutham S and Chang Y (2010) Life cycle assessment modeling of heavy construction activities. *Construction Research Congress 2010: Innovation for Reshaping Construction Practice* (Buwanpura J, Mehamed Y and Lee S-H (eds)). Available at: http://ascelibrary.org/doi/book/10.1061/9780784411094 (accessed 20 August 2013).

Robinson JB (1990) Futures under glass: a recipe for people who hate to predict. *Futures* 22(8): 820–842.

Roussat N, Dujet C and Méhu J (2009) Choosing a sustainable demolition waste management strategy using multicriteria decision analysis. *Waste Management* 29(1): 12–20.

Ryall MJ (2001) Whole life costing, maintenance strategies and deterioration modelling. *Bridge Management* 309–330.

Savic DA, Djordjevic S Cahsman A and Saul A (2008) A whole-life cost approach to sewerage and potable water system management in dangerous pollutants (xenobiotics) in urban water cycle. *NATO Science for Peace and Security* **1**: 3–12.

Simapro (2013) Available at: http://www.simapro.co.uk (accessed 20 August 2013).

Simpson DR (2011) *The Ecosystem Service Framework – A Critical Approach.* UN EP Ecosystem Services Economics Working Paper Series Paper No. 5.

Sterman JD (2002) All models are wrong: reflections on becoming a systems scientist. *System Dynamics Review* **18**: 501–531.

Stern N (2006) *Stern Review on The Economics of Climate Change.* HM Treasury, London.

Stern N (2010) *Blueprint for a Safer Planet: How We Can Save the World and Create Prosperity.* Vintage, London.

Stokes J and Horvath A (2010) Life cycle assessment of urban water provision: a tool and case study in California. *ASCE Journal of Infrastructure Systems* 17(1): 15–24.

Svenfelt A, Engstrom R and Svane O (2010) Decreasing energy use in buildings by 50% by 2050 – a backcasting study using stakeholder groups. *Technological Forecasting and Social Change.* Elsevier, Oxford.

The Natural Step (n.d.) What is Backcasting? Available at: http://www.naturalstep.org/backcasting (accessed 20 August 2013).

Tse R (2001) The implementation of EMS in construction firms: case study in Hong Kong. *Journal of Environmental Assessment Policy and Management* **3(2)**: 177–194.

UKWIR (UK Water Industry Research) (2012) *A Framework for Accounting for Embodied Carbon in Water Industry Assets.* Project CL01/B207. UKWIR, London, Appendix C.

Vukotic L, Fenner RA and Symonds K (2010) Assessing the embodied energy of building structural elements. *Proceedings of the ICE – Engineering Sustainability* **163(ES3)**: 147–148.

Weaver P, Jansen L, Van Grootveld G, Van Spiegle E and Vergragt P (2000) *Sustainable Technology Development.* Greenleaf, Sheffield.

World Green Building Council (2009) Green Building Councils. Available at: http://www.worldgbc.org (accessed 20 August 2013).

Wiedmann T and Minx J (2008) A definition of 'carbon footprint'. In: *Ecological Economics Research Trends* (Pertsova CC (ed.)). Nova Science, Hauppauge, NY, Ch. 1, pp. 1–11.

Sustainable Infrastructure: Principles into Practice
ISBN 978-0-7277-5754-8

Chapter 14
End Words

The discussion of systems in Section 13.11 brings us full circle, and draws this book to its close.

Sustainable infrastructure must properly respect and serve the complex systems – environmental and socio-economic – in which it operates, and will do so with a commitment to stewardship for future generations. We will be really 'inventing a sustainable future' when these *principles* have become embedded as a standard part of our *practice* of 'good infrastructure engineering'.

The book is not, of course, in any way 'complete' – developments in sustainability are moving too fast for that – but it contains the things that seem most important to us now, as we conclude our writing in early 2013. We hope the ideas and evidence that it lays out will stimulate and help you to play a part in that continuing process, and that you will add to them.

Progress will be uneven, and there will be good days and bad days; and sometimes it may feel that we are just muddling through, but we can get there, through the combined decisions and actions by all of us, as individuals and within teams. The most powerful combination of levers for change that we can deploy are excellent technical expertise, applied with a passionate commitment and obvious integrity used at just the right time in the project.

Finally, keep going – be *persistent*. The book *Active Hope* (Macy and Johnstone, 2012) concludes with a story of Tibetan monks, who just keep rebuilding their monastery, in spite of their uncertainty about the future:

> you simply proceed. You do what you have to do. You put one stone on top of another and another on top of that. If the stones are knocked down, you begin again, because if you don't, nothing will be built. You persist. In the long run, it is persistence that shapes the future.

Good luck.

REFERENCE

Macy J and Johnstone C (2012) *Active Hope: How to Face the Mess We're in without Going Crazy*. New World Library, Novato, CA.

Sustainable Infrastructure: Principles into Practice
ISBN 978-0-7277-5754-8

ICE Publishing: All rights reserved
http://dx.doi.org/10.1680/sipp.57548.329

Appendix A
Summaries of common sustainability principles

This appendix provides, in table form, brief summaries of a range of the most popular sustainability principles and methods. They have become available as thinking on sustainability has developed. They are not aimed particularly at infrastructure, but you may come across them. The material will allow you to investigate them further if necessary. They are grouped under our four fundamental principles (see Section 2.2).

Sustainability principle	Comments
A1. Environmental sustainability – within limits	
1. The Precautionary Principle Where there are threats of serious irreversible damage, lack of full scientific certainty shall not be used as a reason for postponing cost-effective measures to prevent environmental degradation (1992 Rio Declaration).	■ Scientific uncertainty is no excuse for inaction on an environmental problem. ■ The best precautionary action will be a 'no regrets' policy where other benefits accrue regardless of whether it helps reduce the environmental threat in question (e.g. reducing energy consumption also lowers fuel bills). ■ The interpretation of the precautionary principle will continue to be different in different circumstances because the different technologies and situations require different degrees of precaution (Azapagic et al., 2004). ■ The Wingspread agreement of 1998 suggested four parts (see the Science Environmental Health Network website): (i) a duty to take anticipatory action to prevent harm (ii) the burden of proof of the harmlessness of an action lies with the proponent (iii) before deploying the action a full range of alternatives should be considered (iv) decisions in applying the principle should be open, informed and democratic.

Sustainability principle	Comments

2. The Natural Step
There are four system conditions in a sustainable society, where nature is not subject to systematically increasing
■ concentrations of substances extracted from the Earth's crust
■ concentrations of substances produced by society
■ degradation by physical means
■ people are not systematically subject to conditions that systematically undermine their capacity to meet their own needs.
	■ Founded in Sweden in 1989, it provides a science-based tool to help organisations build sound programmes.
	■ Recognises that there are thresholds beyond which living organisms are adversely affected by increases in greenhouse gases, contamination of water and metal toxicity.
	■ Provides a 'moral compass' for a focus on sustainability' (Anstey et al., 2009).
	■ The Natural Step and Forum for the Future recommend six steps (Chambers et al., 2008).
	  (i) Adopt the system conditions as sustainability objectives (e.g. Carillion's Sustainability Excellence model).
	  (ii) Identify violations of the system conditions over the full cycle.
	  (iii) Create a vision for success (e.g. Interface FLOR set seven challenges: eliminate waste; achieve benign emissions; use renewable energy; close the loop; use resource-efficient transport, sensitise stakeholders; redesign commerce).
	  (iv) Backcast from success to come up with potential actions.
	  (v) Create an action plan.
	  (vi) Select the right tools for the job (e.g. LCA, EMS, backcasting, energy audits).

Sustainability principle	Comments
3. One Planet Living *Zero carbon* through renewable energy technologies. (i) *Zero waste* with reuse where possible. (ii) *Sustainable transport* and reduce the need to travel. (iii) *Materials* with low embodied energy, sourced locally. (iv) Local, seasonal and organic *food*, and reduce food waste. (v) *Efficient water use*, and protect from flooding and pollution. (vi) Protect/restore *biodiversity* through appropriate land use. (vii) Revive *local identity*, wisdom and culture in the community. (viii) Support fair employment, equity and *local economies*. (ix) Encourage meaningful lives for *health and happiness*.	A global initiative based on ten principles of sustainability, developed by Bioregional and WWF. ■ The London 2012 Olympics and Paralympics used the One Planet Living approach at the bid stage, and applied it to the delivery and legacy strategy. ■ Working with developers around the world, the programme has award-winning endorsed projects in the USA, UK and Portugal, with others planned in South Africa, China, Australia and Canada. ■ Municipal authorities, local residents, voluntary organisations and businesses have put in place initiatives that can be adopted throughout their area to help organisations and individuals to use the ten principles to reduce their environmental footprint (e.g. London Borough of Sutton, Middlesborough). ■ Since 2007, B&Q, the UK's largest home-improvement retailer, has partnered with BioRegional to develop solutions to reduce the environmental impact of its stores, products and supply chains. Key successes include a 20% reduction in the company's carbon footprint and the development of a 3500 strong eco-product range – One Planet Home.

Sustainability principle	Comments

4. Ecological Footprint

The Ecological Footprint (Chambers *et al.*, 2002; Global Footprint Network, n.d.) measures the amount of biologically productive land and sea area that an individual, a region, all of humanity, or a human activity requires to produce the resources it consumes and absorb the carbon dioxide emissions it creates, and compares this measurement with how much land and sea area is available.

Biologically productive land and sea includes area that (1) supports human demand for food, fibre, timber, energy and space for infrastructure, and (2) absorbs the carbon dioxide emissions from the human economy.

Biologically productive areas include cropland, forest and fishing grounds, and do not include deserts, glaciers and the open ocean.

The merit of this approach is that it provides a scientifically objective way (within its many assumptions) of combining a range of key environmental impacts into one overall number, which

- directly expresses the impact term of the $I = P \times A \times T$ equation
- gives a 'footprint' result, as a land area (in hectares) per person, which people can visualise and is easy to communicate
- can be compared to a real available global surface area, expressing the fact that we are exceeding our environmental limit in a factual way – 'We would need three earths if everyone on the planet lived at European levels of affluence'
- can be applied to global averages, countries or regions.

Its disadvantages are that

- it still has to make assumptions – notably by including carbon dioxide emissions by estimating the forest land area needed to absorb them (this is the single largest component in most footprints) – and this oversimplifies a very complex problem
- apart from carbon dioxide, it cannot include pollution from chemicals and radiation, which cannot be turned into this kind of number.

Sustainability principle	Comments

5. Resource Efficiency and the Waste Hierarchy

Resource efficiency means using the Earth's limited resources in a sustainable manner while minimising impacts on the environment. It allows us to create more with less and to deliver greater value with less input (European Commission, n.d.). The *waste hierarchy* classifies *waste management* options in order of their environmental impact, typically with five steps

■ reduce
■ reuse
■ recycle
■ other recovery (e.g. energy)
■ disposal.

WRAP (Waste and Resources Action Programme) in the UK is a main source of information (WRAP, n.d., 2005).

In the 1970s, the Club of Rome warned that there were limits to growth (Meadows *et al.*, 1972). This was based on a world systems model which recognised that the availability of non-renewable resources was finite, and that industrial and agricultural pollution was exceeding the assimilative capacity of the environment. So a critical component in managing our infrastructure, which provides basic needs and services to millions of people, is to use less land and consume resources much more efficiently.

The waste hierarchy remains the cornerstone of most *waste minimisation* strategies. The aim of the waste hierarchy is to extract the maximum practical benefits from products and to generate the minimum amount of waste. Sometimes an additional top level has been added, R for 're-think', implying that the present system may have fundamental flaws, and an entirely new way of looking at waste is needed. *Source reduction* reduces hazardous waste and other materials by modifying industrial production.

Waste minimisation, reuse and recycling can be encouraged by each of the parties involved – clients, designers, and contractors (Keys *et al.*, 2000). Maximum use should be made of renewable resources, minimising use of materials with poor environmental performance. Materials should be used to be beneficial for many generations, allowing structures to be adapted within the shell to suit a variety of potential future uses.

Sustainability principle	Comments

6. Adaptation, Resilience and Adaptive Management

Adaptation: the process or outcome of a process that leads to a reduction in harm or risk of harm, or realisation of benefits associated with climate variability and climate change (United Kingdom Climate Impacts Programme (UKCIP)).

Resilience: the ability of a system to recover from the effect of an extreme load that may have caused harm (UKCIP).

Adaptive management: a structured, iterative process of robust decision-making in the face of uncertainty, aiming to reduce uncertainty over time via system monitoring. In this way, decision-making simultaneously meets one or more resource-management objectives and, either passively or actively, accrues information needed to improve future management (Wikipedia).

Organisations have defined these terms in many slightly different ways according to their own context. Recent use refers particularly to the capability to react to global warming.

The Organisation for Economic Cooperation and Development (OECD) gives a useful summary and discussion of these differences (Levina and Tirpak, 2006). Its conclusion summarises the current situation well: 'The analysis of various definitions of the key adaptation terms and concepts demonstrates that definitions vary across institutions and different groups of stakeholders. The lack of precision is a reflection of a highly dynamic discussion of the adaptation issue, where the lexicon is still evolving, and the relatively young age of these discussions. Once adaptation enters wider circles of policy makers and analytical community it may need to be handled with more care and accuracy.'

The definitions we have quoted are generic, contain the key features, and are relevant to infrastructure. Importantly, the UKCIP definition of *resilience* is wider than some others. It includes the capacity to *recover* from harm, not just the capacity to resist it. This takes a systems approach: for instance, in flooding, it includes the capacity of rescue and response services, and how properties can be designed to be more easily recovered and reused after flooding, not just how big the flood defences are.

Although UKCIP does not define *adaptive management* directly, their website gives a useful summary of how to do it. They include this advice: 'there may be practical, cost-effective options that deliver the required adaptation and also minimise the risks associated with implementation. These options are normally referred to as *no-regrets*, *low-regrets*, and *win–win* options, and should be used where possible, along with a flexible or adaptive management approach.'

Sustainability principle	Comments

A2. Socio-economic sustainability – 'development'

7. Polluter Pays Principle

'Manufacturers and importers of products should bear a significant degree of responsibility for the environmental impacts of their products throughout the product life-cycle, including upstream impacts inherent in the selection of materials for the products, impacts from manufacturers' production process itself, and downstream impacts from the use and disposal of the products.

Producers accept their responsibility when designing their products to minimise life-cycle environmental impacts, and when accepting legal, physical or socio-economic responsibility for environmental impacts that cannot be eliminated by design' (Levina and Tirpak, 2006)

Polluters (including waste producers), rather than society in general, should bear the full costs of the safe management and disposal of waste.
- The principle's immediate goal is that of internalising the environmental externalities of economic activities, so that the prices of goods and services fully reflect the costs of production. When and how much the polluter should pay is often unclear.
- Widely acknowledged in international environmental law, the principle is normally implemented through two different mechanisms:
  (i) based on performance and technology standards
  (ii) uses market-based instruments such as pollution taxes, tradable pollution permits and product labelling.
- The elimination of subsidies is also an important application of the principle. (The Kyoto protocol is an example of this principle –countries have obligations to reduce their greenhouse gas emissions and must bear the cost of reducing such pollution emissions.)

8. Use of Best Available Techniques (BAT)

Best is considered to be the most effective way of achieving a high level of protection of the environment as a whole.

Available means developed on a scale that allows implementation in the relevant sector under technically and economically viable conditions, taking into account the cost and advantages, as long as they are reasonably accessible to the operator.

Techniques include both the technology used and the ways in which any installation is defined, built, maintained, operated and decommissioned (Azapagic et al., 2004).

This concept combines:
(i) an emphasis on technical feasibility and 'state-of-the-art' technology in German pollution regulation, with
(ii) the need for economic feasibility and pragmatic, case-by-case decision-making in UK regulation (Sorrel, 2002).

BAT is a pillar of the Integrated Pollution Prevention & Control Directive of 1996, and is linked to earlier concepts such as BATNEEC (Best Available Technology Not Entailing Excessive Cost) and BPEO (Best Practicable Environmental Option).

BAT will have different interpretations in different countries, depending on the level of economic development. The term also encompasses a moving target on practices, as developing societal values and advancing techniques may change what is currently regarded as 'reasonably achievable', 'best practicable' and 'best available'.

Sustainability principle	Comments
9. BAT Not Entailing Excessive Cost (BATEEC) BATNEEC should consider the following. (i) Process design/redesign changes to *eliminate* emissions and wastes that might pose environmental problems. (ii) *Substitution* of materials (e.g. low-sulphur fuel) by environmentally less harmful ones. (iii) Demonstration of waste *minimisation* by means of process control, inventory control and end-of-pipe technologies. (iv) Use of *energy efficient technologies*, such as combined heat and power generation, when appropriate.	Debate has centred around what is meant by 'excessive cost': whether this is decided by: (i) the environmental cost–benefit approach (which compares the environmental gains that can be achieved with the abatement costs), or (ii) the sectoral affordability approach, which takes what an average company with a sector can afford as a benchmark. How excessive cost is decided is also open to interpretation, being based either on economic information and formal analytical techniques, or expert judgement and negotiation with individual operators. A readable account is given by Environmental Practice @ Work (n.d.), which includes the comment that BATNEEC 'should not to be confused with CATNIP – the Cheapest Available Techniques Not Incurring Prosecution'.
10. The Global Sullivan Principles of Social Responsibility The principles refer to the support for universal human rights (i) equal opportunities (ii) respect for freedom of association, levels of employee compensation, training, health and safety, sustainable development (iii) fair competition (iv) working in partnership to improve quality of life.	The principles provide companies with a framework for addressing issues of human rights and equal opportunity (see Mallenbaker, n.d.).

Sustainability principle	Comments
11. Principles of Social Sustainability (Gates and Lee, 2005; Mitchard *et al.*, 2011)	The principles summarise the idea that individuals should have access to sufficient resources to participate fully in their community and have opportunities for personal development and advancement.
(i) *Equality*	(i) Equal opportunities, impacts and abilities to participate in society for people from all minority and majority groups (e.g. design of transport schemes for vulnerable groups that are operated to be safe, well lit, frequent, direct and accessible).
(ii) *Social inclusion and interaction*	(ii) The strength of unity within communities and how they embrace diverse cultural identities. Fair distribution of resources and the ability of people to engage with wider society in terms of accessing services and amenities and interacting with other community members to facilitate a better quality of life and full participation and collaboration (e.g. social disadvantages can arise from high transport costs (economic exclusion), spatial coverage of a service (geographic exclusion), non-car-owning households and severance and displacement (physical exclusion), concerns about safety (fear-based exclusion), inappropriate design for disability access (spatial exclusion)).
(iii) *Security*	(iii) Individuals and communities have economic security and have confidence that they live in safe, supportive and healthy environments. People need to feel safe and secure in order to contribute fully to their own well-being or engage fully in community life (e.g. sense of identity from local heritage and community aesthetics, and a sense of place that foster a sense of belonging to an area).
(iv) *Adaptability*	(iv) Resiliency for both individuals and communities and the ability to respond appropriately and creatively to change. Adaptability is a process of building upon what already exists, and learning from and building upon experiences from both within and outside the community.
(v) *Health and well-being*	(v) Contributes to the health, happiness and prosperity of people (e.g. reduction in pollution that causes respiratory illness, cancer, premature death, reduction in road traffic injury or death, changes to patterns of walking and cycling that improve health through exercise).

Sustainability principle	Comments

A3. Intergenerational stewardship

12. Delivering value for future generations
(i) Demonstrate enduring value for future generations.
(ii) Clearly outline the future negative impacts (local, regional and cumulative) of the proposal and how they will be managed, and by whom, and how future liability will be managed.
(iii) Hold proponents accountable for commitments (e.g. through mechanisms such as development bonds).
(iv) Demonstrate that the proposal will not impact on the long-term performance of existing significant local or regional land-use activities.

Specific questions that should be asked are (Morrison-Saunders and Hodgson, 2009):
■ Will a particular development be economically viable in the future?
■ Who will have responsibility for managing negative impacts of a development in the future?
■ Will commitments by proponents be acted upon in the future?

A4. Complex systems

13. Biomimicry Principles (Benyus, 2002)
General principles:
(i) Nature runs on sunlight.
(ii) Nature uses only the energy it needs.
(iii) Nature fits form to function.
(iv) Nature recycles everything.
(v) Nature rewards cooperation.
(vi) Nature banks on biodiversity.
(vii) Nature demands local expertise.
(viii) Nature curbs excesses from within.
(ix) Nature taps the power of limits.

It may be difficult to see the application of these general principles to infrastructure projects. Stephen Vogel, Professor of Biomechanics at Duke University, more specifically compares nature to mechanical-engineering-specific principles (Vogel, 2011):
■ Nature's factories produce things much larger, not smaller, than themselves.
■ We use metals, nature never does.
■ Nature makes gradual transitions in structures (curves, density gradients, etc.) rather than sharp corners.
■ We make things out of many components, each of which is homogeneous; nature makes things out of fewer components but they vary internally.
■ We design for stiffness, nature designs for strength and toughness.
■ Our mechanisms have rigid pieces moving on sliding contacts, nature bends/twists/stretches.
■ Nature often uses diffusion, surface tension and laminar flow; we often use gravity, thermal conductivity and turbulence.
■ Our engines are mostly rotary or expansive, nature's are mostly sliding or contracting.
■ Nature's engines are isothermal.
■ Nature mostly stores mechanical work as elastic energy, sometimes as gravitational potential energy.

Sustainability principle	Comments

A5. Cross-cutting principles

14. The Earth Charter

I. Respect and care for the community of life
(i)   Respect the Earth and life in all its diversity.
(ii) Care for the community of life with understanding, compassion and love.
(iii) Build democratic societies that are just, participatory, sustainable and peaceful.
(iv) Secure the Earth's bounty and beauty for present and future generations

II. Ecological integrity
(i)   Protect and restore the integrity of the Earth's ecological systems, with special concern for biological diversity and the natural processes that sustain life.
(ii) Prevent harm as the best method of environmental protection and, when knowledge is limited, apply a precautionary approach.
(iii) Adopt patterns of production, consumption and reproduction that safeguard the Earth's regenerative capacities, human rights and community well-being.
(iv) Advance the study of ecological sustainability and promote the open exchange and wide application of the knowledge acquired.

III. Social and economic justice
(i)   Eradicate poverty as an ethical, social and environmental imperative.
(ii) Ensure that economic activities and institutions at all levels promote human development in an equitable and sustainable manner.
(iii) Affirm gender equality and equity as prerequisites to sustainable development, and ensure universal access to education, healthcare and economic opportunity.
(iv) Uphold the right of all, without discrimination, to a natural and social environment supportive of human dignity, bodily health and spiritual well-being, with special attention to the rights of indigenous peoples and minorities.

IV. Democracy, non-violence and peace
(i)   Strengthen democratic institutions at all levels, and provide transparency and accountability in governance, inclusive participation in decision-making, and access to justice.

The Earth Charter principles provide a comprehensive, multi-dimensional approach to presenting values for worldwide acceptance. Although the principles cover a wide range of sustainability issues, they fall short on practical implementation strategies. Nevertheless, they are an effective expression of a global ethic, which has the following key characteristics (Earth Charter Initiative, n.d.):

- They are about essential human goods alongside concern for the environment, which can guide progress towards a substantially better future.
- They are global in scope with a commitment to global responsibility.
- They are motivational by being emotionally and intellectually engaging.
- They are the product of intercultural agreement and consultation.

The Earth Charter has been criticised by religious groups and others who fear a global super-state, but the Earth Charter Initiative (ECI) has defended the charter as a statement of common ethical values towards sustainability, that recognises humanity's shared responsibility to the Earth and to each other (Earth Charter Initiative, 2008).

Sustainability principle	Comments

(ii) Integrate into formal education and lifelong learning the knowledge, values and skills needed for a sustainable way of life.

(iii) Treat all living beings with respect and consideration.

(iv) Promote a culture of tolerance, non-violence and peace.

15. CERES Principles

Protection of the biosphere – Reduce and make continual progress toward eliminating the release of any substance that may cause environmental damage to the air, water, or the Earth or its inhabitants. Safeguard all habitats affected by our operations and protect open spaces and wilderness, while preserving biodiversity.

Sustainable use of natural resources – Make sustainable use of renewable natural resources, such as water, soils and forests. Conserve non-renewable natural resources through efficient use and careful planning.

Reduction and disposal of wastes – Reduce and, where possible eliminate, waste through source reduction and recycling. All waste will be handled and disposed of through safe and responsible methods.

Energy conservation – Conserve energy and improve the energy efficiency of our internal operations and of the goods and services we sell. Make every effort to use environmentally safe and sustainable energy sources.

Risk reduction – Strive to minimise the environmental, health and safety risks to our employees and the communities in which we operate, through safe technologies, facilities and operating procedures, and by being prepared for emergencies.

Safe products and services – Reduce, and where possible eliminate, the use, manufacture or sale of products and services that cause environmental damage or health or safety hazards. Inform our customers of the environmental impacts of our products or services and try to correct unsafe use.

The CERES (Coalition of Environmentally Responsible Economies) principles came about as a direct result of the Exxon-Valdez environmental disaster in Alaska in 1989, and give large-, medium- and small-sized companies a means to adopt sustainable practices in a low-risk, safe, voluntary and supportive setting (CERES, n.d.). In this respect they establish a forum for sustainability in the business community.

As well as covering protection of the biosphere, energy conservation, environmental restoration, waste reduction and sustainable use of resources, they also focus on employee issues such as personal health and safety and corporate responsibility to the local community.

Companies that endorse these principles pledge to go voluntarily beyond the requirements of the law but, like The Natural Step the principles provide an open-ended framework without a specific methodology.

CERES launched the Global Reporting Initiative (GRI), which has become the international standard by which companies report on environmental, social and economic performance. CERES also founded and directs the Investor Network on Climate Risk (INCR).

Sustainability principle	Comments

Environmental restoration – Promptly and responsibly correct conditions we have caused that endanger health, safety or the environment. To the extent feasible, we will redress injuries we have caused to persons or damage we have caused to the environment, and will restore the environment.

Informing the public – Inform in a timely manner everyone who may be affected by conditions caused by our company that might endanger health, safety or the environment. We will regularly seek advice and counsel through dialogue with persons in communities near our facilities. We will not take any action against employees for reporting dangerous incidents or conditions to management or to appropriate authorities.

Management commitment – Implement these Principles and sustain a process that ensures that the Board of Directors and Chief Executive Officer are fully informed about pertinent environmental issues and are fully responsible for environmental policy. In selecting our Board of Directors, we will consider demonstrated environmental commitment as a factor.

Audits and reports – Conduct an annual self-evaluation of our progress in implementing these Principles. We will support the timely creation of generally accepted environmental audit procedures. We will annually complete the CERES report, which will be made available to the public.

Sustainability principle	Comments
16. The Equator Principles (i) *Review and categorisation* – Categorise projects based on the magnitude of its potential impacts and risks in accordance with IFC environmental and social screening criteria. (ii) *Social and environmental assessment* – For each project a social and environmental assessment will be conducted to identify the relevant social and environmental impacts and risks of the proposed project. The assessment should also propose mitigation and management measures relevant and appropriate to the nature and scale of the proposed project. (iii) *Applicable social and environmental standards* – For projects located in non-OECD countries, and countries not designated as High-Income, the assessment will refer to the then applicable IFC performance standards and the applicable industry-specific EHS guidelines. (iv) *Action Plan and Management System* – The borrower will prepare an action plan which will describe and prioritise the actions needed to implement mitigation measures, corrective actions and monitoring measures necessary to manage the impacts and risks identified in the assessment. (v) *Consultation and disclosure* – The government, borrower or third-party expert will consult with project-affected communities in a structured and culturally appropriate manner to adequately incorporate affected communities' concerns. (vi) *Grievance mechanism* – The borrower will, scaled to the risks and adverse impacts of the project, establish a grievance mechanism as part of the management system. This will allow the borrower to receive and facilitate resolution of concerns and grievances about the project's social and environmental performance raised by individuals or groups from among project-affected communities.	Launched in 2003, these principles guide financial institutions in addressing social and environmental impacts of financing development projects. They are based on the guidelines used by the International Finance Corporation (IFC), the financial division of the World Bank (Equator Principles, n.d.). Like The Natural Step and CERES principles, they highlight the relationship between economic activities, employment and the environment. They set standards with far-reaching effects, and address the challenge of maximising profits within the environmental and social limitations faced by businesses.

Sustainability principle	Comments

(vii) *Independent review* – An independent social or environmental expert not directly associated with the borrower will review the assessment, action plan and consultation process documentation.

(viii) *Covenants* – The borrower will covenant in financing documentation
  (a) to comply with all relevant host country social and environmental laws, regulations and permits
  (b) to comply with the action plan (where applicable) during the construction and operation of the project
  (c) to provide periodic reports prepared by in-house staff or third-party experts, that (i) document compliance with the action plan and (ii) provide representation of compliance with relevant local, state and host country social and environmental laws, regulations and permits, and
  (d) to decommission the facilities, where applicable and appropriate, in accordance with an agreed decommissioning plan.

(ix) *Independent monitoring and reporting* – Require appointment of an independent environmental and/or social expert, or require that the borrower retain qualified and experienced external experts to verify its monitoring information, which would be shared.

(x) *EPFI reporting* – Each organisation adopting the Equator Principles commits to report publicly at least annually about its Equator Principles implementation processes and experience, taking into account appropriate confidentiality considerations.

Sustainability principle	Comments
17. The Hanover Principles (McDonough and Braungart, 2003) 1 Insist on the rights of humanity and nature to co-exist. 2 Recognise interdependence. 3 Respect relationships between spirit and matter. 4 Accept responsibility for the consequences of design. 5 Create safe objects of long-term value. 6 Eliminate the concept of waste. 7 Rely on natural energy flows. 8 Understand the limitations of design. 9 Seek constant improvement by the sharing of knowledge.	These recognise the full-cycle concept, from initial design, to use and eventual disposal. The 'cradle-to-cradle' approach advocated by McDonough and Braungart requires that waste products are reintegrated in the manufacturing process or biodegrade. Designers use resources and design systems that support reuse in new products, and engineered services that support the manufacturing and ecological cycles. The Hanover Principles address the interdependence of humanity and nature, and accept 'responsibility for the consequences of design'. They also emphasise 'long-term value', and challenge engineers to design structures, products and standards that broaden rather than restrict the possibilities for future generations. McDonough states that the Hanover Principles are 'based on ideas of restraint, awareness and concern for solving the world's problems, not hiding them behind a wall of promising machines'.

REFERENCES

Anstey N, Barker V, Clark R *et al.* (2009) The Natural Step Environmental Management System Performance: An evaluation of eight organisations in Canterbury New Zealand.

Azapagic A, Perdan S and Clift R (2004) Sustainable Development in Practice: Case Studies for Engineers and Scientists. Wiley Blackwell, Oxford.

Benyus J (2002) *Biomimicry: Innovation Inspired by Nature*. Harper Perrenial, New York.

CERES (Coalition of Environmentally Responsible Economies) (n.d.) Available at: http://www.ceres.org (accessed 13 August 2013).

Chambers N, Simmons C and Wackernagel M (2002) *Sharing Nature's Interest – Ecological Footprints, as an indicator of Sustainability*. Earthscan, London.

Chambers T, Porritt J and Price-Thomas P (2008) *Wealth Creation Within Environmental Limits*. Forum for the Future, Cheltenham. Available at: http://www.forumforthefuture. org/blog/wealth-creation-within-environmental-limits (accessed 13 August 2013).

Earth Charter Initiative (n.d.) Available at: http://www.earthcharter.org (accessed 13 August 2013).

Earth Charter Initiative (2008) *The Earth Charter Initiative Handbook*. Earth Charter International Secretariat, Costa Rica. Available at: http://www.earthcharterinaction.org/ invent/images/uploads/Handbook%20ENG.pdf (accessed 13 August 2013).

Environmental Practice @ Work (n.d.) *UK Environmental Law*. Available at: http:// www.epaw.co.uk/EPT/law.html (accessed 13 August 2013).

Equator Principles (n.d.) Available at: http://www.equator-principles.com (accessed 13 August 2013).

European Commission (n.d.) *Environment. Online Resource Efficiency Platform (OREP)*. Available at: http://ec.europa.eu/environment/resource_efficiency (accessed 13 August 2013).

Gates R and Lee M (2005) *Policy Report Social Development*. Report to City of Vancouver from Director of Social Planning.

Global Footprint Network (n.d.) Available at: http://www.footprintnetwork.org/en/index.php/GFN/page/frequently_asked_questions/#gen2 (accessed 13 August 2013).

Keys A, Baldwin AN and Austin SA (2000) Designing to encourage waste minimisation in the construction industry. *Proceedings of CIBSE National Conference, CIBSE2000*, Dublin.

Levina L and Tirpak D (2006) Adaptation to Climate Change: Key Terms. OECD, Paris. Available at: http://www.oecd.org/environment/cc/36736773.pdf (accessed 13 August 2013).

Mallenbaker (n.d.) The Global Sullivan Principles. Available at: http://www.mallenbaker.net/csr/CSRfiles/gsprinciples.html (accessed 13 August 2013).

McDonough W and Braungart M (2003) *Cradle to Cradle: Remaking the Way We Make Things*. Rodale Press, Emmaus, PA.

Meadows DH, Meadows D, Randers J and Behrens III WW (1972) *The Limits to Growth*. Universe Books, New York.

Mitchard N, Frost LC, Harris J, Baldrey S and Ko J (2011) Assessing the impacts of road schemes on people and communities. *Proceedings of the ICE – Engineering Sustainability* **164(3)**: 185–196.

Morrison-Saunders A and Hodgson N (2009) Applying sustainability principles in practice: guidance for assessing individual proposals. Presented at: *IAIA09 Impact Assessment and Human Well-being, 29th Annual Conference of the International Association for Impact Assessment*, Accra.

One Planet Living (n.d.) Available at: http://www.oneplanetliving.org (accessed 13 August 2013).

Science Environmental Health Network (n.d.) Available at: http://www.sehn.org/index.html (accessed 13 August 2013).

Sorrel J (2002) The meaning of BATNEEC: interpreting excessive costs in UK industrial pollution regulation. *Journal of Environmental Policy & Planning* **4**: 23–40.

The Natural Step (n.d.) Available at: http://www.naturalstep.org (accessed 13 August 2013).

UKCIP (United Kingdom Climate Impacts Programme) (n.d.) Available at: http://www.ukcip.org.uk (accessed 13 August 2013). For adaptive management: http://www.ukcip.org.uk/adopt/adaptation-options (accessed 13 August 2013).

Vogel S (2011) *Comparative Biomechanics, Life's Physical World*. Princeton University Press, Princeton, NJ.

WRAP (Waste and Resources Action Programme) (n.d.) *Business Resource Efficiency Hub*. Available at: http://www.wrap.org.uk/content/business-resource-efficiency-hub (accessed 13 August 2013).

WRAP (2005) *Quick Wins Guide: Opportunities to Use Recycled Materials in Preliminary Building Works and Civil Engineering*. Available at: http://www2.wrap.org.uk/downloads/Civil_Quick_Wins_Guide_lo_res.459fdc47.1703.pdf (accessed 13 August 2013).

Sustainable Infrastructure: Principles into Practice
ISBN 978-0-7277-5754-8

ICE Publishing: All rights reserved
http://dx.doi.org/10.1680/sipp.57548.347

Index

absolute principles (A), 28, 30, 263, 329–344
  A1 (staying within environmental limits), 6,
    13, 15, 30, 33–44, 263, 329–334
  A2 *see* socio-economic development and
    sustainability
  A3, 30, 57–59, 263, 338
  A4 *see* complex/complicated systems
  linking, 31–33
  sustainability-enabling and measuring tools
    in relation to, 263
academic–practitioner interactions, 237
action (for change/sustainability), 66–69
  for effective change, 231–242
  engineers' monitoring own actions, 21–23
  individual, 245–260
action inquiry, 254–256
activists, 225, 232, 246
adaptation and adaptability, 334, 337
adaptive management, 20–21, 85, 86, 102, 103,
  132, 171, 334
adding evaluation criteria *see* weighting and
  adding
advocacy in conversations, 257
affluence, measures, 31
  see also IPAT equation
Africa, World Bank regional strategy for,
  132–133
agriculture, fertilisers and pesticides, 8, 62, 99
ALARP (as low as reasonably practical),
  157–158
Allen's *How to Save the World*, 9
alliancing and partnering, 123, 128, 129, 132,
  135, 187
analytical hierarchy process (AHP), 310, 311
Anthropocene, 36
appropriate scale (principle of), 43, 101
arguments, choosing your, 237–241
as low as reasonably practical (ALARP),
  157–158

assets
  owner, procurement and, 121
  in project scoping
    developing a future sustainable
      'destination', 96
    planning anchored in strategic
      sustainability objectives, 95–96
atmospheric aerosol load, 35, 80, 154
audit, 341
authority, individual, 247

backcasting, 305–309
Banham, Bryan, 128
Barnes, Martin, 123
Battersea power station, 215
beam design, 169–170
behaviour
  change *see* change
  user, incentivising, 207–208
benchmark sustainable option (BSO)
  methodology, 147–148, 149
benefits transfer method, 287, 289
  see also cost–benefit analysis
best available techniques (BAT), 335
  not entailing excessive (BATECC), 336
Bexley (London Borough), 290, 292
big-bang change, 230
biodiversity (and its loss), 34, 153, 293
  in detailed design landscaping, 180–181
  global trends, 5
  setting limits to biodiversity impacts, 80
biomass fuel, 82
biomimicry, 60, 318, 338
biosphere, protection, 340
bottom-up approach to change, 223, 232, 233,
  234, 239, 241, 242, 247
brainstorming, 129, 142, 147, 161, 239
BREEAM, 168, 191, 205, 206, 235, 269, 279,
  281, 282–283, 284, 285, 297